5G/6G Advancements in Communication Technologies for Agile Management

Ramesh Kumar Miryala
Mahatma Gandhi University, India

Young-Chan Lee
Dongguk University, South Korea

Marica Mazurek
University of Zilina, Slovakia

Madhavaiah Chendragiri
Pondicherry University, India

IGI Global
Scientific Publishing
Publishing Tomorrow's Research Today

Vice President of Editorial — Melissa Wagner
Managing Editor of Acquisitions — Mikaela Felty
Managing Editor of Book Development — Jocelynn Hessler
Production Manager — Mike Brehm
Cover Design — Phillip Shickler

Published in the United States of America by
IGI Global Scientific Publishing
701 East Chocolate Avenue
Hershey, PA, 17033, USA
Tel: 717-533-8845
Fax: 717-533-8661
E-mail: cust@igi-global.com
Website: https://www.igi-global.com

Library of Congress Cataloging-in-Publication Data

CIP PENDING

ISBN13: 9798369367254
Isbn13Softcover: 9798369367261
EISBN13: 9798369367278

British Cataloguing in Publication Data
A Cataloguing in Publication record for this book is available from the British Library.

Table of Contents

Detailed Table of Contents

Mobile Communication technology is no longer serving the needs of Smart Phone users, rather it has made it smart in meeting the needs of Enterprises. With the advent of 5G Cellular network, Enterprise Automation and transformative opportunities across the managerial functions in various sectors has been seen. This chapter gives a brief overview of evolution of 5G and highlights the usage and applications of 5G technology. Further, this chapter explores the impact of 5G Technologies through real times cases studies. By examining the deployments in diverse settings like Health Care, Manufacturing, Logistics and Transportation, this chapter highlights the unique challenges and innovative solutions encountered with 5G Technology implementation. This synthesis of real time experiences offers a crucial resource for policymakers, industry leaders, and technology managers navigating the complexities of 5G deployment in business.

The continuous evolvement of new formative 5G and evolution towards 6G are pertinent to changing the dynamics of the development of the principles of the agile

data from customers through online can be considered as a big challenging task for both online businesses partners and consumers. So, it is tedious and time consuming task for buyers to go through all these reviews to make a correct decision. Hence, this research work focuses on developing a prediction model based on machine learning technique which can help in suggesting valuable products to buyers based on the real-world data. As a result, buyers can make effective decision while purchasing products and helps businesses organization to improvise their business process and their websites.

Qazi Saima Syed, Islamic University of Science and Technology, India
Shakir Hussain, Islamic University of Science and Technology, India
Irfan Bashir, Islamic University of Science and Technology, India

This chapter provides a comprehensive review of the technological shifts, highlighting AI's and machine learning's transformative potential in communications while acknowledging the challenges they present. This chapter will delve into the development of various mobile technologies over time and the future of 5G and 6G communication networks. Furthermore, this chapter provides an overview of the use of AI and machine learning in upgrading 5G and 6G networks, concentrating on critical topics such as its evolution, significance, influencing factors, and the repercussions of its deployment in network capabilities.

Swati Sharma, Amity University, Noida, India
Deepti Bhatt, Amity University, Noida, India
Narendra Kumar, Amity University, Noida, India
Levent Karadağ, Muğla Sıtkı Koçman University, Turkey

This study explores the effect of Chatbots & Virtual Assistants (CVA) on Guest satisfaction (GS). Although several studies have been published that investigate guest satisfaction in Jordan, only a few have reported on how AI influences Guest satisfaction in Delhi NCR. Therefore, this study addresses the gap. It uses a structured questionnaire to collect data from 155 employees working in five-star hotels. The data is analysed using structure equation modelling (SEM). Smart-PLS is applied to examine the relationship between the independent and dependent factors- Chatbots & Virtual Assistants, Fraud Detection & Security, Operational Efficiency, Personalised Experience, Predictive Maintenance, Revenue Management. This study finds that

sources of data are collected. It tries to have a good impact on the local communities of tea estates leading to economic development in the region.

This chapter addresses the cooperative relay communications in 6G radio networks for high reliability and low latency communications. In this regard, we consider two-hop buffered relay network. To achieve these goals, advanced protocols need to be developed that are state-of-art in the cooperative relaying networks. To mitigate the delay caused by data packets arrival and scheduling at the relay, the advanced random-access protocol and non-random-access protocol are investigated for buffer state information (BSI). Following this, an average queueing delay (AQD) is formulated in closed-form which is minimized by satisfying certain queueing stability of buffer and maintain the constant power level by exploiting the Markov reward process (MRP) at the relay for RA protocol. Furthermore, the closed-form expression of the AQD is also, derived for NRA protocol. Our proposed numerical results perform better as compared to existing schemes available in the literature. In addition, this chapter provides the research directions in the high-speed mobility scenarios for 6G radio system.

The book chapter is focused on the specific cases of smart city applications in the selected Slovakian city, which were implemented to improve the competitiveness as well as to create better co-operation of the chosen city representatives (municipalities) with the city users (citizens, entrepreneurs) and the main focus will be the principle of co-creation and new forms of communication among these entities, which were influenced by new technologies and smart decisions. The digitalization has enormously influenced the new trends of communication and the process of co-creation has been fully applied in the contemporary municipal managerial practices. The specific case study will be focused on the city of Banska Bystrica and the primary and secondary research in this city will be a base of our research and discussion.

Preface

Rapid advancements in communication technologies, especially in the realm of 5G and 6G, have significantly redefined the scope and dimensions of industries and society. These next-generation networks enhance connectivity and offer unprecedented speed, reliability, and real-time responsiveness, paving the way for more agile management and operational flexibility across sectors. As organizations and industries gear up to harness the full potential of these technologies, it is essential to explore their implications for management, business agility, and overall productivity.

This edited volume, "5G/6G Advancements in Communication Technologies for Agile Management," brings together leading researchers, scholars, and industry practitioners to provide in-depth insights into the transformative power of 5G and 6G technologies. The chapters in this book cover a wide array of topics, from the technical underpinnings of 5G/6G systems to their application in management practices that require agility and swift adaptability. The contributions delve into how businesses can leverage these innovations to enhance efficiency, competitiveness, and strategic decision-making in a rapidly changing global environment.

With the increasing complexity of business processes and the need for agile responses to market demands, this book aims to serve as a comprehensive resource for academics, industry professionals, and students to understand how cutting-edge communication technologies can drive organizational agility. It also highlights the future possibilities as 6G technologies continue to evolve, potentially redefining the boundaries of communication and management further.

Chapter 1, "5G Communication Technologies in Business: Applications and Real-World Cases", authored by Swetha Suram, Mr. Raam Sai Bharadwaj Miryala and Ramesh Kumar Miryala, gives a brief overview of evolution of 5G and highlights the usage and applications of 5G technology and explores the impact of 5G Technologies through real times cases studies in Health Care, Manufacturing, Logistics and Transportation highlighting the unique challenges and innovative solutions encountered with 5G Technology implementation.

Chapter 2, "Future Proofing Agile Management with 5G/6G Advancements" authored by Dankan Gowda V, Kirti Rahul Kadam, Manoj Kumar S B, Srinivas D and Prasad KDV examines how advanced communication technologies increase flexibility to facilitate quick shifts in line with market trends and improve efficiency and innovation. They also demonstrate how 5G/6G increase the connectivity, reduce the latency, and enhance the bandwidth which enable the IoT integration, the AI adoption, and real-time analytics into an adaptable platform.

Chapter 3, "Bridging the Digital Divide- Accessibility and Inclusivity Strategies in 5G/ 6G Technology" authored by Fatima Farooq and Irfan Bashir discusses methods to improve the accessibility and inclusivity of 5G and 6G technologies to reduce the digital gap. It underscores the importance of developing affordable infrastructure, cost-effective devices, and comprehensive digital literacy programs while considering ethical implications.

Chapter 4, "Evaluating the economic impact of 5G-6G implementations across diverse industry sectors" authored by Vetrivel SC, Vidhyapriya P and Mr. Arun VP evaluates the economic impact of implementing 5G and 6G networks on diverse industry sectors, including healthcare, manufacturing, transportation, agriculture, and entertainment.

Chapter 5, "Transition from 5G to 6G Communication Technologies: Workforce Evolution and Skill Development Needs" authored by Suryanarayana Alamuri, Ravi Aluvala and Ramesh Kumar Miryala explores the workforce evolution necessitated by the shift from 5G to 6G and identifies the most essential skill sets required to thrive in this new era.

Chapter 6, "5G to 6G-Empowering the Workforce Skills for Next-Gen Connectivity" authored by Anurag A. S, Swathi A, Rohith Ramesh and Johnpaul M. examines the technological transformations resulting from the shift from 5G to 6G, particularly in the context of empowering workforce skills. Utilizing existing literature, the authors identified several workforce changes and proposed strategies to adapt to this evolving landscape.

Chapter 7, "Transformative Potential of 5G and 6G Wireless Communication Technologies and IOT in Intelligent Transportation Systems and Smart Cities: IOT in Intelligent Transportation Systems and Smart Cities" authored by Santhakumar Govindasamy, Vairavel Madeshwaren, Siva Brindha G and Sumathy R examines the transformative impact of 5G, 6G, and IoT on Intelligent transportation systems (ITS) and smart cities. It delves into the fundamentals of IoT within ITS and emphasizes how advancements in 5G and emerging 6G technologies are crucial for enhancing transportation sustainability, safety, and efficiency.

Chapter 8, "Developing and Deploying a Recommendation System for Predicting Trends in Users" authored by Sathiyamoorthi V, Vidhya Gopal, Murali Kalipindi, Prakash B S and Somasekar J focuses on developing a prediction model based on machine learning technique which can help in suggesting valuable products to online buyers based on the real-world data, so that the buyers can make effective decision while purchasing products and helps businesses organization to improvise their business process and their websites.

Chapter 9, "Artificial Intelligence and Machine Learning as Pioneers in Advancing 5G/6G Network Capabilities: Evolution, Introduction, Antecedents and Consequences" authored by Qazi Saima Syed, Shakir Hussain and Irfan Bashir provides a comprehensive review of the technological shifts, highlighting AI's and machine learning's transformative potential in communications while acknowledging the challenges they present.

Chapter 10, "Exploring the Impact of Artificial Intelligence on the Tourism & Hospitality Industry A Comprehensive Analysis" authored by Swati Sharma, Deepti Bhatt, Narendra Kumar and Levent Karadağ explores the effect of Chatbots & Virtual Assistants (CVA) on Guest satisfaction (GS) from the employees working in five-star hotels through a questionnaire.

Chapter 11, "Revolutionizing Tourism and Hospitality Industry: The Impact of 5G/6G Technology" authored by Prashant Kumar Yadav, Jayaprakashnarayana G and S. Rajadurai adopted quantitative approach to assess the potential of 5G/6G technologies to enhance the tourist experience through immersive technology applications, improve operational efficiencies with IoT integrations, and boost safety protocols through advanced, real-time monitoring systems.

Chapter 12, "Assessing the Economic Impact of 5G/6G Integration on Tea Tourism Sector in Barak Valley, Assam" authored by Jayeeta Bhattacharjee and K.N. Lokesh Kumar focuses on the 5G and 6G integration on the technological advancements in the digital era. It tries to analyze the economic impacts of 5G/6G integration on the tea-estates of Barak Valley in the state of Assam leading to tourism development in the region.

Chapter 13, "Cooperative Relaying for Hyper Reliable Low Latency Communications in 6G Radio Networks and its Potential Research Directions: Cooperative Relaying for Hyper Reliable Low Latency Communications in 6G Radio Networks" authored by Rajeev Kumar, Sunandita Debnath and Ashraf Hossain addresses the cooperative relay communications in 6G radio networks for high reliability and low latency communications. In addition, this chapter provides research directions in the high-speed mobility scenarios for 6G radio system.

Chapter 14, "Smart cities practices applied in the selected city of Slovak Republic" authored by Marica Mazurek focuses on the specific cases of smart city applications in the selected Slovakian city, which were implemented to improve the competitiveness as well as to create better co-operation of the chosen city representatives (municipalities) with the city users (citizens, entrepreneurs) and the main focus will be the principle of co-creation and new forms of communication among these entities, which were influenced by new technologies and smart decisions.

This volume will inspire readers to explore the intersection of technology and management and contribute to shaping a future where communication technologies are fully integrated into agile management systems.

Chapter 1
5G Communication Technologies in Business: Applications and Real-World Cases

Swetha Suram
Mahatma Gandhi University, India

Raam Sai Bharadwaj Miryala
https://orcid.org/0009-0006-8319-8164
Sri Venkateswara Engineering College, India

Ramesh Kumar Miryala
https://orcid.org/0000-0003-2282-1558
Mahatma Gandhi University, India

ABSTRACT

Mobile Communication technology is no longer serving the needs of Smart Phone users, rather it has made it smart in meeting the needs of Enterprises. With the advent of 5G Cellular network, Enterprise Automation and transformative opportunities across the managerial functions in various sectors has been seen. This chapter gives a brief overview of evolution of 5G and highlights the usage and applications of 5G technology. Further, this chapter explores the impact of 5G Technologies through real times cases studies. By examining the deployments in diverse settings like Health Care, Manufacturing, Logistics and Transportation, this chapter highlights the unique challenges and innovative solutions encountered with 5G Technology implementation. This synthesis of real time experiences offers a crucial resource for policymakers, industry leaders, and technology managers navigating the complexities of 5G deployment in business.

DOI: 10.4018/979-8-3693-6725-4.ch001

INTRODUCTION

Mobile networks marked the beginning of widespread telecommunication by enabling basic voice calls via analog technology, which is where the advent of the first generation (1G) has begun. The first generation of cellular systems used analog radio technology that consisted of three elements: a mobile telephone (mobile radio), cell sites, and a mobile switching center (MSC) (Harte, 2002). However, unlike later generations, 1G did not support internet connectivity. Though the internet connectivity has begun the advent of 2G, the improvement in the speed of the connection, bandwidth, technological up gradations has led to 3G, 4G and 5G in the last two decades. The world is yet to experience 6G in the coming years. The Table 1 presents few of the applications of 2G, 3G, 4G, and 5G.

Table 1. Applications of Generations of mobile networks

Generation	Applications
2G	Voice call, Short Messages
3G	Video Conferencing, Mobile TV, GPS
4G	High speed applications, Wearable devices
5G	High resolution Video streaming, Remote control of vehicles, robots

This chapter gives a brief overview of the evolution of 1G to 5G and the importance of 5G and the anticipated advancements towards 6G and the importance of telecommunications advancements in shaping digital transformation. This chapter rides you through the deployment of 5G and its applications in various sectors showcasing the benefits and highlighting few cases where organizations have implemented 5G technologies to achieve operational efficiencies and improve customer experiences.

HISTORICAL PERSPECTIVES FROM 1G TO 5G

It all started with the man's innate need to communicate. Early humans used to send smoke signals and signalling through instruments like drums to communicate over long distances. Later with the invent of telegraph, it has revolutionized the communication system. The telegraph was the fastest and most efficient mode of communication in the 1800s. However, it was subsequently replaced by the telephone, fax, and internet.

Moving from the wired telephone to wireless communication was a big revolution. This technology was first called as mobile radio telephone or called as pre-Cellular later after the invent of cellular technology which was named as zero generation (0G) (Chauhan D, et. al.). Pioneer for this technology was Motorola and Bell Systems (See Figure 1). It started in 1940s. The technology used in this 0G was Push to talk mobile. It later got improved as Mobile Telephone Service (MTS), Improved Mobile Telephone Service (IMTS), and Advanced Mobile Telephone System (AMTS). The telephone of this generation are too big to be carried. They were mounted in the cars with Antennas on the car and later the briefcase version of the system has come into practice. Though it was a big change in technology, it was not used by many.

Figure 1. Motorola Car Telephone

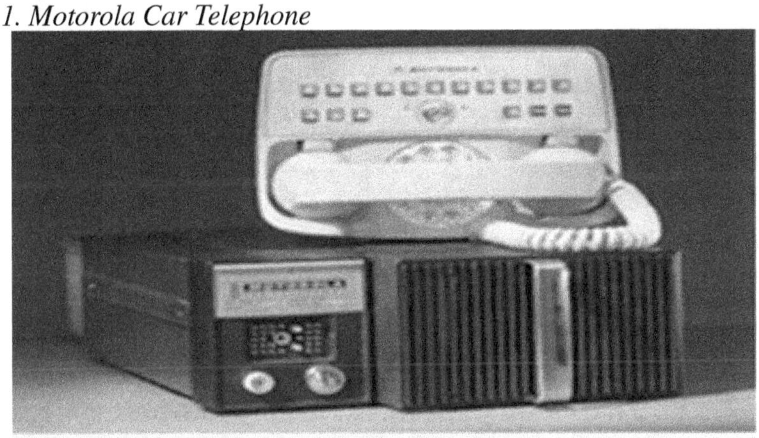

Motorola Improved Mobile Telephone Service (IMTS) car telephone, 1964.

Source:*https://www.motorolasolutions.com/en_us/about/history/explore-motorola-heritage/cell-phone-development.html*

First Generation (1G): In 1979, the first generation (1G) mobile Technology has opened doors to wireless communication. It was first commercially released by Nippon Telegraph and telephone(NTT) in Tokyo, Japan and in 5 years, Japan was first Nation to fully connected wirelessly. In 1981, Nordic Mobile Telephone (NMT) introduced 1G in Denmark, Finland, Norway, and Sweden. It also started first international roaming facility which led to people from different countries to speak wirelessly on phone. In 1983, Motorola Dyna TAC mobile phone was used in Chicago, USA and slowly the whole world began experiencing wireless communication.1G used analog signals to transfer the information which is only limited to voice calls with maximum speed of 2.4 kilobits per second(Kbps). The very low frequency of 150 Megahertz(MHZ) was used which led to a high coverage but also high latency and battery consumption. Though the people at that time used to carry the 1G mobile phone but when compared to the present generation it looks like impractical to carry in the present times. The below figure (Figure 2) shows how a 1G mobile looks.

Figure 2. 1G Mobile

Motorola DynaTAC 8000X portable cellular phone, 1984.

Source:*https://www.motorolasolutions.com/en_us/about/history/explore-motorola-heritage/cell-phone-development.html*

Second Generation(2G): Having tasted the mobile technology, people felt the need for more. Over the next 10 years, Technology also got advanced. The outcome was 2G Technology. The second generation(2G) of mobile technology was a big achievement opening the doors to the current advancement. The biggest change in 2G is using digital signals rather than analog signals. The advances in the Technology has given possibility for short message services along with the increase in quality over voice calls (Acker, 2014). The size of the mobile phones has been decreased and the phones are now more secure. And as the speed further increased it led to the introduction of Multimedia Message Service(MMS) which gave the user the capability to send media like pictures. It also supports Internet Protocol (IP) to connect to the internet.

Third Generation (3G): In less than 10 years, the third generation(3G) mobile technology was introduced, where the network speed was increased. The first commercial release of 3G was done in South Korea. 3G uses Code Division Multiple Access (CDMA) Technology (Prasad & Ojanpera,1998). The maximum speed was depended on the device mobility. For moving devices like a mobile phone in the car or on the train, the maximum speed was limited to 384 Kbps and for non-moving devices like for people on foot, the maximum speed was 2 Mbps. With the big increase in transmission speed, 3G has led to the introduction of video calling facilities, mobile internet and streaming services (Elsen et.al., 2001).

Fourth Generation(4G): With the efforts to increase the data speed, the quality and security, the new steps towards 4G Technology has been taken. The fourth generation(4G) of wireless technology is also known as LTE (long term evolution) (Martín-Sacristán et.al, 2009). Here data transmission was possible with low latency. Having supported by high speed and low latency, people experienced new features like Voice over Internet Protocol (VoIP) that allows users to make phone calls over the internet instead of a traditional phone line. With 4G, 3D television support, video conferencing with multiple active participants, HD mobile TV, gaming and cloud computing was possible (Agarwal & Agarwal, 2014).

Fifth Generation (5G): 5G is a broadband mobile communication technology with high speed, low latency and extensive connectivity. 5G's low latency and high throughput signal a shift from just serving the needs of smartphone users towards meeting the needs of enterprises. 5G provides minimal interference and fast data transmission, making it perfect for enterprise automation. 5G gives hope for the metaverse in the coming years. The usage of 5G around the world is presented in Exhibit 1.

Exhibit 1. 5G availability worldwide

Regarding availability of 5G:
As of June 2023, South Korea and Puerto Rico are leading in the availability of 5G Network. Then, it is followed by US with 31% availability rate. In European continent varies between different countries with Finland and Bulgaria leading in Europe with about 24%, where as UK is lagging with 10.1% availability. In Asian countries Singapore and Taiwan are at 30% and Malaysia at 20.5%. In India, though 5G started in October 2022, number of 5G Users exceeds 100 million which is one-sixth of the 600 million phone users of the country. In China, though initial setup was started in 2019 around 560,000 5G base stations were deployed by November 2021. In August 2023, China Broad Net has announced that it has reached nationwide 5G Coverage.

Source: https://www.thalesgroup.com/en/markets/digital-identity-andsecurity/mobile/inspired/5G

The Future: Sixth Generation (6G): 6G is the name given for sixth generation of cellular networks. These networks are envisaged for delivering truly omnipresent wireless intelligence. They are expected to be available by 2030. The vision of 6G networks is to build seamless integration of digital and physical worlds. This merged reality of the future will provide new ways of meeting and interacting with other people, new possibilities to work from anywhere and new ways to experience faraway places and cultures. (Ericsson, n.d.). The following Figure 3 shows the evolution of the network technologies.

Figure 3. Evolution of 5G/6G

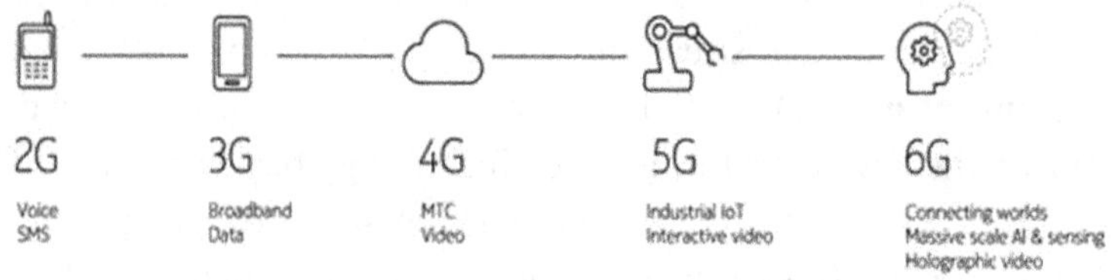

Source:*https://www.nokia.com/about-us/newsroom/articles/6g-explained/*

IMPORTANCE OF 5G AND 6G

5G network is designed to connect virtually everyone and everything together including machines, objects, and devices. 5G wireless technology is meant to deliver higher multi- gigabits per second (Gbps), with peak data speeds, ultra-low latency, more reliability, massive network capacity, increased availability, and a more uniform user experience to more users. Higher performance and improved efficiency empower new user experiences and connects new industries. 5G provides the fastest wireless internet connection to date. 5G is faster than 4G and more flexible and reliable. The impact of internet of Things(IoT) on 5G is huge as IoT is one of the most common use cases for 5G technology helping 5G expand its reach in the digital world. 5G allows for the support of a large number of static and mobile IoT devices (Palattella et.al. 2016). For example, it provides enhanced operational capabilities for Enterprises as IoT devices grow and become more prevalent. 5G is a viable solution in terms of its speed and quality of service. 5G offers higher levels of reliability, security and system performance. With the help of 5G, manufacturing plants can improve their performance. 5G allows maximum efficiency to be achieved. Even Health Care practitioners can also benefit from the implementation of 5G IoT devices. High-speed Service delivery is critical to the healthcare industry as even a few seconds can save a person's life. 5G can guarantee this speed and allow it to reach unknown Heights much earlier. If we consider smart cities, 5G can enhance traffic management by enabling efficient connectivity between various traffic IoT devices including surveillance cameras and traffic signal sensors. 5G can also improve Energy Efficiency in cities by efficiently monitoring energy consumption and identifying ways to conserve resources. 5G can efficiently collect and process all the data collected from home IoT devices to provide insight into efficient home management (Weixian Li et.al., 2018).

With high speeds, superior reliability and negligible latency, 5G will expand the mobile ecosystem into new realms. 5G impacts every industry, making safer transportation, remote healthcare, precision agriculture, digitized logistics and more. 5G technology is a real game changer for the current digital world.

As the world might experience 6G very soon, it brings new possibilities. 6G networks are expected to have a significant impact on many aspects of life, including healthcare, education, and sustainability.6G networks are expected to be much faster than 5G networks, with the ability to provide hundreds of gigabits per second (Gbps). This will make streaming more enjoyable with stable connections even when traveling to remote locations. 6G networks might enable immersive digital experiences, such as holographic communication. With 6G, Wearables can also be used to monitor vital parameters 24/7. 6G networks might enable immersive and inclusive hybrid learning for everyone. 6G networks might provide limitless connectivity everywhere, including rural land, sea, and air areas.

USAGE AND APPLICATIONS OF 5G

While 6G was mainly in the research and development stage with no operational deployments, 5G's deployment and real-world applications were rapidly growing across a variety of industries. 5G applications in management contexts can be found in Manufacturing, Logistics and Transportation, Health care management and infrastructure management. Significant improvements in a range of managerial tasks across diverse industries are anticipated with 5G technology. This section presents the possible uses and applications of 5G Technology:

- **Real-time Data Analytics:** Managers are able to make decisions and respond to situations more quickly by having instant access to and analysis of real-time data stream (Rikkeisoft, 2024). Real-time inventory tracking and customer analytics, for example, can improve supply chain management and customer experience in the retail industry (Sahay, B. S., & Ranjan, J.,2008).
- **Remote Monitoring and Control:** Industrial processes and equipment can be remotely monitored and controlled through 5G's low latency and high reliability (Cheng, Chen, Tao, & Lin, 2018). This feature allows managers to monitor production lines from a distance and take quick action when problems occur (Ansari et al., 2022).
- **Virtual reality (VR) and augmented reality (AR):** 5G enables smooth AR/VR apps that improve remote collaboration, design visualisation, and train-

ing (Vega et al., 2020). Supervisors have the ability to run virtual meetings, provide employees with simulated training, and view intricate data in real time. Wearable technologies impact the way human resources can be managed. The integration of wearable technologies carries profound implications for HR practices (Suram & Miryala, 2023)

- **Integration of the Internet of Things (IoT):** 5G enables extensive data collection and management by supporting a large number of IoT devices with low latency (Loghin et al., 2020). IoT sensors can be used by managers for energy management, asset tracking, and predictive maintenance (Teoh, Gill, & Parlikad, 2023).
- **Supply Chain Optimisation:** 5G's improved connectivity and dependability make effective supply chain management possible. Managers are able to guarantee on-time delivery, dynamically optimise logistics routes, and track shipments in real-time (Taboada & Shee, 2020).
- **Smart Cities Management:** Applications like utility monitoring, public safety, and traffic management are made possible by 5G (Liu et al., 2017). IoT sensors and AI-powered analytics can be used by Public administrators to enhance resource allocation and urban planning (Rathore, Ahmad, Paul, & Rho, 2016).
- **Enhanced Customer Engagement:** Managers can provide individualised and engaging customer experiences with faster and more dependable connectivity (Rane, 2023). 5G, for instance, makes individualised recommendations based on real-time data and augmented reality tours possible in the tourism and hospitality industry (Pradini, Awaloedin, Agustiani, & Kelly, 2024).
- **Telemedicine and healthcare management:** 5G enables remote medical diagnostics and high-definition video conferencing, which expands access to healthcare services(Liu, 2019). Healthcare administrators are able to supervise telemedicine consultations and manage remote patient monitoring systems (Rahman et al., 2024).
- **Environmental Monitoring:** 5G makes it possible for IoT devices to monitor the environment extensively, giving managers real-time access to data on water levels, air quality, and other important parameters. This data supports informed decision-making for environmental management (Su et al., 2021).
- **Security and Surveillance:** 5G's increased bandwidth and connectivity boost security systems' efficacy. In order to detect and respond to threats in real time, managers can implement sophisticated video analytics and surveillance systems (Dutta & Hammad, 2020).
- **Fleet Management:** 5G makes it possible to track cars in real time, optimise routes, keep an eye on driver behaviour, and guarantee on-time delivery in the transportation and logistics sectors (Apruzzese, Bruni, Musso, & Perboli,

2023). Predictive analytics is a tool that managers can use to lower maintenance costs and increase fleet productivity (Karimzadeh & Shoghli, 2020).

- **Energy Management:** By utilising 5G, utilities can more effectively manage grid operations, keep an eye on real-time energy consumption, and seamlessly integrate renewable energy sources. This lowers operating costs and optimises the distribution of energy (Porcu et al., 2022).
- **Smart Agriculture:** 5G enables IoT sensors to monitor crop health, conditions of the soil, and patterns of weather in real-time, facilitating precision farming. To maximise yields and resource utilisation, farmers and agricultural managers can make data-driven decisions (Tang et al., 2021). IoT concepts are also adopted in Agriculture and Breeding (Khanna A, Kaur S, 2019).
- **Financial Services:** 5G facilitates real-time market analysis, expedites and secures transactions, and improves mobile banking capabilities. Managers are better able to reduce fraud risks and offer individualised financial services (Tian et al., 2019).
- **Education and Training:** Augmented Reality (AR) /Virtual Reality (VR) apps, virtual classrooms, and remote collaboration tools provide immersive learning experiences (Meccawy, 2022). Globally, educational institutions can provide employees and students with interactive, high-quality content and training programmes.
- **Event Management:** By facilitating interactive attendee experiences, real-time crowd control, and high-quality live streaming, 5G improves event planning and management (Neuhofer, Magnus, & Celuch, 2020). AR overlays can be used by event managers to provide personalised event content and facilitate navigation enhancing customer service and satisfaction.
- **Media and Entertainment:** 5G can be deployed by content producers to stream high-quality videos, create immersive gaming environments, and host interactive live broadcasts. Media managers are able to create unique experiences for audiences and deliver content promptly (Perez et al., 2022).
- **Urban Planning:** By utilising AI algorithms and 5G-enabled IoT sensors, city planners can enhance urban infrastructure planning, analyse traffic patterns, and optimise public transportation routes (Alahi et al., 2023). This makes cities more sustainable and comfortable.
- **Insurance Services:** By utilising 5G, insurers can expedite the processing of claims, assess risks in real time, and engage with clients in a personalised manner (Chauhan, Chirputkar, & Pathak, 2022). IoT data can be used by insurance managers to improve customer satisfaction and provide usage-based policies.

REAL-TIME CASE STUDIES

To better understand the applications of 5G in the various sector of industries, the real time cases are to be studied. The cases from the domains of Health Care, Manufacturing, Logistics and Transportation are presented in the sections below:

Real-Time Case Studies in Healthcare Sector

The healthcare sector has undergone major changes with the advent and integration of 5G technology. The technology has helped the Hospital administration in terms of operational efficiency and provided enhanced services to the patients. Following are the few real-time case studies in this regard:

1. University Hospital in Düsseldorf, Germany

With the collaboration of Vodafone, University Hospital in Düsseldorf has implemented 5G technology in its operations. This integration has helped the hospital to provide variety of applications including the remote surgeries. This is possible only because of high-speed internet connectivity with low latency. Thus, the technology has allowed surgeons to perform complex operations with the assistance of remote experts. (Vodafone, 2020)

2. **Ping An Good Doctor in China**

The leading telemedicine provider from China "Ping An Good Doctor" has effectively utilised the power of 5G Internet Connectivity in its services. The 5G technology has facilitated real-time consultation to the patients, especially to the ones living in remote and underserved areas. The 5G enabled High Speed & High Quality video consultations has helped the patients to connected to health care providers with timely and accurate medical support (Xie, 2020).

3. **King's College Hospital in London, UK**

The King's College Hospital, London with the support of BT Network has explored the applications of 5G Technology in health care industry. The 5G Internet Connectivity with real-time data exchange has streamlined the operations of the hospital, improved the patient monitoring and enabled it to bring effective use of telemedicine (BT, 2021).

Real-Time Case Studies in Manufacturing Sector

The manufacturing sector also has effectively integrated 5G technology. The high-speed internet connectivity with low latency has provided the ability to the firms to connect and manage large number of devices. This has resulted in operational efficiency, improved quality in production and provided real-time monitoring and maintenance of equipment & systems. Following are the few real-time case studies in this regard:

1. Ericsson's Smart Factory in Lewisville, Texas

Ericsson in its plant at Lewisville, Texas deployed 5G applications to connect its wide range of machines and devices. These interconnected devices has created a manufacturing environment on which real-time data analytics and predictive maintenance of equipment is possible. This arrangement has significantly helped the maintenance team to anticipate and address maintenance needs before they become critical.

The firm also uses augmented reality (AR) for training and troubleshooting. The AR devices connected with 5G internet connectivity are helping the workers with real-time information and guidance thus leading to efficiency and accuracy in operations. Further, this technology improves the safety to the workers letting them to just focus on tasks. The firm also installed various machine learning applications to optimize production process and improve quality of product (Ericsson, 2021).

2. BMW Group's Plant in Dingolfing, Germany

BMW Group in its plant at Dingolfing, Germany has successfully integrated 5G technology to enhance the flexibility and efficiency in production process. The applications integrated at this plant enables the firm to seamlessly integrate autonomous guided vehicles (AGVs) for movement of materials. These AGVs will communicate with central control system in real time and ensures precise and efficient operations at the plant.

Along with AGVs, advanced robotics also contribute for the real-time quality control systems of the plant. Robots on the production line make instant decisions based on the real-time data. All these technologies enhance the quality, precision and reduces the wastages in the production. This is all possible only with the support of 5G high-speed connectivity.

Apart from operational efficiency, the applications installed in the plant also monitors the workplace safety. The real-time control system prevents accidents by ensuring that machines and vehicles operate within safe parameters. The worker's health is also monitored by using wearables (BMW, 2020).

3. Foxconn's Smart Manufacturing in Taiwan

Foxconn in its Taiwanese facilities have integrated 5G technologies to monitor real-time production lines, equipment and address the potential issues. The 5G Connectivity has allowed them to implement advanced automation including the use of robots and IoT devices to optimise the production process.

The firm has also implemented cloud-based applications for data analysis and operational insights. The voluminous data generated from production lines are analysed by these applications and the insights are drawn for improving performance and efficiency. Certain times these insights are helping the firm to take data-driven decisions.

With the technology implemented, the firm has the capability to adapt and reconfigure production lines to changing demands of the markets. This flexibility is helping the firm to combat the challenges of the markets (Foxconn, 2021).

Real-Time Case Studies in Logistics and Transportation Sector

The applications of 5G Technology in Logistics and Transportation sector industries have disrupted the operations in the sector. The technology has revolutionized Supply Chain Management, Freight Transportation and Port Operations. With 5G Technology the firms can achieve real-time positioning, tracking and monitoring of their assets improvising logistics operations. Following are the few real-time case studies in this regard:

1. Port of Rotterdam: Optimizing Port Operations with 5G

The Port of Rotterdam, one of the busiest ports of Europe has adapted 5G technology-based applications to enhance their efficiency in operations. With the aid of IoT devices the port can manage the movement of containers in real-time. The technology also helps the ports in planning of movement of containers in real-time.

Further, the technology helps in scheduling loading and unloading processes, optimizing turnaround times and improvising the capacity to handle volume of goods. Additionally, automated cranes, smart containers and real-time inventory management systems have remarkably reduced the bottlenecks in operations and improved efficiency in operations. (Clemente et al., 2023)

2. **Deutsche Bahn: Enhancing Railway Operations**

The German railway company Deutsche Bahn has significantly integrated 5G technology in its operations. This implementation is helping them to monitor real-time movement of trains, track conditions and passengers' mobility. This integration is also helping them to efficiently schedule the trains, reduce delays and improvising service reliability.

Further, the technology also aids in safety systems by providing solutions in terms of automated braking system and collision avoidance system. Thus, the technology enhances efficiency in operations, brings reliability and safety in railway operations. (Rückert, 2023)

3. **China Mobile and SAIC Motor: 5G-Connected Vehicle Solutions**

China Mobile, in collaboration with SAIC Motor has developed solutions to enhance transportation efficiency and safety. Their solution includes deployment of 5G network-based applications along major highways and urban areas. The vehicles fitted with 5G solutions will help the agencies to communicate with real-time vehicle-to-vehicle (V2V) and vehicle-to-infrastructure (V2I) data. These vehicles share data pertaining to traffic, road hazards and other critical information, allowing the entities to optimize the route dynamically and also helps in effective traffic management.

The integration of technology in connected vehicles will significantly helps in reducing congestions in traffic and enhances safety in transportation. (Ren, J., & Xia, D., 2023)

OVERALL IMPRESSION ON 5G TECHNOLOGY BASED ON THE CASE STUDIES

The 5G Technology has revolutionized the health care industry in many folds. The hospital operations have significantly improved with the technology. The patients are not only be monitored seamlessly but remote experts help can also be provided at times of need without any difficulty. Telemedicine has become reality only because of the technology helping the patients located at remote places. However, there are concerns of data security, privacy concerns and need for robust regulatory frameworks to derive more resilient and responsive healthcare systems.

The manufacturing sector is evidently benefitted by the use of 5G technologies in their operations. The advanced applications clubbed with Augmented Reality, IoT, Machine Learning, Robotics and Wearables have helped the firms to not just monitor the production lines, predict the challenges but enhances workplace safety.

Furthermore, the ability to reconfigure the production lines to the changing needs of the demand and possibility to take data driven decisions is another remarkable utility of the technology.

The Logistics and Transportation sector got significantly disrupted with the use of 5G technologies. The agencies can now exercise real-time monitoring, optimized scheduling and efficient resource allocation resulting in efficiency and effectiveness in logistics and transportation operations.

CONCLUSION

In conclusion, 5G technology, through facilitating faster data processing, improved connectivity, and cutting-edge applications like IoT, AR/VR, and real-time analytics, offers transformative opportunities across managerial functions. Across a range of industries, 5G integration can greatly increase productivity, decision-making, and customer satisfaction. 5G is predicted to have a huge impact on society and business as it develops further.

REFERENCES

Acker, A. (2014). The Short Message Service: Standards, infrastructure and innovation. *Telematics and Informatics*, 31(4), 559–568. DOI: 10.1016/j.tele.2014.01.004

Agarwal, A., & Agarwal, K. (2014). The Next Generation Mobile Wireless Cellular Networks – 4G and Beyond. *American Journal of Electrical and Electronic Engineering*, 2(3), 92–97. DOI: 10.12691/ajeee-2-3-6

Alahi, E. E., Sukkuea, A., Tina, F. W., Nag, A., Kurdthongmee, W., Suwannarat, K., & Mukhopadhyay, S. (2023). Integration of IoT-Enabled Technologies and Artificial Intelligence (AI) for Smart City Scenario: Recent Advancements and Future Trends. *Sensors (Basel)*, 23(11), 5206. DOI: 10.3390/s23115206 PMID: 37299934

Ansari, J., Christian, A., de Bruin, P., Farkas, J., Grosjean, L., Sachs, J., Torsner, J., Varga, B., Harutyunyan, D., König, N., & Schmitt, R. (2022). Performance of 5G Trials for Industrial Automation. *Electronics (Basel)*, 11(3), 412. DOI: 10.3390/electronics11030412

Apruzzese, M., Bruni, M. E., Musso, S., & Perboli, G. (2023). 5G and Companion Technologies as a Boost in New Business Models for Logistics and Supply Chain. *Sustainability (Basel)*, 15(15), 11846. DOI: 10.3390/su151511846

BMW. (2020). *BMW Group uses 5G for flexible, efficient production.* Retrieved from https://www.bmwgroup.com/en/news/general/2020/bmw-group-5g-production.html

BT. (2021). BT and King's College London to deploy 5G at King's College Hospital. Retrieved from https://www.bt.com/news/2021/may/bt-and-kings-college-london-to-deploy-5g-at-kings-college-hospital

Chauhan, D., Hadia, S. K., Desai, D., Parekh, D., & Patel, K. (2019). Evolution of mobile communication network: From 0G to 5G. *International Journal of Research and Analytical Reviews*, 6(2), 278–281. https://ijrar.com/upload_issue/ijrar_issue_20543738.pdf

Chauhan, R., Chirputkar, A., & Pathak, P. (2022, March 23). Blockchain and IoT in developing Fintech Ecosystem- An assistance to Insurance Industry. *2022 International Conference on Decision Aid Sciences and Applications (DASA)*. https://doi.org/DOI: 10.1109/DASA54658.2022.9765026

Cheng, J., Chen, W., Tao, F., & Lin, C.-L. (2018). Industrial IoT in 5G environment towards smart manufacturing. *Journal of Industrial Information Integration*, 10, 10–19. DOI: 10.1016/j.jii.2018.04.001

Clemente, D., Cabral, T., Rosa-Santos, P., & Taveira-Pinto, F. (2023). Blue seaports: The smart, sustainable and electrified ports of the future. *Smart Cities*, 6(3), 1560–1588. DOI: 10.3390/smartcities6030074

Dutta, A., & Hammad, E. (2020). 5G Security Challenges and Opportunities: A System Approach. *2020 IEEE 3rd 5G World Forum (5GWF)*.https://doi.org/DOI: 10.1109/5GWF49715.2020.9221122

Elsen, I., Hartung, F., Horn, U., Kampmann, M., & Peters, L. (2001). Streaming technology in 3G mobile communication systems. *Computer*, 34(9), 46–52. DOI: 10.1109/2.947089

Ericson. (n.d.). 6G – Preparing for the next decade of connectivity. Retrieved July 12, 2024, from https://www.ericsson.com/en/6g

Ericsson. (2021). *Ericsson smart factory: 5G-enabled manufacturing*. Retrieved from https://www.ericsson.com/en/news/2021/5/ericsson-smart-factory-5g-enabled -manufacturing

Foxconn. (2021). *Foxconn integrates 5G for smart manufacturing*. Retrieved from https://www.foxconn.com/news/2021/foxconn-integrates-5g-for-smart-manufacturing

Harte, L., Kitka, R., & Levine, R. (2002). *3G Wireless Demystified*. McGraw-Hill.

Karimzadeh, A., & Shoghli, O. (2020). Predictive Analytics for Roadway Maintenance: A Review of Current Models, Challenges, and Opportunities. *Civil Engineering Journal*, 6(3), 602–625. DOI: 10.28991/cej-2020-03091495

Khanna, A., & Kaur, S. (2019). Evolution of Internet of Things (IoT) and its significant impact in the field of Precision Agriculture. *Computers and Electronics in Agriculture*, 157, 218–231. https://doi.org/. DOI: 10.1016/j.compag.2018.12.039

Li, W., Logenthiran, T., Phan, V.-T., & Woo, W. L. (2018). Implemented IoT-Based Self-Learning Home Management System (SHMS) for Singapore. *IEEE Internet of Things Journal*, 5(3), 2212–2219. DOI: 10.1109/JIOT.2018.2828144

Liu, D. (2019). 5G and intelligence medicine—How the next generation of wireless technology will reconstruct healthcare? *Precision Clinical Medicine*, 2(4), 205–208. DOI: 10.1093/pcmedi/pbz020 PMID: 31886033

Liu, J., Wan, J., Jia, D., Zeng, B., Li, D., Hsu, C., & Chen, H. (2017). High-Efficiency Urban Traffic Management in Context-Aware Computing and 5G Communication. *IEEE Communications Magazine*, 55(1), 34–40. DOI: 10.1109/MCOM.2017.1600371CM

Loghin, D., Cai, S., Chen, G., Dinh, T. T. A., Fan, F., Lin, Q., Ng, J., Ooi, B. C., Sun, X., Ta, Q.-T., Wang, W., Xiao, X., Yang, Y., Zhang, M., & Zhang, Z. (2020). The Disruptions of 5G on Data-Driven Technologies and Applications. *IEEE Transactions on Knowledge and Data Engineering*, 32(6), 1179–1198. DOI: 10.1109/TKDE.2020.2967670

Martín-Sacristán, D., Monserrat, J. F., Peñuelas, J., Calabuig, D., Garrigas, S., & Cardona, N. (2009). On the Way towards Fourth-Generation Mobile: 3GPP LTE and LTE-Advanced. EURASIP Journal on Wireless Communications and Networking, 2009(1). DOI: 10.1155/2009/354089

Meccawy, M. (2022). Creating an Immersive XR Learning Experience: A Roadmap for Educators. *Electronics (Basel)*, 11(21), 3547. DOI: 10.3390/electronics11213547

Motorola Solutions. (n.d.). Cell phone development. Motorola Solutions. https://www.motorolasolutions.com/en_us/about/history/explore-motorola-heritage/cell-phone-development.html

Neuhofer, B., Magnus, B., & Celuch, K. (2020). The impact of artificial intelligence on event experiences: A scenario technique approach. *Event Management*, 31(3), 601–617. DOI: 10.1007/s12525-020-00433-4 PMID: 38624486

Palattella, M. R., Döhler, M., Ruta, M., Rizzo, G., Torsner, J., Engel, T., & Ladid, L. (2016). Internet of Things in the 5G Era: Enablers, Architecture, and Business Models. *IEEE Journal on Selected Areas in Communications*, 34(3), 510–527. DOI: 10.1109/JSAC.2016.2525418

Pérez, P., Corregidor, D., Garrido, E., Benito, I., González-Sosa, E., Cabrera, J., Berjón, D., Díaz, C., Morán, F., García, N., Igual, J., & Ruiz, J. (2022). Live Free-Viewpoint Video in Immersive Media Production Over 5G Networks. *IEEE Transactions on Broadcasting*, 68(2), 439–450. DOI: 10.1109/TBC.2022.3154612

Porcu, D., Castro, S. M., Otura, B., Encinar, P., Chochliouros, I. P., Ciornei, I., Hadjidemetriou, L., Ellinas, G., Santiago, A. R., Grigoriou, E., Antonopoulos, A., Cadenelli, N., di Pietro, N., Betzler, A., Prieto, I., Battista, F., Brodimas, D., Rumenova, R., & Bachoumis, A. (2022). Demonstration of 5G Solutions for Smart Energy Grids of the Future: A Perspective of the Smart5Grid Project. *Energies*, 15(3), 839–839. DOI: 10.3390/en15030839

Pradini, G., Awaloedin, D. T., Agustiani, F., & Kelly, A. (2024). 5G Technology Transformation In Tourism. *International Journal of Economics, Management, Business, and Social Science*, 4(1), 523–536. DOI: 10.59889/ijembis.v4i1.376

Prasad, R., & Ojanpera, T. (1998). An overview of CDMA evolution toward wideband CDMA. *IEEE Communications Surveys and Tutorials*, 1(1), 2–29. DOI: 10.1109/COMST.1998.5340404

Rahman, A., Wadud, M., Islam, M. J., Kundu, D., Bhuiyan, T. M. A.-U.-H., Muhammad, G., & Ali, Z. (2024). Internet of medical things and blockchain-enabled patient-centric agent through SDN for remote patient monitoring in 5G network. *Scientific Reports*, 14(1), 5297. Advance online publication. DOI: 10.1038/s41598-024-55662-w PMID: 38438526

Rane, N. L. (2023). Enhancing Customer Loyalty through Artificial Intelligence (AI), Internet of Things (IoT), and Big Data Technologies: Improving Customer Satisfaction, Engagement, Relationship, and Experience. *Social Science Research Network*.https://doi.org/DOI: 10.2139/ssrn.4616051

Rantcell. (n.d.). Comparison of 2G, 3G, 4G, 5G. Rantcell. Retrieved July 12, 2024, from https://rantcell.com/comparison-of-2g-3g-4g-5g.html

Rathore, M. M., Ahmad, A., Paul, A., & Rho, S. (2016). Urban planning and building smart cities based on the Internet of Things using Big Data analytics. *Computer Networks*, 101, 63–80. DOI: 10.1016/j.comnet.2015.12.023

Ren, J., & Xia, D. (2023). Introduction to 5G C-V2X. In *Autonomous driving algorithms and Its IC Design* (pp. 283–294). Springer Nature Singapore. DOI: 10.1007/978-981-99-2897-2_13

Rikkeisoft. (2024, April 4). How real-time data analysis empowers your business. Retrieved July 12, 2024, from https://rikkeisoft.com/blog/how-real-time-data-analysis-empowers-your-business/

Rückert, R. (2023). *Disruptions in public transport: from dispatching assistance to improved robustness* (Doctoral dissertation, Dissertation, Halle (Saale), Martin-Luther-Universität Halle-Wittenberg, 2023).

Sahay, B. S., & Ranjan, J. (2008). Real time business intelligence in supply chain analytics. *Information Management & Computer Security*, 16(1), 28–48. DOI: 10.1108/09685220810862733

Su, X., Liu, X., Motlagh, N. H., Cao, J., Su, P., Pellikka, P., Liu, Y., Petäjä, T., Kulmala, M., Hui, P., & Tarkoma, S. (2021). Intelligent and Scalable Air Quality Monitoring With 5G Edge. *IEEE Internet Computing*, 25(2), 35–44. DOI: 10.1109/MIC.2021.3059189

Suram, S., & Miryala, R. K. (2023). Wearable technologies in human resource management: A comprehensive analysis of adoption and its impact. ANU Journal of Commerce and Management, 15(2), 89-96. https://doi.org/DOI: 10.6084/m9.figshare.25484752

Taboada, I., & Shee, H. (2020). Understanding 5G technology for future supply chain management. *International Journal of Logistics*, 24(4), 392–406. DOI: 10.1080/13675567.2020.1762850

Tang, Y., Dananjayan, S., Hou, C., Guo, Q., Luo, S., & He, Y. (2021). A survey on the 5G network and its impact on agriculture: Challenges and opportunities. *Computers and Electronics in Agriculture*, 180, 105895. DOI: 10.1016/j.compag.2020.105895

Teoh, Y. K., Gill, S. S., & Parlikad, A. K. (2023). IoT and Fog-Computing-Based Predictive Maintenance Model for Effective Asset Management in Industry 4.0 Using Machine Learning. *IEEE Internet of Things Journal*, 10(3), 2087–2094. DOI: 10.1109/JIOT.2021.3050441

Tian, M.-W., Wang, L., Yan, S., Tian, X., Liu, Z., & Rodrigues, J. J. P. C. (2019). Research on Financial Technology Innovation and Application Based on 5G Network. *IEEE Access : Practical Innovations, Open Solutions*, 7, 138614–138623. DOI: 10.1109/ACCESS.2019.2936860

Vega, M. T., Liaskos, C., Abadal, S., Papapetrou, E., Jain, A., Mouhouche, B., Kalem, G., Ergüt, S., Mach, M., Sabol, T., Cabellos-Aparicio, A., Grimm, C., De Turck, F., & Famaey, J. (2020). Immersive Interconnected Virtual and Augmented Reality: A 5G and IoT Perspective. *Journal of Network and Systems Management*, 28(4), 796–826. DOI: 10.1007/s10922-020-09545-w

Vodafone. (2020). Vodafone and University Hospital Düsseldorf: First 5G remote surgery live. Retrieved from https://www.vodafone.com/news/press-release/vodafone-and-university-hospital-dusseldorf-first-5g-remote-surgery-live

Xie, D. (2020). Ping An Good Doctor launches first 5G telemedicine service in China. Retrieved from https://www.pingan.com/news/ping-an-good-doctor-launches-first-5g-telemedicine-service-in-china

Chapter 2

Future Proofing Agile Management With 5G/6G Advancements

Dankan Gowda V.
https://orcid.org/0000-0003-0724-0333
Department of Electronics and Communication Engineering, BMS Institute of Technology and Management, Bangalore, India

Kirti Rahul Kadam
Management Department, Bharati Vidyapeeth University, India

Manoj Kumar S. B.
https://orcid.org/0000-0002-3037-2564
Department of ECE, BGS Institute of Technology, Adichunchanagiri University, India

Srinivas D.
https://orcid.org/0000-0002-6446-6649
School of Business, SR University, Warangal, India

K. D. V. Prasad
https://orcid.org/0000-0001-9921-476X
Symbiosis Institute of Business Management, Symbiosis International University, Pune, India

ABSTRACT

The continuous evolvement of new formative 5G and evolution towards 6G are pertinent to changing the dynamics of the development of the principles of the agile management and open new horizons of development for enterprises. This chapter

DOI: 10.4018/979-8-3693-6725-4.ch002

examines how these advanced communication technologies increase flexibility to facilitate quick shifts in line with market trends and improve efficiency and innovation. We demonstrate how 5G/6G increase the connectivity, reduce the latency, and enhance the bandwidth which enables the IoT integration, the AI adoption, and real-time analytics into an adaptable platform. Through the discussion of various cases and industry examples, the reader is offered practical guidelines on the use of the discussed technologies for gaining a competitive edge.

INTRODUCTION

Overview of Agile Management

Agile Management is a term used to refer to a set of principles and practices that are used in the management of organizations with an aim of increasing efficiency and effectiveness as well as increasing responsiveness and flexibility. originated in software engineering, but more and more organizations apply them in different fields to improve projects and work processes' management. Kanban management is flexible and works in progress, has short cycles of developing and continuous feedback and an enhanced focus on teamwork. It means that the primary objectives of using agile are to provide the maximum value to the customer, to welcome the change, to support the teams' self-organization, and to avoid burnout.

The concept of agility is especially significant in today's business environment since the technological transformations and market conditions change rather quickly (Al-Fuqaha, A., Guizani, M., Mohammadi, M., Aledhari, M., and Ayyash, M. 2015).. The environmental turbulence with regards to the tastes and desires of consumers, rules and regulations, and market players' characteristics makes adaptiveness a critical competitive advantage. Agile management also sustains that flexibility is achieved by dividing the large tasks into smaller parts which can be done in short time periods referred to as sprints (Alhassan, A. H., Rahman, T. A., Abd-Alhameed, R. A., et al. 2021). This approach helps teams to constantly redefine and reorient themselves according to the feedback and the changing circumstances thus reducing the chances of having large p_projects fail and in equal measure ensuring that the final product meets or exceeds the customers expectations.

Also, Agile management supports the concept of the continuous improvement of the organizational culture. It is common that coaches apply the principle of evaluation on a frequent basis by such procedures in the team, for instance, retrospectives where teams note highs, lows, and what specifically they should obviate to improve organizational dynamics (Akyildiz, I. F., Kak, A., and Nie, S. 2020). This iterative learning process keeps organizations relevant and thus will ensure that the process and products are constantly improving. Also, it has been established that agile management promotes the notion of the team made up of individuals from different functional areas that can share knowledge and experience towards overcoming the challenges and coming up with new ideas.

Need for Future-Proofing

As the global environment evolves from being relatively stable towards the VUCA (volatile, uncertain, complex, ambiguous) type, the necessity of planning the prospects of the agile management as a potential development type grows. Future proofing is a set of measures and procedures aimed at the preparation and adaptation of the various organizations for the changes in the future environment over the course of doing business (Basar, E., Di Renzo, M., de Rosny, J., Debbah, M., Alouini, M.-S., and Zhang, R. 2019). To the agile methods, future proofing is translation of the basic agility principles to new levels of practices and tooling up with progressive technologies.

Technological advancement remains one of the chief stimuli for the preparation of management for the future agile development. New technologies like Artificial Intelligence, IoT, big data and analytics are changing business disciplines and generating new approaches. If these technologies cannot be fully adopted in agile frameworks of respective organizations, they are often left behind where competitors are more advanced in the use of such technologies (Bennis, M., Debbah, M., and Poor, V. 2018). Future-proofing makes it possible to consider management practices that must be adopted today as future-oriented to support organizational needs resulting from the future technological environment.

The other imperative characteristic of future-proofing is the capacity to preempt, counter alterative currents in the marketplace. Consumers' behavior, rules and regulations, and competitors' activities are unpredictable and flexible, and an organization need to adapt quickly (Choi, J., Lee, B., and Jindal, N. 2021). The management of such processes is most effectively done as agile processes since these propose the flexibility and continuous progression. Still, to develop truly future-proof agile practices, it is also necessary to apply causal, predictive tools and scenario modeling, which implemented organization's ability to forecast future changes and adjust the necessary practices accordingly.

Moreover, the future-proofing work on the organisational culture that is based on innovation and learning. Employees and their companies also need to create the conditions in which it is safe to create awareness of new trends and technologies and to work out changes to practice (Cotton, S. L., Bultitude, R. J. C., and Scanlon, W. G. 2011). This culture of innovation is critical for keeping adaptability on the foreground and guaranteeing to the organization to take advantage of new opportunities whenever they appear. By advocating an Agile ready approach within their frameworks, organisations are able to lock-in their market position and fuel growth.

Role of 5G/6G Technologies

The continued extension in the 5G ability and the progressively developing 6G technologies has a tremendous impact on the management flexibility. Technology in the fifth generation has LTE connectivity that offers higher speed in data transfer, low latency and high number of connections hence is promising to revolutionize industries by providing reliable means of communication. With the increase of 5G networks, the infrastructure for real-time processing of data and more efficient interface of IoT and remote collaboration are major elements of agility.

5G's advanced data transfer speeds enable resilient and adaptive teams to collect and share big data within a short span of time thus enhancing the decision-making process and removing workflow congestion. The 5G networks have near-zero latency that makes it possible to receive and transmit feedback within the shortest time possible (Ghosh, A., Ratasuk, R., Mondal, B., Mangalvedhe, N., and Thomas, T. 2010). This capability is quite useful now days due to increasing usage of cross-abilities or decentralized structures of work-teams which means agile working-teams that require ongoing communication and few iterations with customers. Further, enhanced richness of band width in 5G greatly contributed for incorporating IoT devices that will enable organisation to gather information from a variety of sources hence improving on surveillance and monitoring of processes in real time.

With regards to the rest of the benefits, one can expect the same to be scaled even beyond by 6G technology in the future. Although still being in the early stages of the process, 6G may provide greater transmission rates, even smaller delays, and considerably better connectivity options (Govardhana Reddy, G., and Raghavendra, K. 2022). With these enhancements, there will be a possibility of the advanced application of artificial intelligence and the increased usage of machine learning, the more extensive coverage of IoT, and more automation than is ever possible. For agile management, 6G will offer the enablers to achieve highly dynamic environments to work and adapt feedback loops consequently which help organizations save time for planning, implementing, and delivering the innovations on the market.

However, the part of 5G and even of 6G in improving agile management is not limited to technical points of view. They also facilitate the requisite cultural and organisational changes for agility (Han, S., I, C.-L., Xu, Z., and Rowell, C. 2015). Its accessibility to empower a global and remote workspace means that everyone, regardless of their location, is considered equally valuable, thus making the usage of 5G and soon 6G networks an instrumental factor for organizations' diversity and inclusiveness (Liu, Y., Liu, X., Mu, X., et al. 2021). Such function to link and share information across geography is imperative to establish durable and flexible cross geographical teams that can manage the current business environment dynamics.

To sum up, the use of the 5G and 6G technologies and changes in the organization's management procedures based on the principles of task agility are the foundation for building organizational resilience for the future. These technologies give the needed connectivity, speed and data in order to support, adapt and innovate at a pace that is faster due to increasing volatility. In this way, through the use of 5G, and getting ready to go 6G, organisations can ensure the continuous build-up of their agile management frameworks and practices as the future unfolds.

ADVANCEMENTS IN 5G AND 6G TECHNOLOGIES

5G Technology Overview

5G is an innovative telecommunications system, the fifth generation of mobile networks that allows increasing the connection speed and ensuring the clamor for the higher data rates. However, 5G is much more significant than its predecessors in terms of performance increase such as data transfer rates of 5G are much higher than that of 4G, latency is lower and it can support a large number of connected devices (Park, J., Park, S., and Ji, H. 2020). Another characteristic of 5G that has been widely discussed is one that relates to the speeds of transfer, which it is expected could go as high as 10 GBPS, or even 100 times the speed of 4G networks. This improvement enables the easy transfer of extensive files as well as capability to handle data-hungry applications including, HD video streaming, virtual and augmented realities.

Another important parameter of the 5G is low latency specified to be as low as 1 ms of response time. The time it takes to communicate the signals is almost negligible and is crucial in systems that need fast feedback and control actions by an operator in areas such as vehicle control with no drivers involved, robotic surgeries, and auto manufacturing plants (Peng, B., Li, Z., and Li, S. 2021). The low latency of 5G improves customers' satisfaction replacing the idea of waiting and keeps consumer transition between different services seamless.

The availability of service can also be enhanced by 5G, which also features better network dependability and capability. From generational technologies like massive MIMO and beamformig, 5G networks provide for higher density of connected users. This capability is significant in making way for the Internet of Things IoT since billions of devices produce streams of data continuously (Popovski, P., Trillingsgaard, K. F., Simeone, O., and Durisi, G. 2018). However, with strong 5G structure, industries like smart cities, intelligent transportation system, and superior manufacturing system that need an unbroken connection of communication and data transfer.

Also, 5G networks are very dynamic and adaptive in that they are very easy to deploy for a various number of purposes. Operators can utilise a feature in 5G known as network slicing, which enable them to build multiple virtual networks for different applications or users. Such customization provides for the best results and resource utilization for various services which range from emergency response channels to amusement facilities for consumers (Qin, Y., Gacanin, H., Saad, W., et al. 2021). The current state of 5G implementation is diverse with many countries have deployed the technology in urban centres and are now extending it to Suburban and rural areas (Rappaport, T. S., Xing, Y., MacCartney, G. R., et al. 2017). The ever-evolving and enhancing of 5G networks future the potential of opening up new doors of opportunity for creativity and connectivity.

6G Technology Prospects

Although now 5G is under the process of deployment all over the world, there is a research going on for the development of 6G technology to support the future communication scenario. 6G, which is considered to be the sixth generation of systems of mobile networks, is believed to be constructed on the progress of 5G, which will bring capabilities that are unparalleled and will help to speed up the interactions between human beings and technologies (Reddy, A. M., and Tanguturi, R. 2022). Future trends of 6G study show that it entails the transmission of very high rates of data and speeds of up to 1 Tbps and latency of less than microseconds.

Perhaps one of the most exciting areas that are possible with 6G is AR/VR/MR where 6G will offer new possibilities for the advancement of the related applications. These technologies will leverage the usa and low latency connectivity that is expected with the 6G to provide real-time interactions within the virtual world (Saad, A., Hossain, M. S., Muhammad, G., Alhamid, H., Guizani, M., and Al-Dhelaan, A. 2020). It will change industries including education, healthcare and entertainment, and make users more meaningfully and actively engaged with the presented content.

6G is also expectated to improve the implementation of artificial intelligence and machine learning in the network platform making networks intelligent and self-sustainable networks. The proactive character of this approach based on artificial intelligence will bring to enhancement of the requirements of effective interaction networks with the condition and the demand of users in the real time (Shekhar, R., and Ashreetha, B. 2023). Furthermore, new application areas and use-cases that can be supported with the help of advanced sensing technologies in 6G include environmental monitoring, public safety, and smart agriculture because of high resolution data and information.

The future trends of 6G also include exploration of the terahertz (THz) frequency bands that are characterized by large bandwidths and data rates than present spectrum licenses (Shruthi, M., and Naveen K., B. 2020). This expansion into the higher frequencies will cater for the sophisticated applications which require large amounts of data and more connected devices. Furthermore, 6G is likely to address network sustainability and energy consumption, with an intention of including green aspects in all operations of the network.

In other cases, advanced technologies such as 6G technology will conquer the imperfections that current communication systems have, and open a new horizon for service and application. These notions indicate that cooperation between academia, industry, and governmental institutions will be paramount for defining the goals, demands, and directions for development of the new generation of the wireless networks – 6G.

Key Improvements

The development that has been introduced by the 5G and the further prospective enhancements by 6G are the great enhancement in the aspects of connectivity, latency and bandwidth which are indispensable for supporting the new generation of digital services and applications. A clear improvement that may easily be pointed out is the great enhancements in data transfer rates. 5G networks can reach up to 10 Gbps and 6G is going to break that barrier and go up to 1 Tbps. This increase in the speed helps in the exchange of massive amounts of data within a short period, a requirement for applications like high definition video streaming, cloud gaming, and remote working.

Another area of a large enhancement is latency, or the time required for data to transfer from one particular location to another. The low latency is as low as 1ms enabling real-time communication and interaction through 5G. This ultra-low latency is essential for any application that relies on real-time response, the level of which is not possible with the current networks as exemplified by self-driving cars, tele-surgery, and smart industries. The next generation will be 6G which will reduce latency to the microsecond level paving way for more detailed interaction in digitally related things.

Also the ability of a network to carry traffic or data otherwise referred to as bandwidth has also received major boost with the 5G and with 6G further improvements will also be realized. Bandwidth capability also has increased over the years and more devices can now connect concurrently without any drop in performance (Tataria, H., Shafi, M., Molisch, A. F., et al. 2021). This is especially so in the IoT context whereby a large number of devices in the billions continually originate and transfer data. The new features like massive MIMO and beamforming in 5G make efficient use of the bands required and guarantees the connectivity. In terms of frequency it is expected that 6G will operate in even higher frequencies bands, with potential to use terahertz, to meet the demand of more bandwidth for the upcoming data hungry applications.

Besides those improvements of the communication technology, 5G and 6G have also introduced new changes in network architecture and management. Another key concept closely associated with 5G is network slicing which lets establish several virtual networks based on the applications or users' needs (Yang, G., Yang, Y., Li, K., et al. 2020). As mentioned earlier, this flexibility is required to allow for a wide range of applications from core infrastructure to various consumer services. This capability is set to improve in 6G with artificial intelligence and Machine Learning to improve the management and well as optimisation of the network in real time.

In general, the envisaged enhancements in terms of connectivity, latency, and bandwidth associated with 5G and 6G will open up opportunities for various new and advanced applications in various industries thus kicking starting a number of technological revolutions. Such improvements will help to create a basis for a better connected, smooth, and timely oriented digital environment for present and future businesses and consumers.

ENHANCING AGILITY WITH 5G/6G

Real-time Connectivity

Real time connection, due to the low latency and high speeds in 5G and 6G connections also improves the agility processes greatly. One core success factor related to agile development methodologies refers to fast feedback loops with users as well as the C.I /C.D process undertaken iteratively. 5G latency can be as low as 1ms, while 6G standard is expected to provide microsecond latency, and such SA allows for real-time communication between team members, devices, and systems (You, X., Wang, C.-X., Huang, J., et al. 2021). It is also an affordance that is very important to agile teams since it enables teams to give and receive feedback instantaneously and to change their strategies when necessary.

In software development for instance, real time connectivity makes it possible to integrate and test changes in code. It is highly possible for developers to simply push updates to the cloud, get feedbacks from continuous integration systems, and then fix issues immediately if there are any. This capability does not only compress time but also increase the quality of the final output of a development cycle since more test can be performed in a shorter span of time (Zhang, R., Cui, Y., Cui, Y., et al. 2021). In addition and more importantly, real time communication that includes video conferencing, instant messaging, collaborating platform are critical to Agile teams especially when the teams are distributed. These tools run seamlessly in high-speed and low latency environments of 5G and 6G networks making remote work and decision making possible.

Figure 1. Latency Reduction Over Time

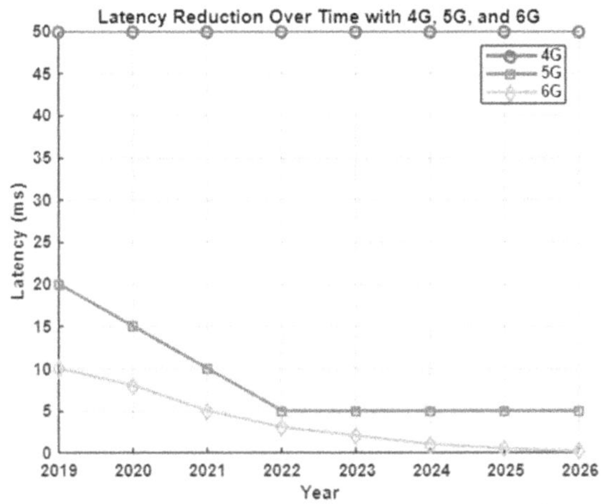

The line graph in figure 1 illustrates the decline in latency as 5G evolved and the estimated potential of 6G. The horizontal axis reflects years from 2019 up to 2026 while the vertical axis displays latency in milliseconds (ms). The graph used by Akintoye has three lines labelled as 4G, 5G and 6G technologies. The 4G line stability is hold at 50 ms of latency line while the 5G latency exhibits a progressive decline from 20 ms in 2019 to 5 ms in 2022 with the subsequent stabilization. The actual 6G line seems to have a further drop to as low as 0% infection rate as projected on the below graph. 2 ms by 2026. Here, we have a depiction of the growth in communication technology and how the area of latency is of essence in real time connectivity in agile management.

As with internal processes, real-time connectivity also improves customers' experiences with a company. One of the things about agile frameworks are that it focuses a lot on customer feedback and further improvement, and being able to communicate with users in real-time is a true asset. The 5G and 6G networks make it possible for companies to capture data from customers' engagement, process the information, and plug it back into a product cycle in a short span (Zhang, Z., Xiao, Y., Ma, Z., et al. 2019). The above responsiveness to customer needs not only enhance the satisfaction of the customers, but also acts as an outstanding basis of competition and innovation. For industries such as retail, health care, finance where information plays a crucial role the concept of real time connectivity proves to be even more advantageous.

Also, IoT devices in relation to the concept of agility are seamlessly implemented in real-time by the use of 5G and 6G. Connected devices have the capability to churn out large volumes of data that can be used in monitoring and even enhancing several processes. The low latency and HSS can then facilitate the real time transmission of data from IoT sensors and enable the subsequent real time analysis to support the operational dynamics of agile teams. For example, in manufacturing, IoT devices used in operations, can provide instantaneous feedback of the manufacturing performance and this informs the teams to rectify problems as they occur thus boosting the operational flexibility and cutting out losses.

Bandwidth and Data Handling

The key aspect of having higher bandwidth from 5G, and the future improvements with the 6G technology is the role they have in overlaying the data-intensive applications which are a significant feature of contemporary agile development. Bandwidth is a term that relates to the amount of data that can be transmitted at a go through a network and high bandwidth data can handle large quantity of data without compromising on time. This capability is crucial to the kinds of teams involved in real-time analytics, the use of machine-learning algorithms, big data and the like.

As for the capabilities of the bandwidth, they have increased with the advancing of the generations and, by now, it is possible to reach the 10 Gbps. This degree of increase ensures that HD video streaming, distribution of big data sets, and the use of analytically-enabled, real-time applications are possible. It is thus important for agile teams to capture these capabilities in order to gain greater understanding and awareness of their operations, customer and market environment. For instance, marketing departments would use real-time analytics to monitor the outcomes of current campaigns, modify strategies with data from the campaigns, and design better approaches to engage the customers. Likewise, the product development teams have an opportunity to monitor the user feedback and usage patterns in real-time for the purpose of continuously refining the products and specific features.

Figure 2. Bandwidth Capabilities Comparison

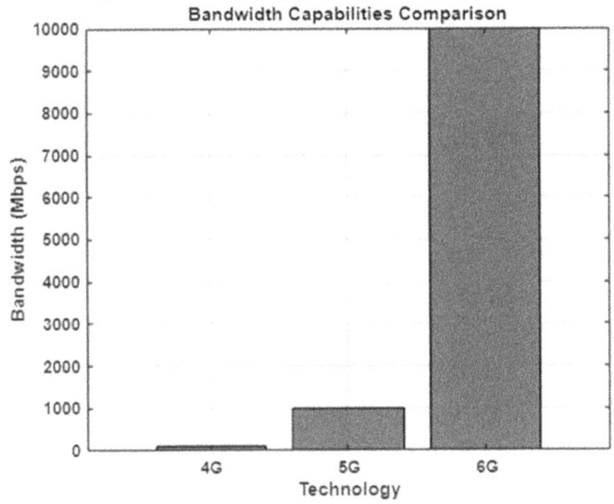

This is further depicted in figure 2 below, which shows a bar graph of the bandwidth capacity of 4G, 5G and 6G technologies. The catheter is also referred to as the x-axis; the technology type the y-axis has bandwidth in megabits per second (Mbps). The bars illustrate that there was enhancement of bandwidth from 100 Mbps of 4G to 1000 Mbps of 5G, and 10,000 Mbps of 6G. This figure also signifies the great improvement that 5G, and particularly 6G, consists of in the ability to process large amounts of data, foster data-intensive applications and improve agile processes.

The promised increase in the bandwidth in the 6G networks, which may range from 500 Gbps to 1 Tbps, will reinforce these possibilities and enable more and more data-hungry services. This will be especially helpful for less-mature technologies like VR & AR because these technologies demand a lot of bandwidth to provide adequate immersion. For the more dynamic VR/AR teams that work in an agile environment, this will create a better bandwidth boost by rendering the complicated graphics and interactive environment in real time through making the necessary models, thus allowing for user trials. Moreover, with the new 6G networks having even higher bandwidth, there will be the realisation of enhanced AI and machine learning models that demand a lot of computations and transmission rates.

Another important area that contributes to the aspects of resilience in agile processes is handling of data. This is due to the fact that IT agile teams need to work through vast amounts of data and being able to make proper analysis and subsequent decisions. 5G and 6G network proprieties, in cooperation with Edge computing and cloud solutions let process it effectively (Shruthi, M., and Naveen K., B. 2020). The idea of edge computing which take the processing of data near to the place of origin helps in reducing latency and bandwidth utilization for analytics.

This is especially useful for IoT devices since data logged by sensors and connected devices can be analyzed at the client's premises thereby avoiding unnecessary data transmission to servers.

Reliability and Ubiquity

Maintaining consistent and reliable connections is a basic prerequisite for agile operations and with the improvements in 5G and 6G technology, this is achieved in an all-round manner. Dependability in networks means that one is capable of achieving guaranteed communication and data transfer that is essential to support the operations and fast-growing integrated software development teams while reducing the level of potential delays in project completion. In contrast, there is also a strict meaning of coverage, which is the availability of Network coverage so that self-organizing teams can always be reached.

5G networks are characterized with reliability that is based on characteristics like network slicing, under which several virtual networks can be built on one physical network. An individual slice can be optimized for different needs so as to achieve the optimum performance and robustness for the intended use. For instance, the network slice that is used for the critical communication such as Emergency services or Industrial automation, the ULLR is an essential resource that can be adapted to guarantee that specific verticals applications which require this high resources utilization are met. This level of reliability is good for the agile teams that are working on the mission critical projects since any delay in the project has major implications.

A further extension of High speed connectivity is also made possible by the massive implementation of 5G. 5G has invested in network coverage for both urban, suburban and rural areas meaning that dynamic teams are always guaranteed a network connection. It can be even more crucial for distributed workers which are increasingly present in the contemporary working environment. Due to fast and steady 5G connection everyone can work remotely and have access to the same and fast connection, so proximity doesn't impact the possibility to contribute to the agile processes.

As we move further in the future, another generation, known as 6G, is being set to provide even more secure and almost context-aware connectivity. Towards the creation of the 6G network, the goal is to achieve almost complete reliability and coverage and to use such technologies as satellite communications and intelligent networks. In a satellite-based 6G networks, areas that are so hard to reach and have very slow internet connections will be hooked onto the network, making it possible for the agile teams to communicate cross borders. Furthermore, self-automated intelligent network management systems using Artificial Intelligence and machine learning algorithms will proactively adjust the network and foresee or prevent an disruption that may affect users.

For agile operations, this means 5G and 6G networks are more reliable and more accessible which results in better efficiency, flexibility and robustness. Users can always count on a reliable communications and data transfer some of which can be seen as: Continual integration and deployment practices can also be reliably achieved on this framework since constant updates and iterations are necessary to sustain the level of flexibility. In addition, the availability of fast connection means that even distributed teams can work effectively, thus leading to a more diverse workforce.

In conclusion, one can state that improvements in the 5G and soon the 6G technology fascinatingly contribute to the augmentation of agile processes through the support for real-time connection and increased bandwidth together with a reliable and global network coverage. These enhancements facilitate economically viable processes of agile teams, quick and effective change, and data analysis and the exploitation of new technologies. As more firms incorporate 5G technology and look toward 6G, it will be easier for firms to predict and compete in today's increasingly complicated digital environment.

INTEGRATION OF IOT, AI, AND REAL-TIME ANALYTICS

IoT Integration

The process of integrating IoT devices into business processes is rapidly changing in organizations' ways of functioning, and the achievements in the 5G and 6G fields are critical in the organization of IoT device connectivity and control. IoT is the connection of physical objects to the Internet so that they can operate and gather data that may be used to get meaningful information. The higher speed of network-5G and expected network-6G, low latency, and large spectrum bandwidth improve the functionality and operation of IoT systems.

With an ability to support up to one million devices per sq. km., 5G technology guarantees that IoT networks can grow without the prospects of a bottleneck. This high device density is important in smart city infrastructure, industrialization and especially in areas which need many devices to be responsive at once such as the healthcare industry. Due to the nature of 5G to allow much response time of milliseconds, it is possible to have real-time interaction between IoT gadgets and central hubs. For example, in smart production, sensors on equipments are able to detect flaws or a failure in the machinery and interface with the control system in real-time, critical to avoiding potential delays in manufacturing lines. In figure 3, we find pie chart representing IoT devices connected over 4G 5G & 6G technologies. The chart can be, therefore, divided into three segments to represent the proportion of IoT devices connected through the mentioned technologies. As for the connectivity schematics it supposes that 30% of devices are connected to 4G, 50% connected to 5G, and only 20% of devices connected to 6G. This visualization highlights the manner in which organizations are using 5G and 6G technologies to manage more IoT devices and improve organizational agility in the networks they run.

Figure 3. IoT Device Connectivity Distribution

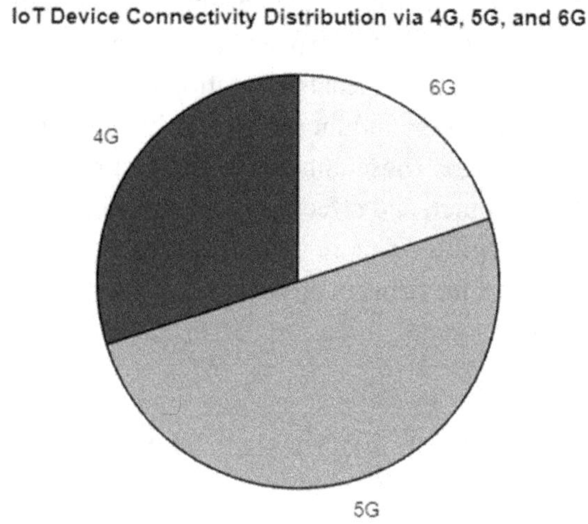

Further, with the improved bandwidth of the 5G network, the flow of enormous data produced by the IoT devices is enabled. This capability is key especially for business such as self-driving cars which require real-time inputs from multiple sensors to operate optimally. Real-time high-definition video and environmental data along with telemetry enables users to have a more precise control system. 5G networks also guarantee dependable and stable connections for IoT devices which are important for applications with no interruption such as the use of IoT in the healthcare, remote patient monitoring.

AI and Machine Learning

AI and ML's incorporation into modern agile frameworks is what changed the way companies work, respond, work faster and innovatively. AI and ML are very important for predictive analysis of data and for decision making the improvement of 5G and 6G technologies help a lot. The integration and operations of the AI/

ML models are easily manageable in the actual working sphere due to high-speed connection, low latency, and vast bandwidth of these networks.

Figure 4. Data Speed vs. Decision Accuracy

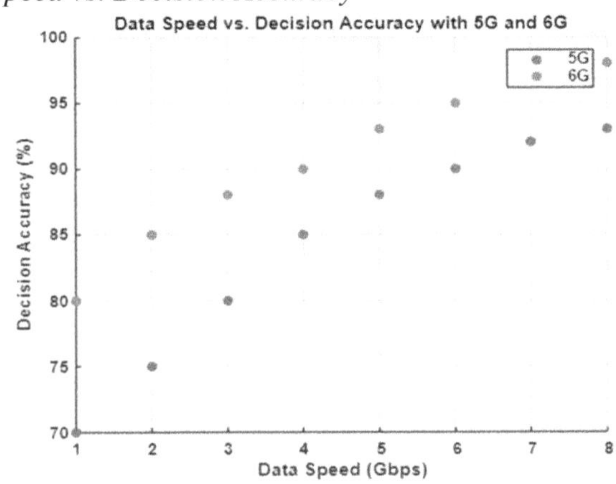

The analysis shown in Figure.4. is a scatter plot that analyzes the impact of data processing speed on the decision's accuracy while using the 5G and 6G technologies. Measures: While the x-axis plots data speed which was at a range of 1Gbps to 8Gbps, the y-axis depicts decision accuracy which was in percentage (%). Two sets of data points are plotted: It goes without saying that some of them are aimed at 5G implementation and others are meant for the 6G networks. According to the plot as the data speed rises, the decision gets enhanced and the plot also shows that for a given speed, 6G delivers enhanced decision levels than 5G. This figure compares how 6G will create better support systems for the AI and ML applications, which are imperative for predictive analysis and fast decision-making. With low latency, 5G networks also prove to be effective to the machine learning models to process data immediately and simultaneously in return.

Real-time Analytics

Real-time analytics is a very important concept of the agile decision-making process that is beneficial to the organization as it gives an opportunity to make decisions based on the data in the production process. Here it is important to note that the progress of 5G and 6G technology is essential in improving real-time analytics to meet the need for fast responses as the condition changes and opens up new opportunities. Real time analytics means real time data analysis which is used to gain insights for decision making in real time and for controlling 'agile' processes.

Figure 5. Growth in Real-time Data Processing Capabilities

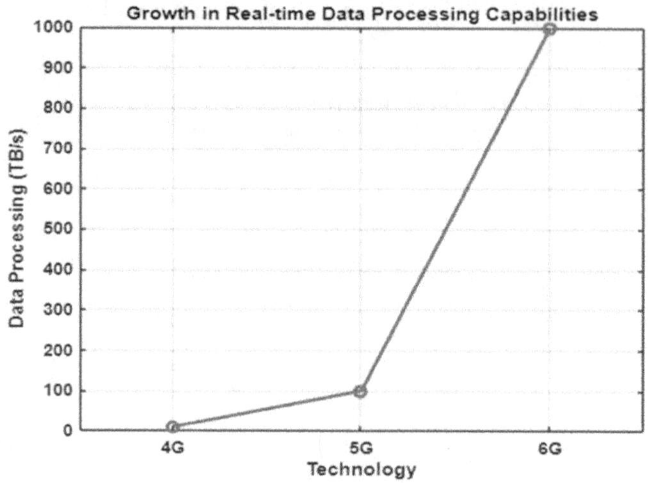

The increase in capabilities of real-time data processing of 4G to 6G is shown in figure 5 in the form of a line graph. The x-axis is Categorically while the y-axis is Proportional, where data processing capabilities are indicated in terabytes per second (TB/s). In this line graph, 4G network has 10 TB/s, 5G network has 100 TB/s and 6G, network has 1000 TB/s. This kind of visualization emphasizes the kind of growth in the data processing capability, which has a positive impact on the quality of real-time analysis and increase in the flexibility of business processes. One of the other features of the 5G network is the availability of a much broader spectrum of frequencies, as well as such advantages as the ability to develop more complex models based on the data received in real time and process broader datasets. It is especially relevant to industries that still create humongous quantities of data, like telecommunications, retail, and healthcare.

CASE STUDIES AND INDUSTRY APPLICATIONS

Manufacturing

5G technology also changes the traditional manufacturing industry by bringing the concept of smart manufacturing to create more efficiency and real-time control. One of the most well-known cases is Siemens' Amberg Electronics Plant in Germany, which has transformed into a reference example for the digital manufacturing. Currently deploying 5G networks, the plant links several thousand sensors and pieces of equipment, thereby making it possible to continuously monitor and analyze the obtained data. This connectivity also enables constant check on production processes with a view of identifying any disparity or deviation and promptly correct it. This leads to transformed prospect of product quality, decrease in periods of downtime for maintenance and average expenses for maintenance.

Figure 6. Impact of 5G Across Industries

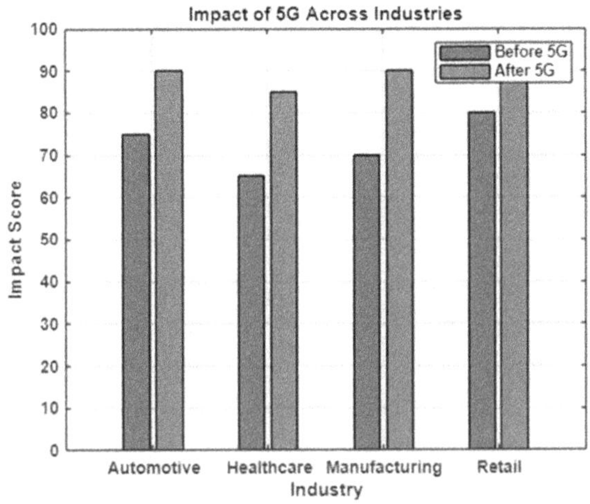

Figure.6. illustrates a grouped bar chart that depicts the benefits of the 5G technology in manufacturing, healthcare, retail and manufactures industries. The horizontal axis shows the industries while vertical axis demonstrates the impact ratio ranging from 0 to 100. Before and after the 5G is in the chart with two bars each for every industry. The data indicates significant improvements in impact scores post-5G implementation: they range from 70-90 to manufacture, 65-85 to health care, 80-95 to retail and between 75-90 in automotive industry.

Another area where real-time monitoring enabled by 5G has significant importance is the preservation of the workers' safety as well as the optimization of operations. Using wearable devices connected through 5G network, health and safety conditions of the worker can be monitored and immediate danger indications can be provided to the worker as well as the management. With this real time data it is possible to take the necessary precautions to minimize the occurrence of some accident as well as to improve working conditions. In addition, remote-controlled vehicles such as the Autonomous Guided Vehicle (AGV) and robots access via 5G; control supply chain traffic with accuracy to cut down manual interferences in factory floors.

Healthcare

Telemedicine as one of the areas of development in the healthcare industry has become possible due to the implementation of the 5G network. As shown in the case of telemedicine, 5G with its ultra-reliable, high-speed connections has impacted telemedicine that surged due to COVID-19 greatly. Policies like Teladoc and Amwell are benefiting from 5G to perform video consulting between the patient and the healthcare service provider regardless of area. This way helps in coordinating the patient's medical needs, advice, prescriptions and follow up thereby enhancing the patient's access to healthcare needs especially in the remote areas.

Retail

The retail industry is one of the industries that have deployed 5G technology to deliver real-time information and AI-powered solutions to the end-user. The most evident is probably the usage of AI-based chatbots and virtual assistants that beneficially use high speeds to promptly reply to customers' inquiries. Today's retail behemoths such as Amazon and Walmart apply these technologies to implement individual approaches to consumption, suggest products based on the previous purchases of a certain client, and enable safe purchasing processes.

Automotive

The automotive industry will be drastically impacted by the 5G technology, especially in the aspects of Self-Driving cars and Intelligent Traffic Systems. Real-time data transfer and data processing are very central components in AVs so that they can successfully and efficiently perform their functions. These include low latency, high speed, which makes the AVs to communicate with each other and with other structures of traffic rapidly. Manufacturers such as Tesla, Waymo and Uber are currently working towards adopting this technology to enhance how the autonomous cars adapt to different situations on the roads and avoid accidents.

STRATEGIC ADVANTAGES OF 5G/6G IN AGILE MANAGEMENT

Market Adaptability

The applications of 5G and soon the 6G in agile management are immense in terms of the increase in the market flexibility. Indeed, dynamism is a critical area of concern in the current world economy for business since the market is continuously changing and so are the customer preferences. 5G technology helps in the availability of real-time data and analysis and thus supports quick decision making so that businesses become more competitive. The tight interconnectivity of IoT devices and complex analytics with the sophisticated frameworks is especially valuable when monitoring the market trends on a regular basis, collecting customers' feedback in real time, and promptly adapting the strategies to make them more responsive to the constantly evolving environment.

Figure 7. Strategic Advantages Radar Chart

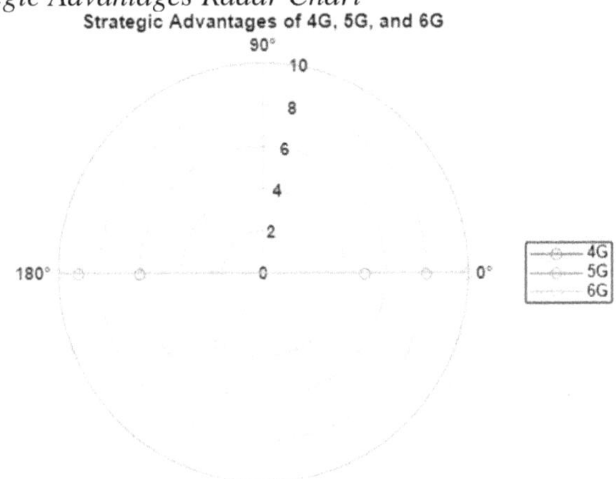

It is a radar chart presented in figure.7, which indicates how 4G, 5G and 6G is advantageous in terms of market adaptability, increase in work efficiency and innovations. Each of the 4G, 5G, and 6G portfolios includes radar charts showing relative scores of up to 10 for each strategic advantage. By comparing the absolute score of 4G and 5G plot, the moderate scores depicts on all the four dimensions and 5G plot shows significant scores. The graph showing the 6G also depict the maximum scores in order to signify its enhanced efficiency. It highlights the manner in which developments in communication technologies improve some important strategic elements of agile management that would enable organisations to remain relevant and create more value propositions. For example, in the retail sector 5G supported systems can monitor stock movement, consumer buying behavior and sales trends to forecast the future needs. This capability enables the retailers to constantly modify the stocking density, having those products that are in high demand in their stores, at the same time, they avoid having many stocked goods. Likewise in financial industry 5G provides real time monitoring of market environment and trading operations. Such information will be beneficial for financial institutions in making and organizing trading strategies, avoiding possible risks, and making considerable profits at the correct time.

Workflow Optimization

Another major disruptive opportunity 5G and 6G has in the concept of Agile Management is the improvement of work flow. These technologies operating to boost the efficacy of conveying messages and automation of tasks as well as processes. With 5G's high-speed connections and low latency, it becomes possible that team members, systems and devices can always communicate in real-time, which means that there can never be those unnecessary stages that tend to slow down work while tasks are being completed.

For instance, connected devices and sensors in a manufacturing environment form part of the large IoT space leveraging on 5G to continuously feed data on equipment and manufacturing processes. They are utilized in real-time with the aim of identifying areas of underperforms, identifying potential maintenance issues and perhaps, setting up an efficient schedule for production. In integrating 5G into their operations, manufacturers will be able to cut on time lost, work more on the output and also get better quality products. Also, the innovative aspect of 5G involves optimizations of the material handling through the autonomy of guided vehicles and robots in intralogistics.

Innovation

From the 5G and 6G technologies point of view, the most important value and impact of those solutions are connected with their perfect data and connectivity opportunities which are the important factors of business innovations facilitating various new product, service, and business models. 5G's data transfer speed and low latency enable the implementation of fresh technologies like augmented reality (AR), virtual reality (VR) as well as advanced robotics in dynamic frameworks. These technologies provide fresh opportunities for introducing changes to different fields.

For example, 5G revolutionizes entertainment industry to bring high quality Augmented Reality (AR) and Virtual Reality (VR) experiences to the creation and consumption of the content. This is a great way through which companies can create new experiences that can captivate users in ways that was not possible before, such as games, concerts and other events. 5G's relatively low latency and high bandwidth help these experiences to be seamless to the end-user and generate additional revenues.

In the automotive industry 5G is used for the construction of self-driving cars and also it improves smart traffic control. Real time transmission and processing of data is very important as it is found in the autonomous vehicles. The 5G and vehicle-to-everything allows cars to be connected to each other and to traffic signals so that they are informed about the road traffic and conditions, and possible risks. This real-time communication makes efforts to improve safety, decrease traffic density and consequently increase the efficacy of transport networks. Thus, automotive companies can develop new ideas in possession of better self-driving car technologies and build smarter cities where human beings and smart machines integrate.

CHALLENGES AND BEST PRACTICES FOR IMPLEMENTATION

Implementation Challenges

The increased usage of 5G networks and future emergence of 6G technologies create a great potential for the development of effective agile management practices but it is important to recognize a number of implementation issues related to the technological framework of these networks. Infrastructure upgradation is one of the major challenges here as it entails a significant amount of capital expenditure. The graphical representation of this challenge and best practice is presented in the form of the stacked bar chart present in the figure.8. below. On the horizontal axis there is placed the challenges and on the vertical axis, there is placed the percentage of difficulty. Two bars with a side by side hexadecimal comparison of the difficulty of each challenge before and after the best practice implementation are provided. The data shows a reduction in difficulties: it has reduced the financial challenges from 80% to 30%, technological challenges from 70% to 20% and security challenges from 60% to 10%. This figure illustrates why following best practices is needed to manage the challenges with the implementation of 5G.

Figure 8. Implementation Challenges and Best Practices

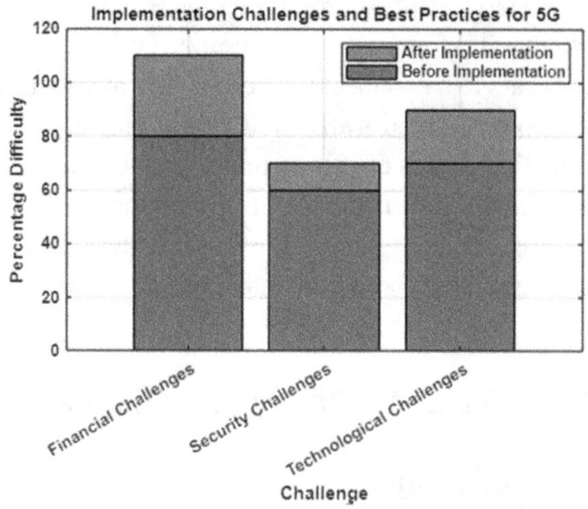

Moreover, the shift from 5G to the advanced 6G Technology needs a lot of integration and compatibility at the technological level. The new networks may not easily integrate into already existing systems or devices in the organization's network and this requires a massive overhaul of IT assets. This can often take quite a while to implement and can often involve significant planning and co-ordination to get everything right. The problem is further complicated by the fact that the transition must be done with minimal disruption of business as this may lead to costly impacts such as loss of business and customer dissatisfaction.

Security Concerns

As the connectivity advances and the overall data capabilities grow along with 5G and 6G solutions, the cybersecurity threats do as well, forcing organizations to be ready. More and more IoT devices are connected to the network, and the size of network perimeters themselves is increasing – all this contributes to the fact that the number of potential points of cyber-attacks also increases. These devices especially with basic security features are prone to hacking and therefore open to acts of burglary in regard to data and access to vital systems. Protection of these devices and the information that they produce is a big concern here.

Best Practices

Based on the reviewed literature and the research findings, it is crucial for implementing 5G and 6G technologies in agile management to follow a number of recommendations, which would cover technological and organisational factors. Essentially, the following measures should be followed in implementation phases: The process of integrating acquisitions is to do so step-by-step, including the improvement of the infrastructure and systems and using pilot projects in order to fine-tune the integration strategy. The rollout of 5G can be done in phases, which will help to mitigate risks, know customers' feedback and update the strategy, if needed.

Another strategic best practice is interaction with technology partners and vendors. Outsourcing implementation to external providers could also be an advantage in overcoming challenges in implementing 5G and to make sure that the organisation implements the best, most current technologies available. It also helps to have strong partnerships as this would enable one to gain access to other resources and help in making the transition easier.

Change Management

Change management and employees' training should be regarded as one of the most essential tasks in the process of implementing the 5G and 6G. These networks present a drastic leap from the traditional networks in business operations, and therefore, call for efficient management of change processes to enhance the achievement of the intended goals. Communication is one of the major pillars in dealing with change since people need to be informed about what is happening. It is crucial for organisations to define how 5G and 6G advantages contribute to its operations and the potential that these technologies will bring to the future growth. Communicating the change vision and the justification for change is another way of gaining commitment from the workforce and minimize any against change.

FUTURE OUTLOOK

Evolving Technologies

Finally, maintaining its focus at the future, the rates of innovative technologies' development will further contribute to the transformation of the environment of agile management. With the current advancement in the 5G technology still being in its initial deployment process globally, the ideas of 6G and even the next further versions are already giving a lot of anticipations. It is predicted that the next wireless communication generation, or 6G, will have data transfer rates up to one terabyte per second and latency in microseconds. Such capabilities are to open even higher-level applications of artificial intelligence, machine learning, and IoT, thereby making the world significantly more interconnected, whereby real-time data processing and real-time decision-making are to become the new normal. For the agile management, it means even higher flexibility, shorter response time, improved operational efficiency and ability to come up with new ideas. Other possibilities for the future developments are quantum computing, edge computing, as well as enhancing the usage of augmented reality which all can contribute their part in making the businesses more flexible and effective. These are presented in figure 9 below in the form of a line graph to show the expected trend in technology over the next decade and its implication on the agility management. The horizontal axis is the years from the fiscal year 2020 to the fiscal year 2030 while the vertical axis is a measure of advancement and the impact on agility on an arbitrary scale. Two lines illustrate the trends: One of the two could be dedicated to technological improvements; the other could concern agility. In this graph, both metrics are rising steadily and at 2030 tech advancements hit 100 on the scale and the same thing goes for the impact on agility. This visualization brings out the continuing progression of technology and its impact of making moves in business processes more revolutionary.

Figure 9. Technology Advancements and Agile Impact

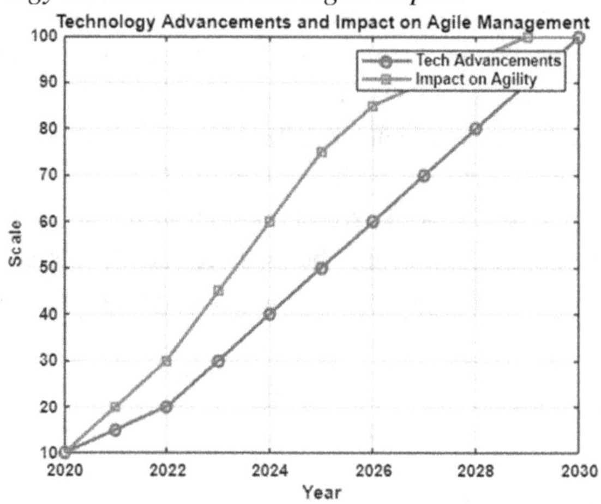

Long-term Strategies

Thus, the increase of the technological pace and changes of technologies require the owing business strategies that ensure further development of the technologies. The essence of this is the continuous investment in the technology base and the human capital. Employees should be well trained ways of using the emerging technologies and methodologies that are adopted by the businesses. Other sources of funding can also be obtained with technological vendors and research organizations that can reveal the new technologies to businesses and also help the organizations to understand how the technologies can be effectively deployed.

Sustainability Considerations

Having noted the roles of 5G and the progress toward the launching of 6G, it is important that its utilization should be responsible and green. Environmental factors that are associated with the construction and continued running of enormous structures that connect tower, base station and data centers must be regulated. Head of companies and moguls should extend their efforts by using energy efficient technologies and renewable energy sources in order to apply strategies on reducing the emissions of carbon. The ideas of the circular economy which presupposes the recycling and reuse of materials should be implemented in the technological equipment's lifecycle.

CONCLUSION

This chapter has discussed the influence that has been occasioned by 5G and looming 6G on the subject of agile management. We have seen how 5G/6G makes the needed connectivity in real time enhancing the bandwidth and the established networks to realize higher market agility, efficiency and innovation. Real-time IoT, AI, and analytics help in managing devices' connectivity and enable high quality of decisions at the same time. Application samples across industries including manufacturing, healthcare, retail and automotive industries show how these technologies work and the available advantages. Further, we descuscussed the issues of implementation, security, practices, and management needed for proper adoption.

In the context of 5G and with regards to the 6G era, the management agility prospects are vast. The future trends of agile management include an increased level of connectivity, elaborate speed, and new data features which will allow an organization to function more effectively and in a more flexible manner. The continuous advancement of these technologies also opens up newer and improved ways to enable business in coming up with better optimization of these technologies. Nevertheless, to derive the said benefits organizations must have the willingness on the part of the organization to engage in learning, planned investment and embracing of sustainable strategies.

Since 5G and 6G present such possibilities and openings, companies are urged to adapt and incorporate these connected possibilities into their operations the entire concept of the paper targets entrepreneurs as they should adopt the innovations offered by the new generations of connectivity technology. It is within this regard that adopting the agile frameworks that harness the advanced means of communication can help organizations to remain pertinent in the market as well as enhance operational efficiency while promoting creativity.

REFERENCES

Akyildiz, I. F., Kak, A., & Nie, S. (2020). 6G and Beyond: The Future of Wireless Communications Systems. *IEEE Access : Practical Innovations, Open Solutions*, 8, 133995–134030. DOI: 10.1109/ACCESS.2020.3010896

Al-Fuqaha, A., Guizani, M., Mohammadi, M., Aledhari, M., & Ayyash, M. (2015). Internet of Things: A Survey on Enabling Technologies, Protocols, and Applications [Fourthquarter.]. *IEEE Communications Surveys and Tutorials*, 17(4), 2347–2376. DOI: 10.1109/COMST.2015.2444095

Alhassan, A. H., Rahman, T. A., & Abd-Alhameed, R. A.. (2021). Comprehensive Survey on 6G Wireless Networks: Key Technologies, Challenges, Standardization, and Future Directions. *IEEE Access : Practical Innovations, Open Solutions*, 9, 148191–148230.

Basar, E., Di Renzo, M., de Rosny, J., Debbah, M., Alouini, M.-S., & Zhang, R. (2019). Wireless Communications Through Reconfigurable Intelligent Surfaces. *IEEE Access : Practical Innovations, Open Solutions*, 7, 116753–116773. DOI: 10.1109/ACCESS.2019.2935192

Bennis, M., Debbah, M., & Poor, V. (2018). Ultra-Reliable and Low-Latency Wireless Communication: Tail, Risk, and Scale. *Proceedings of the IEEE*, 106(10), 1834–1853. DOI: 10.1109/JPROC.2018.2867029

Choi, J., Lee, B., & Jindal, N. (2021). 6G Era: Challenges and Solutions for Future Communications and Networking. *IEEE Transactions on Wireless Communications*, 20(2), 1160–1170.

Cotton, S. L., Bultitude, R. J. C., & Scanlon, W. G. (2011). Millimeter-Wave Technology in Wireless PAN, LAN, and MAN: From IEEE 802.15.3c to IEEE 802.11ad and Beyond. *IEEE Communications Magazine*, 49(4), 68–74.

Ghosh, A., Ratasuk, R., Mondal, B., Mangalvedhe, N., & Thomas, T. (2010). LTE-advanced: Next-generation wireless broadband technology [Invited Paper]. *IEEE Wireless Communications*, 17(3), 10–22. DOI: 10.1109/MWC.2010.5490974

Govardhana Reddy, G., & Raghavendra, K. (2022). Vector space modelling-based intelligent binary image encryption for secure communication. *Journal of Discrete Mathematical Sciences and Cryptography*, 25(4), 1157–1171. DOI: 10.1080/09720529.2022.2075090

Han, S., i, C., Xu, Z., & Rowell, C. (2015). Large-Scale Antenna Systems with Hybrid Analog and Digital Beamforming for Millimeter Wave 5G. *IEEE Communications Magazine*, 53(1), 186–194. DOI: 10.1109/MCOM.2015.7010533

Liu, Y., Liu, X., Mu, X., Hou, T., Xu, J., Di Renzo, M., & Al-Dhahir, N. (2021). Reconfigurable Intelligent Surfaces: Principles and Opportunities [Thirdquarter.]. *IEEE Communications Surveys and Tutorials*, 23(3), 1546–1577. DOI: 10.1109/COMST.2021.3077737

Park, J., Park, S., & Ji, H. (2020). Dynamic Spectrum Sharing for 5G and Beyond: A Survey. *IEEE Wireless Communications*, 27(5), 12–18.

Peng, B., Li, Z., & Li, S. (2021). A Survey on 5G and 6G Technologies: Opportunities, Challenges, and Solutions. *IEEE Communications Magazine*, 59(1), 38–43.

Popovski, P., Trillingsgaard, K. F., Simeone, O., & Durisi, G. (2018). 5G Wireless Network Slicing for eMBB, URLLC, and mMTC: A Communication-Theoretic View. *IEEE Access : Practical Innovations, Open Solutions*, 6, 55765–55779. DOI: 10.1109/ACCESS.2018.2872781

Qin, Y., Gacanin, H., & Saad, W.. (2021). Integrating 6G with AI: Vision, Challenges, and Opportunities. *IEEE Network*, 35(2), 20–26.

Rappaport, T. S., Xing, Y., MacCartney, G. R., Molisch, A. F., Mellios, E., & Zhang, J. (2017). Overview of Millimeter Wave Communications for Fifth-Generation (5G) Wireless Networks—With a Focus on Propagation Models. *IEEE Transactions on Antennas and Propagation*, 65(12), 6213–6230. DOI: 10.1109/TAP.2017.2734243

Reddy, A. M., & Tanguturi, R. (2022). "Improved Secure Communication Mechanism for IoT Platform Devices," *2022 International Conference on Smart Generation Computing, Communication and Networking (SMART GENCON)*, Bangalore, India, pp. 1-6.

Saad, A., Hossain, M. S., Muhammad, G., Alhamid, H., Guizani, M., & Al-Dhelaan, A. (2020). The Road to 6G: A Comprehensive Survey [Fourthquarter.]. *IEEE Communications Surveys and Tutorials*, 22(4), 2192–2235.

Shekhar, R., & Ashreetha, B. (2023). Performance Analysis of Energy Efficiency and Security Solutions of Internet of Things Protocols. *IJEER*, 11(2), 442–450. DOI: 10.37391/ijeer.110226

Shruthi, M., & Naveen, K. B. (2020). "Internet of Things: A Survey on Devices, Ecosystem, Components and Communication Protocols," *2020 4th International Conference on Electronics, Communication and Aerospace Technology (ICECA)*, pp. 611-616.

Shruthi, M., & Naveen, K. B. (2020). "Internet of Things: A Survey on Devices, Ecosystem, Components and Communication Protocols," *2020 4th International Conference on Electronics, Communication and Aerospace Technology (ICECA)*, pp. 611-616.

Tataria, H., Shafi, M., Molisch, A. F., Dohler, M., Sjöland, H., & Tufvesson, F. (2021). 6G Wireless Systems: Vision, Requirements, Challenges, Insights, and Opportunities. *Proceedings of the IEEE*, 109(7), 1166–1199. DOI: 10.1109/JPROC.2021.3061701

Yang, G., Yang, Y., & Li, K.. (2020). A Survey on Advanced Techniques for Spectrum Sharing in 5G and Beyond Mobile Networks. *IEEE Wireless Communications*, 27(1), 17–25.

You, X., Wang, C.-X., Huang, J., Gao, X., Zhang, Z., Wang, M., Huang, Y., Zhang, C., Jiang, Y., Wang, J., Zhu, M., Sheng, B., Wang, D., Pan, Z., Zhu, P., Yang, Y., Liu, Z., Zhang, P., Tao, X., & Liang, Y.-C. (2021). Towards 6G wireless communication networks: Vision, enabling technologies, and new paradigm shifts. *Science China. Information Sciences*, 64(1), 1–74. DOI: 10.1007/s11432-020-2955-6

Zhang, R., Cui, Y., & Cui, Y.. (2021). AI-Empowered 6G: A Technology Review and Framework. *IEEE Communications Magazine*, 59(6), 26–32. DOI: 10.1109/MCOM.2012.6211482

Zhang, Z., Xiao, Y., Ma, Z., Xiao, M., Ding, Z., Lei, X., Karagiannidis, G. K., & Fan, P. (2019). 6G Wireless Networks: Vision, Requirements, Architecture, and Key Technologies. *IEEE Vehicular Technology Magazine*, 14(3), 28–41. DOI: 10.1109/MVT.2019.2921208

Chapter 3
Bridging the Digital Divide:
Accessibility and Inclusivity Strategies in 5G/6G Technology

Fatima Farooq
https://orcid.org/0009-0008-1073-2957
Islamic University of Science and Technology, India

Irfan Bashir
https://orcid.org/0000-0003-0000-1065
Islamic University of Science and Technology, India

ABSTRACT

The chapter discusses methods to improve the accessibility and inclusivity of 5G and 6G technologies to reduce the digital gap. It underscores the importance of developing affordable infrastructure, cost-effective devices, and comprehensive digital literacy programs while considering ethical implications. Collaboration among governments, network providers, and technology developers is crucial to implementing inclusive policies ensuring widespread access to transformative technologies. Examining the transition from 4G to 5G and the vision for 6G addresses the socioeconomic factors influencing ICT adoption and emphasizes the role of SDN in network efficiency. By proposing solutions to overcome existing barriers and supporting strategic initiatives, this chapter aims to promote diversity in technology design and enhance accessibility. Ultimately, fostering an equitable digital landscape will allow societies to harness the full potential of information and communication technology, reshaping consumer behavior and elevating global standards in both industrialized and emerging nations.

DOI: 10.4018/979-8-3693-6725-4.ch003

INTRODUCTION

Information and communication technology are rapidly evolving with new software platforms and business ecosystems constantly being created (Matinmikko-Blue et al., 2020). Many of these are market-driven and might significantly impact our daily lives (Saarnisaari et al. 2020). Fifth-generation (5G) and sixth-generation (6G) broadband cellular networks are expected to impact modern civilization significantly. While commercial markets are hyping 5G, it also has the potential to analyze and influence consumer behavior, both online and offline. The definition of 6G remains unknown, but it serves as a conceptual model for the future path to 2030 (Yrjola et al. 2020). Universal access to information and communication technology (ICT) is expected to create a worldwide community for contact, business, and learning, leading to increased living standards and enhanced social welfare. In the 1990s, scholars and policy professionals debated the existence of a ''digital divide'' between people with access to ICT (e.g., PCs and the internet) and those without. A recent research by the Pew Internet & American Life Project revealed that yearly income is the biggest predictor of individual internet use, regardless of other characteristics. Large firms are more likely to implement innovative ICT solutions compared to smaller organizations. ICT expenditure correlates with the degree of development and leads to increased productivity in developed countries. However, it is not yet as productive in underdeveloped countries. Despite a decade of discussion among specialists in public policy, communications, philosophy, social sciences, and economics, there is still no agreement on the definition, scope, and effect of the digital divide.

The Internet of Things (IoT), smart homes and cities, automated passenger cars, autonomous driving, intelligent manufacturing, remote surgery, cloud service robotics, asset tracking and management, and cloud service robotics are some potential uses for 5G technology (Oinas-Kukkonen, Karppinen, and Kekkonen 2021). Furthermore, 5G provides a universal connectivity solution that can efficiently handle all existing traffic profiles, increasing the economic viability of smart logistics (Clancy et al., 2023). The use of smart logistic applications, such as object tracking, automated guided vehicles, drones, remote assistance, machine and robot monitoring and control, instruction and/or hazard surveillance, and ambient sensors, is made easier by Massive Machine Type Communication (mMTC), Ultra Reliable Low Latency communications (URLLC), and Enhanced Mobile Broadband (eMBB) (Khatib and Barco2 2021).

This chapter examines accessibility and inclusion opinions for 5G and 6G technology. The chapter will analyze how impoverished and marginalized groups are denied equitable access to contemporary telecommunications networks. The chapter examines these concerns to underscore the need to emphasize accessibility and inclusiveness in 5G/6G infrastructure and service design, development, and deployment. Furthermore, the chapter will examine global best practices and successful attempts to increase 5G/6g rollout accessibility and inclusivity. In many cases, these case studies may help stakeholders find practical and imaginative solutions to accessibility concerns and digital divides. Additionally, the chapter will highlight innovative ideas and new technology that might improve accessibility and inclusion in evolving telecommunications standards.

OVERVIEW OF THE DIGITAL DIVIDE

With the rise of the digital economy, the topic of the digital divide has gained substantial and dynamic attention. When the idea began to appear in several United Nations (UN) reports, it became well-known and widely acknowledged. Since then, organizations, decision-makers, and scholars from many fields have found it to be a serious concern (Van Dijik., 2020). The difference between those who have adequate access to information and communication technology (ICT) and others who don't or have restricted access to ICT is referred to as the "digital divide" (Soomro and YUlhui., 2020). Diverse manifestations of digital disparities may exacerbate social inequalities by either constraining or enhancing individuals' social and economic resources, as well as their capacity to engage in societal activities. The digital gap is seen as a crucial issue for social equity in the modern era. Certain researchers have even labeled it as a cause of poverty, referring to it as ''digital poverty'' (Setthasuravich and Kato., 2020).

The continuous digital revolution does not afford equal opportunities to all persons, resulting in socioeconomic inequality. ICT is seen as an effective instrument for accomplishing the United Nations' sustainable development goals (SDGs). Its efficient use has been identified as a manner of contributing to SDG objectives, notably goal 10, which aims to alleviate disparities (Guterres 2020). Acquiring a thorough comprehension of the notion of the "digital divide," in addition to its diverse interpretations and expressions, facilitates the determination of particular technological needs. This, in turn, can facilitate the creation of a more cohesive framework and policies aimed at addressing these needs. Additionally, goal 10 is heavily influenced by ICT since it can promote equality by giving citizens access to vital information and enabling them to actively participate in their economy. This understanding can also spur increased efforts to close and close the gaps in digital access and usage. Utilizing ICT effectively is linked to opportunities for training, education, and employment.

The digital gap still exists despite efforts to close it. In addition, the rise in prominence of this problem has been greatly impacted by the coronavirus. The epidemic has forced people to work, learn, access services, and socialize from home, making the digital divide an urgent problem. Some governments, groups, and the media have brought attention to this concern. The UN Secretary-General stated that the "digital divide" has emerged as a crucial problem that might mean the difference between life and death during the pandemic. The COVID-19 epidemic has compelled individuals to make a remarkable transition to digital platforms in their daily routines (Livari, Sharma, and Ventä-Olkkonen 2020). This has led to the emergence of new protocols where the global community strongly depends on the internet and digital gadgets. Nevertheless, the epidemic has intensified the problem by exacerbating the isolation and disadvantages faced by those who lack resources and effective internet access (De', Pandey, and Pal 2020).

The growing reliance on digital technology makes it appropriate to look at how the digital gap has changed in the growing digital economy. Emphasis on empirical research, in particular, has been restored in recent literature (Dutton & Reisdorf, 2019; Reddick et al., 2020). However, comprehensive overviews of the digital divide since 2017 have been limited (Aissaoui, 2022; vassilakopoulou & Hustad, 2023). This study aims to address this gap by (1) identifying factors influencing the digital divide in recent research and (2) exploring potential new forms or dimensions of the divide. Through a systematic review and structural content analysis of 50 studies published since 2017, this research contributes to understanding the digital divide, its evolving nature, and the factors contributing to it. The objective of this study is to provide insights into the creation of efficient methods to close the gap and advance digital inclusion by identifying new trends and aspects of the digital divide. Ultimately, it seeks to contribute to global efforts to achieve sustainable development goals by highlighting the importance of digital skills and competencies. Consequently, this research is intended to encourage further academic exploration of the digital divide and its complex manifestations. It is envisaged that these results will help governments, organizations, and legislators create plans that effectively address the digital gap and promote a more inclusive digital society.

IMPORTANCE OF ACCESSIBILITY AND INCLUSIVITY IN TECHNOLOGY

The field of wireless technology has advanced significantly, offering tools that can improve community engagement, participation, and independent living (Moon, Baker & Goughnour, 2019). The IoT has emerged with the development of intelligent, interconnected devices such as wearable computers and environmental sensors.

However, the design and development of these devices often overlook the needs of individuals with disabilities. People with disabilities have diverse psychological, sensory, and physical characteristics. Wearables and IoT devices have the potential to accommodate specific needs, which aligns with the principles of universal design. In this context, physical disabilities include limitations in mobility, while sensory disabilities encompass visual and hearing impairments, as well as communication difficulties. Cognitive impairments include developmental or cognitive challenges, as well as memory or processing difficulties resulting from accidents.

The COVID-19 epidemic has brought attention to the growing significance of accessibility due to an increased reliance on the internet. In the first quarter of 2020, the United States had a 47% increase in global internet usage (Cohen, 2020), with nearly 50% of Americans considering it essential during the pandemic (Vogels et al., 2020) the pandemic has exacerbated existing accessibility challenges for the disabled community, as face-to-face interactions become limited. Consulting firm Knowbility emphasizes the need for businesses and organizations to integrate accessibility into their COVID-19 response efforts and future technology plans (Rush et al., 2022).

FACTORS CONTRIBUTING TO THE DIGITAL DIVIDE

Telecommunications technology may assist in narrowing the digital divide. Mobile technology and m-commerce have the potential to bridge the global digital gap with appropriate policies in both public and commercial sectors (Buchanan, Ricky, and Natasha Layton., 2019). Even with telephones, the separation has been explored and shown to be persistent. According to one study, racial disparities exist in the United States. In 2000, around 6.3 million American homes lacked access to basic telephone services. This issue is more severe in underdeveloped countries. Approximately 80% of the world's population lacks a telephone connection. Access to PCI computers, internet, telephone, and mobile phone technologies may quantify the ''digital divide''.

THE EVOLUTION FROM 4G TO 5G/6G

Mobile communications have played a crucial role in the development of our modern interconnected civilizations. The progression from earlier analog mobile systems to more advanced long-term evolution (LTE) networks has significantly transformed how individuals get and share information via wireless devices. Presently, wireless communications find themselves at a critical juncture (Network 2019). Undoubtedly, the growing need for more capacity and widespread use of intelligent devices, which rely on high-speed applications, needs the development of more advanced networks to achieve significant performance improvement (Rajatheva et al., 2020).

The rapid emergence of fifth-generation (5G) mobile communication systems is effectively addressing the escalating demand for advanced wireless communication in our present and future societies. With the proliferation of connected devices and the increasing volume of data traffic, 5G faces the daunting task of accommodating diverse service requirements and handling unprecedented traffic loads. The global commercialization of 5G has sparked significant interest and investment in research initiatives focused on the development of the sixth-generation (6G) within academic and industrial domains. These comprehensive research endeavors aim to identify the key drivers propelling the evolution of 6G technology. They also seek to establish the exact performance requirements and technological innovations essential for the realization of the future generation of mobile communication networks. Notably, the International Telecommunication Union (ITU) has taken a proactive step by establishing a ''Focus Group on Technologies for Network 2030'' (FG-NET 2030) to specifically address the development of 6G technology, often referred to as ''Network 2030''. This concerted effort underscores the significance of advancing research and innovation to meet the complex demands and expectations of future mobile communication networks.

In the United States, the Federal Communications Commission (FCC) has taken many actions. Firstly, they have chosen to make the Terahertz spectrum available for 6G (Huq et al. 2019). Additionally, they have proposed the use of blockchain-based technologies to enable flexible and dynamic sharing of spectrum. Additionally, satellite communication is being seen as a crucial aspect of the next 6G technology to provide extensive and global communication. This will include the establishment of integrated networks that encompass space, air, and ground, therefore ensuring comprehensive coverage throughout the globe. Similar to the United States, South Korea is engaged in extensive research on Terahertz and "space-air-ground" communication networks.

The data presented in Figure 1 illustrates the exponential rise in the use of smart devices and the subsequent rapid increase in data traffic. This surge in data consumption will exert significant pressure on the existing 5G network infrastructure. It is noteworthy that data rates for wireless connections have been doubling approximately every 18 months, highlighting the relentless growth in data usage. As the commercialization of 5G technologies gains momentum, both academic institutions and businesses are actively embarking on research initiatives to pave the way for the future of 6G communication. The anticipated leap from 5G to 6G is expected to not only meet but also surpass the expectations that 5G has yet to fully satisfy. According to the data depicted in Figure 1, the growth rate of mobile traffic, excluding machine-to-machine (M2M) traffic, is projected to be approximately 54% per year. When including machine-to-machine (M2M) traffic, the compound annual growth rate of mobile traffic is expected to surpass 55% from 2020 to 2030. It is crucial to emphasize that the current 5G technology deployment will not suffice to meet the escalating demands of surging traffic and expanding user base. Therefore, significant enhancements are imperative to cater to the high volume of traffic, especially in terms of high data transfer rates, uninterrupted service, and minimal latency.

Figure 1. Estimation of global mobile traffic from 2020 to 2030

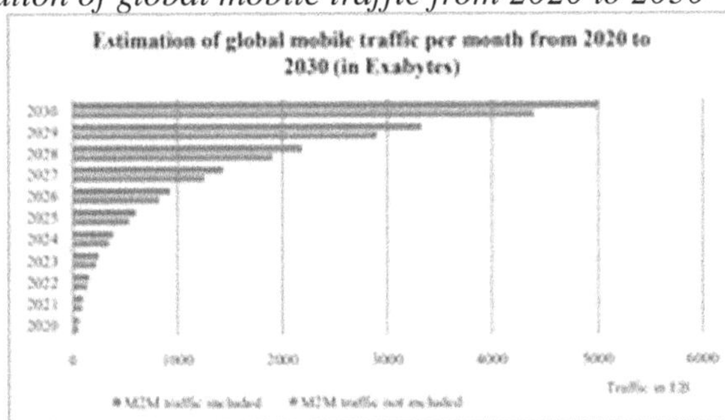

The illustration in Figure 2 depicts the evolution of telecommunications technology across different generations. The progression from first-generation (1G) to second-generation (2G) technology marked the shift from analog to digital technologies and the widespread adoption of mobile phones by telecom companies, making mobile devices accessible to a broader customer base. The transition from 2G to 3G entailed an upgrade from low-speed phone and data services to high-speed multimedia services, significantly enhancing the capabilities of mobile communication. Moving from 3G to 4G aimed to substantially improve data transfer speed through notable advancements in design and the development of Voice over IP (VoIP) and Smartphones. The advent of 5G has introduced three distinct service classifications. The first category is known as enhanced mobile broadband (eMBB), enabling very fast data transmission, which is crucial for bandwidth-intensive applications. The second category, massive machine-type communication (mMTC), facilitates the connection and communication of a large number of devices, supporting IoT and other interconnected systems. The third category, ultra-reliable and low latency communications (uRLLC), enables the achievement of very low latency and exceptionally high dependability, making it suitable for mission-critical applications such as autonomous vehicles and remote surgery. The new 5G network infrastructures will use virtualized technology. A contemporary design will replace physical nodes with software-based network protocols (NFV) on available servers (SDN). This new paradigm allows for more adaptability, scalability, and expertise, resulting in more agile and efficient services with lower operating expenses (OPEX) and investment (CAPEX) compared to existing technology. It also makes it easy to upgrade and fix software. Failure to include security in software development might increase the vulnerability of network services to cyber-attacks. Virtualization enables network slicing to adapt services to specific devices or sectors (e.g., internet access, industrial 4.0, corporate networks), but it also increases risk exposure. Mobile edge computing decreases latency and adds intelligence to a capital-intensive 5G access network. 5G technology provides coverage and processing capability for thousands of linked devices (IoT), reducing the need for security measures in today's networks.

Figure 2. Evolution of technologies and services from 1G to 6G (Docomo 2020)

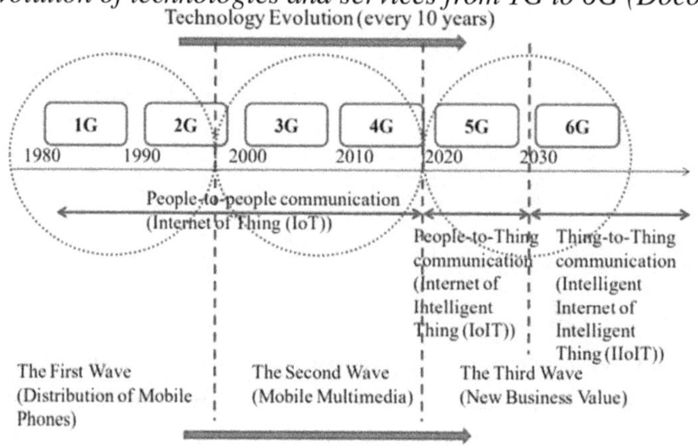

The incorporation of emerging 5G technologies extends the reach of existing networks, amplifying the sensitivity and susceptibility of components to emerging threats and risks, such as integrity, availability, and confidentiality (Tripathi and Reed, 2019). These core components will empower 5G technology to deliver more accurate and targeted 5G devices. Additionally, it is important to mention that version 15 offers a strong foundation for enhancing network performance and expanding mass service offerings (Farhadi Zavleh and Bakhshi 2021). It is worth noting that 3GPP is now engaged in ongoing efforts to further enhance this framework, as seen in Figure 3.

Figure 3. Evolution path of technology

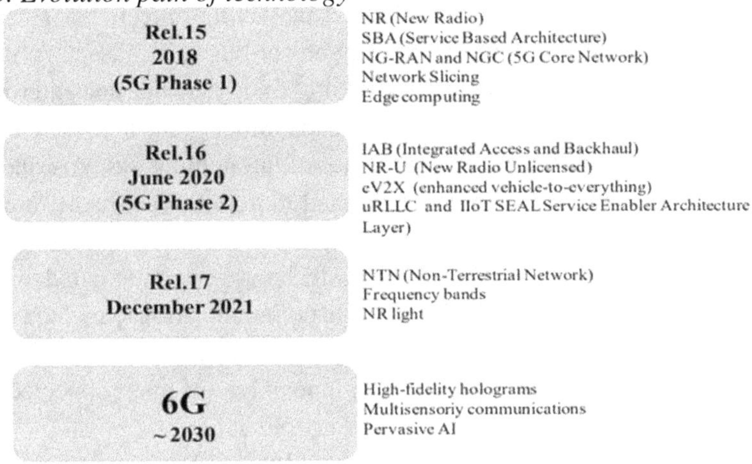

EMERGING TECHNOLOGIES AND THEIR POTENTIAL

The advent of the 5G network is poised to usher in the era of the Internet of Everything (IoE), enabling seamless integration of wired and wireless communications on a unified infrastructure. Simultaneously, a fundamental change in approach is required to develop advanced broadband cellular network frameworks for the future generation. This is necessary to provide a dependable, universally available, very low-latency broadband connection that can support and oversee vital and highly demanding applications and services. The IoE is the future advancement of 5G apps and use cases, consolidated into a single hub (Prakasam, Sayeed, and Ajayan 2021). The provision of infrastructure "as a Service" results in added value. 'Infrastructure as a Service' (IaaS) vendors include Microsoft Azure and Amazon Web Services. These suppliers make available their processing power, memory, and disk space for the use of virtual networks. They are regarded as role models for emerging IaaS companies. Though the IaaS approach and network resource provision share a similar business model, the latter poses a fundamentally different difficulty due to spectrum constraints. The system's capacity is determined by the mobility patterns of the users, including the increasing trend of nomadic user behavior. Additionally, service level agreements may impose more stringent criteria on the perceived quality of the customer experience.

5G technology is anticipated to bring about substantial socioeconomic advantages. Unlike its predecessors, which were primarily designed to meet the needs of mobile network operators and end consumers, the fifth-generation broadband cellular networks are expected to prioritize and target socioeconomic benefits in addition to technical telecommunication and marketing objectives. This shift in focus is expected to have a significant impact on various vertical sectors, including manufacturing, the automobile industry, eHealth, energy, media, and entertainment. These industries are expected to leverage the capabilities of 5G networks to develop sophisticated applications that utilize automation, autonomous systems, and various types of sensors, combined with advanced and more intelligent software and information systems. This convergence has the potential to enable a diverse range of novel software applications. Furthermore, 5G technology is anticipated to facilitate an expansion of connections beyond traditional communication to include machine-type communication. This expansion will lead to the seamless integration of people, data, devices, and sensors, paving the way for innovative and interconnected systems that can revolutionize industry practices and everyday life.

The objective of 6G wireless networks is to address societal needs and contribute to the attainment of the Sustainable Development Goals (SDGs). Anticipated to adhere to rigorous standards, Sixth-Generation wireless networks are expected to deliver the following key attributes: 1) Exceptionally high data speeds, reaching up to 1Tbps. 2) Remarkably low latency, at less than 1 ms. 3) Enhanced mobility and extensive coverage. 4) Efficient and adaptable connectivity for a myriad of devices, reaching a trillion-level scale. 5) Optimized spectrum efficiency at 60 b/s/Hz. 6) Superior system dependability. 7) Augmented network security measures.

The discussion encompasses the emerging technologies anticipated to propel the development of 6G. Specifically, it delves into the potential impact of cell-free massive MIMO, Artificial Intelligence (AI), and Reconfigurable Intelligent Surfaces (RIS). Notably, the deployment of RIS on building doors and windows is being contemplated to facilitate signal reflection without interference. Furthermore, the text expounds on the advantages of RIS technology over existing relays. In the context of 5G networks, the denser distribution of access points (APs) has facilitated the incorporation of massive MIMO technology. Building upon this, 6G advances this paradigm by integrating a cell-free network, thereby significantly enhancing the spectral efficiency of communication networks.

The following four tiers of connection:

1) "Not Connected" describes a situation in which no one has access to the Internet. This usually goes hand in hand with inaccessible rural places that have little road, electricity, transportation, and communications infrastructure because of how difficult it is to get there.
2) People with limited and/or intermittent connections are represented by the "under-connected" scenario. They may, for instance, have 2G phone and text services and, on occasion, spotty WiFi from a nearby network. One example is

a low-speed local mesh network that is linked to a gateway via a VSAT. Another is a local network that has limited Internet access, via delay-tolerant networking. With coverage gaps in certain places with too low a population density, the data rates that may be achieved by this category are below 256 kb/s, which is the limit set by the International Telecommunication Union (ITU) for defining broadband access.

3) "Connected" describes a situation in which users have access to GPRS or 3G networks. They can connect to the web at moderate download rates. Additionally, they may be linked to a reasonably fast backhaul network at a gateway via WiFi (Yaacoub and Alouini., 2020). Possible feasible data rates for this category often fall between 256 kb/s and 10 Mb/s, with little chance of coverage gaps in certain places.
4) Hyper-connected people have access to the latest and greatest in connection, whether it's high-speed fixed DSL, fiber-to-the-home (FTTH) wireless-to-network (FTTN) networks, or cellular networks with 4G and 5G capabilities.

Because of all the obstacles we've already mentioned, it's next to impossible to locate these subscribers in rural regions; they tend to reside in cities. People may be said to be hyper-connected if they are always online and their data rates are over 10 Mb/s.

ARCHITECTURAL INNOVATIONS

The intersection of sustainability, technology, and socio-cultural considerations in modern architecture and environmental design. It highlights the IDP, a collective method that aligns environmental goals with user needs, as seen in the Bullitt Center in Seattle (Wang et al. 2024). Biophilic Design, which integrates natural elements into architecture, enhances human health and productivity, as seen in Singapore's Khoo Teck Puat Hospital. Adaptive Reuse, a sustainable strategy that balances cultural heritage conservation with modern environmental needs, is also discussed, as seen in London's Bankside Power Station transformation into the Tate Modern art gallery. Green Building Technologies, Digital Design and Simulation Tools, and Smart Building Systems are explored, highlighting their role in optimizing building performance and contributing to environmental sustainability. The paper also emphasizes the importance of community engagement and participation in designing, as demonstrated in the participatory design process of the High Line in New York City. Cultural Sensitivity in Design is presented as a necessary approach to respect and integrate local cultural nuances, while Inclusive Design is advocated for creating accessible spaces. In architectural design, it is imperative to take into account the social and cultural composition of communities.

6G NETWORK ARCHITECTURE

The forthcoming 6G technology is poised to deliver significantly enhanced throughputs, reduced latency, higher component density, and pervasive integration of artificial intelligence across all network segments, surpassing the capabilities of 5G technology. As we progress towards the next-generation 6G mobile radio, several challenges about the cohesive functioning of its various components will need to be addressed. Notably, the future 6G wireless network will comprise numerous small mobile radio cells, facilitating efficient and expeditious transmission of substantial data volumes. Consequently, the network will necessitate the capability to transmit tens or even hundreds of gigabits per second over a single channel, thereby requiring exploration of the uncharted Terahertz band frequencies for telecommunications. Moreover, the seamless integration of wireless transmission lines with fiber-optic networks will amalgamate the strengths of both technologies, particularly in terms of capacity, reliability, mobility, and adaptability.

A satellite is a kind of radio-electric repeater that is placed in space. It is designed to receive signals that are created on Earth, amplify them, and then transmit the signals back to Earth, either to the same location or to a new one. A 6G architecture sample is shown in Figure 4. Figure 5, illustrates how many coverage kinds, from extensive coverage to worldwide coverage, are possible with 6G technology. To provide deep coverage, 6G technology makes use of visible light bands, Terahertz waves, and millimeter waves (mmWave). 6G makes use of satellite communications for worldwide coverage and cellular mobile communications for medium and wide coverage.

Figure 4. 6G Architecture

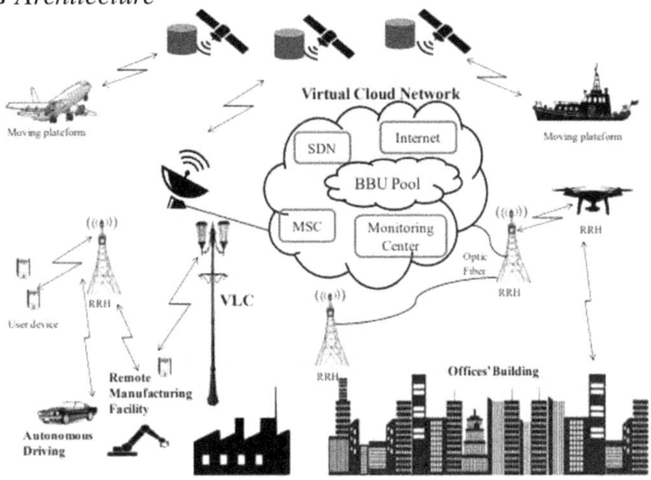

5G NETWORK ARCHITECTURE

The system model presents a network framework for 5G mobile systems, utilizing an all-IP-based approach for seamless wireless and mobile network connectivity. The system comprises a user terminal and multiple independent radio access technologies, each functioning as an IP link to the broader Internet. The initial two OSI levels define radio access technologies, offering Internet access with quality of service (QoS) support mechanisms. The network layer, encompassing OSI-1 and OSI-2 layers, is governed by the Internet Protocol (IP), specifically IPv4 or IPv6, to ensure efficient routing of IP packets for specific application connections (Tudzarov and Janevski, 2019).

Internet sockets serve as endpoints for data communication between clients and servers on the Internet, involving local and target IP addresses, and transport protocols. To enable interoperability across diverse networks and seamless transition between radio technologies, the local and destination IP addresses need to remain constant, ensuring continuity in Internet connectivity, particularly when at least one end of the connection involves a mobile user.

Figure 5. 5G Architecture

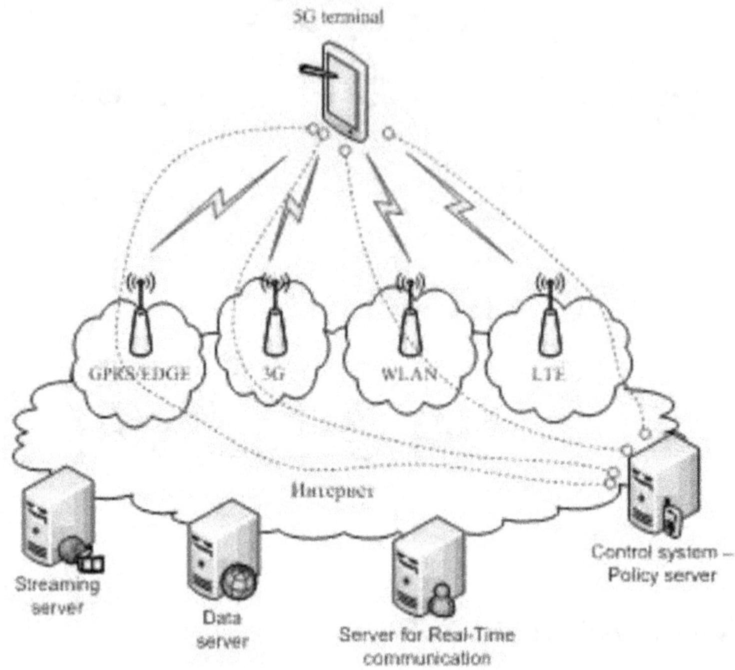

The proposed architecture for the 5G mobile system is based on using an all-IP model for wireless and mobile network interoperability. This system includes a user terminal and various independent radio access technologies (RATs) that act as IP links to the wider Internet. Each RAT should have its unique radio interface in the mobile terminal. The first two OSI levels define radio access technologies, providing access to the Internet with support for Quality of Service (QoS) mechanisms. The network layer, covering OSI-1 and OSI-2 layers, is the Internet Protocol (IP), which can be either IPv4 or IPv6, ensuring sufficient control data for proper routing of IP packets. Internet sockets serve as endpoints for data communication flows between clients and servers on the Internet, consisting of local IP addresses, target IP addresses, and transport protocol. To ensure compatibility between different networks and a smooth transition between radio technologies, the local IP address and destination IP address should remain fixed and unchanged. Routing to the target destination and vice versa should be unique and use the same path. Each radio access technology has an appropriate IP interface, characterized by its IP address, netmask, and parameters associated with IP packet routing across the network.

SOFTWARE-DEFINED NETWORKING (SDN)

The architectural concept of software-defined networking (SDN) involves the segregation of the network control plane from the forwarding plane, allowing a singular control plane to oversee numerous devices (Eissa, Bozed, and Younis, 2019). SDN is utilized to manage traffic flows independently from the systems routing traffic and the underlying infrastructure. This separation and integration of the control and data planes enable a single control program to govern multiple data plane components. The delineation between the control plane and the data plane is defined by the application programming interface (API) connecting the network device to the SDN controller. One notable API is the OpenFlow protocol, which facilitates communication between the controller and the switch, enabling the establishment of rules using a programmable interface switch. The OpenFlow switch exhibits the functionalities of a network address translator, router, switch, and firewall.

Figure 6. Traditional and software-defined architectures

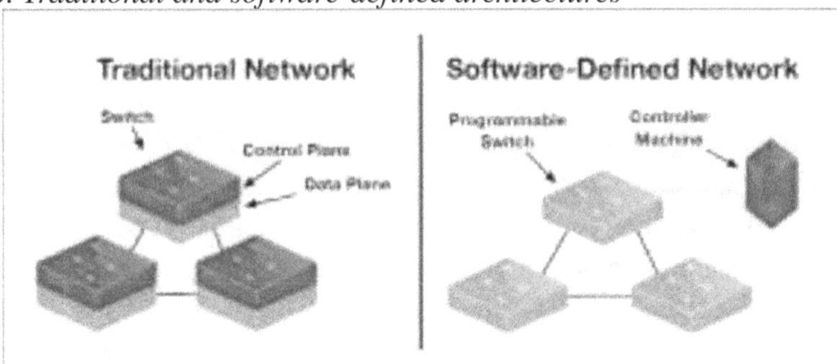

SDN brings fresh perspectives to network management and is rapidly emerging as the top choice for those encountering challenges with traditional networking (Rana, Dhondiyal, and Chamoli 2019). SDN allows for centralized control and management of hardware through a separate software application. This separation involves isolating the control plane, responsible for determining traffic routes, from the data plane, which carries out these instructions and forwards traffic.

NETWORK SLICING AND EDGE COMPUTING

To improve resource usage, a physical network may be divided into virtual networks. This technique is known as network slicing, and the article presents 5G technology. But there are also intricate security issues with this technology, especially when it comes to preserving network integrity (Gantumur 2024). This study investigates the technological foundation of 5G, including its essential elements, and advancements like SDN and NFV. It then concentrates on the security elements of network slicing, underlining the significance of safeguarding these technologies and possible attacks. The study discusses the difficulties posed by these security-related issues and suggests fixes. A summary of the results and their implications is included in the conclusion.

CORE AND EMERGING TECHNOLOGIES IN 5G/6G

Spectrum Management

The planned frequencies for 5G vary from 600-700 MHz to 3-4 GHz. These frequencies can achieve ranges of 26-28 and 38-42 GHz, and potentially up to 86 GHz. This would result in a significant boost in transmission capacity compared to existing technology. For 5G telecommunication to operate within the 700 MHz band, it will necessitate a new retuning procedure. This process would likely need governments to provide reimbursement for a portion of the costs.

The millimeter-wave technology pertains to the correlation between frequency and wavelength, as expressed by the equation: the speed of light equals the product of frequency and wavelength. The speed of light remains constant; therefore, as the energy increases, the wavelength becomes shorter. Contrary to, as the energy decreases, the wavelength becomes longer. With the increasing frequency of 5G, the wavelengths of these radiations will be shorter compared to those previously utilized. Considering the speed of light at 300,000,000 meters per second and the frequencies under discussion being on the order of 10,000,000,000 Hertz (tens of GHz), division of the speed of light by the frequency yields a wavelength of approximately 30 millimeters, commonly referred to as millimetric waves, mmWaves, or mmW.

Elevating the frequency over 30 GHz affects antenna coverage and communication. Despite the term "electromagnetic radiation," visible light and FM radio waves have many similarities. However, their unique qualities (energy, wavelength, and frequency) result in additional properties. At 5G frequencies, the radiation beam is directed like a pipe, concentrating energy in a small area, improving connection quality, and reducing radiation from antennas. This eliminates the need for traditional "traditional" coverage. Radio spectrum management is crucial for the deployment of new and old wireless communications technologies (Matinmikko-Blue, Yrjölä, and Ahokangas 2020). National regulators' choices on spectrum management aim to maximize the optimal use of limited natural resources. This approach promotes fairness among stakeholders, including incumbent mobile network operators and service providers, while adhering to ITU principles. Each nation has its laws for using the radio-electric spectrum to deliver services, such as licenses and authorizations. Next-generation 6G research highlights the importance of the Terahertz band in mobile communication networks.

In the future, 6G networks may continue to use the frequency bands in the spectrum from earlier generations (2G-5G) in certain areas. The wide range of frequency bands for 6G will require multiple radio spectrum management systems to handle differences between moderate, high, and extremely high bands. 6G technology needs to be flexible to operate across numerous frequency bands and spectrum management methods. Local networks will be crucial in the next 6G generation, necessitating the implementation of various spectrum management strategies.

Figure 7. Spectrum for future 6G

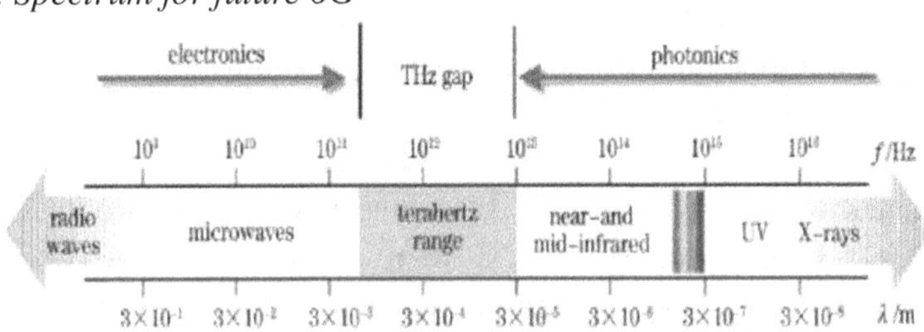

Security and Privacy Considerations

Every network generation has flaws. Despite efforts to reduce exploitation, vulnerabilities persist due to the difficulties of updating essential protocols. Table 2, outlines known security attacks and privacy breaches against previous generations' security architectures and authentication methods, along with their state of being repaired. Attacks are classified based on security criteria (e.g. confidentiality, integrity, and availability). Several threats, including signaling DoS, DDoS against authentication servers, battery depletion assaults, and user tracking, will pose challenges for 6G security architecture and applications. This derives from the fact that problems in the underlying protocol designs (e.g., poor authentication) and the restrictions of radio resources are general difficulties across all network generations, making it difficult to repair them fully. The following are the major lessons from security concerns and legacy network enhancements:

1) New applications pose security concerns, necessitating security enhancements. New apps have enhanced network standards compared to previous generations. However, these changes may introduce new vulnerabilities. One such security flaw in the VoLTE protocol is the ability to reuse keystreams in subsequent sessions. An attacker may use this vulnerability to decode and eavesdrop on encrypted VoLTE calls (Rupprecht et al. 2020). According to research by Saad, Bennis, and Chen., (2019); and Pandi et al., (2024), 6G technology may enable novel applications like mixed reality and autonomous driving. However, these systems may be vulnerable to impersonation and DoS attacks. Improving technology security before deployment is crucial.
2) When a new protocol deployment supports a legacy protocol, it may expose outdated vulnerabilities. The main issues include securing devices on outdated network segments with insufficient security capabilities and the incompatibility of essential security functions between two network standards. To address this incompatibility, the new standard typically requires the old architecture to authenticate older devices. However, this access control paradigm may reveal existing flaws in an older standard. An attacker could potentially use a downgrade attack to force 4G-LTE devices to connect to 2G/3G networks. The 2G/3G standards lack mutual verification between User Equipment (UE) and authentication servers, allowing attackers to collect International Mobile Subscriber Identity (IMSI) and monitor UE position. If 6G supports older 4G/5G devices, careful consideration must be given to the security aspects of dual network access authentication and identity management.
3) Fewer protocol design changes but greater protocol implementation changes reduce new vulnerabilities and fix existing problems faster. This is because fixing security architectural and protocol flaws in AKA and subscriber identity management requires major core equipment modifications, even to end-user devices.
4) Many operators and consumers may not accept the finance load. To avoid real-world vulnerabilities, a new architecture and protocol design requires time to be security tested. A good strategy is to prioritize protocol implementation security patches or endpoint intrusion prevention system upgrades that address vulnerabilities. Ultimately, a design upgrade that eliminates the old architecture's flaws and weaknesses is necessary.
5) End-to-end encryption and mutual authentication remain challenging and need a breakthrough. The lack of these two properties leads to famous attacks like eavesdropping, phony operators, and traceability.
6) Even 5G may not achieve security standards due to increased computing and communication demands. A 6G need for robust end-to-end encryption and mutual authentication may affect latency-sensitive applications without im-

provements in processing capacity and management methods. Any delay in adding these functionalities to 6G might undermine efforts to address current security concerns.

7) Every generation of cellular networks aims to improve security architecture, including authentication and key management, to solve new application and business model concerns. This section covers the 6G roadmap and developments in supporting technologies across three levels (physical, connection/network, and service/application) based on current research and recent white papers (Mahmood et al. 2020). Thus, this study identified security needs and solutions, especially in security architecture, via adjustments to these components.
8) 5G networks prioritize high-quality experience, including numerous radio access technologies (RAT) (Ramezanpour, Jagannath, and Jagannath 2023). Spectrum scarcity is a significant barrier for applications requiring varying RF propagation ranges. Re-allocating unused spectrum bands is a time-consuming operation that may disrupt important military and government uses due to opposition from existing users. Spectrum sharing is a possible solution for spectrum scarcity and a key motivator for B5G networks to achieve a high-quality experience.

FUTURE PROSPECTS OF 5G/6G

The upcoming 5G and 6G broadband cellular networks are anticipated to bring about significant changes in modern society. While 5G is currently a major focus in the business world, its impact on both online and offline behavior is noteworthy. Although the specifics of 6G are yet to be determined, it serves as a blueprint for developments leading up to 2030. Envision a world where 5G is utilized for various applications such as asset tracking, management, intervention planning, remote surgery, robotics in the cloud, remote monitoring, IoT, smart cities and homes, smart medication, automated passenger cars, autonomous driving, intelligent manufacturing, and the Internet of Things (IoT) (Oinas-Kukkonen, Karppinen, and Kekkonen 2021). Furthermore, 5G enables smart logistics to be more cost-effective and is the first global connection solution capable of handling all current traffic profiles. Smart logistic applications such as object tracking, AGVs, drones, remote assistance, machine and robot monitoring and control, intrusion and hazard surveillance, and ambient sensors are made possible by Massive Machine Type Communications (mMTC), Ultra Reliable Low Latency Communications (URLLC), and Enhanced Mobile Broadband (eMBB). The next-generation broadband cellular networks promise even more impressive developments in the coming decade, building on technology advancements over the last decade. Both 5G and 6G will need to consider applications for potential major use cases. This piece covers the benefits and drawbacks of 5G and 6G, aiming to address IT and persuasive systems in general research and design. Mobile technologies will continue to play a significant role in Internet access, with advancements like 5G and 6G paving the way for communities and people to adopt more sustainable practices. Modern technology in fisheries and agriculture will expand opportunities for local populations, including indigenous people in the United States. In addition, for those in remote areas such as the urban-rural split in India and China, technology will also lead to better healthcare, more efficient schools, and increased opportunities to connect with others online. Efforts to improve communication standards to bring wireless broadband internet access to underserved areas have been severely underfunded or completely disregarded. The sixth generation (6G) of wireless networks is still in its early stages, however, it is already addressing the gaps left by earlier generations and will be designed with the specific needs of rural and remote areas in mind, with the end goal of reaching digital inclusion and connecting everyone (Chaoub et al. 2022). This vital market segment must not be disregarded by the research community in favor of more alluring fields like terahertz communications, 3D augmented and virtual reality, haptics, and artificial intelligence (AI).

To improve remote connectivity, we need to address the critical issue of spectrum availability. The licensed spectrum below 1 GHz is expensive, so we may need innovative methods of frequency reuse to meet the unique needs of remote areas. Better network coverage in these locations can be achieved through advanced spectrum-sharing models, improved co-existence strategies, and prudent use of unlicensed bands, which can reduce overall costs. To encourage new participants, such as community micro-operators, to build and operate local networks, we may need new regulatory and commercial models. A potential solution for a thriving remote market is to have flexible and diverse licensing.

Figure 8. A summary of the challenges for providing wireless connectivity to remote areas with proposed solutions

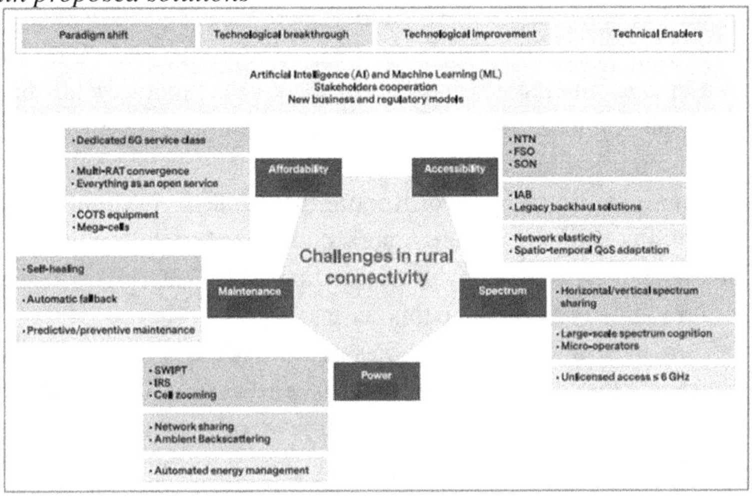

ARTIFICIAL INTELLIGENCE (AI) AND THE INTERNET OF THINGS (IOT)

With the advent of the IoT, the globe has become more accessible, secure, available, scalable, confidential, and interoperable in terms of the connection of devices. Unfortunately, assaults may target IoTs because of their many attack surfaces along with the fact that they are still relatively new, meaning there aren't yet any security standards or criteria in place. Cybercriminals may employ a variety of methods to hack Internet of Things devices, depending on their objectives and the area of the system they're trying to compromise. Therefore, there is a great deal of study on cybersecurity as it pertains to the Internet of Things. Included in this category are methods that use Artificial Intelligence (AI) to safeguard IoT systems against intruders, often by identifying suspicious activity that may point to an assault. Kuzlu, Fair, and Guler., (2021) on the other hand, cybersecurity professionals have several targets to secure, giving cyber-attackers an advantage.

The emergence of Industry 4.0, characterized by the integration of advanced technologies such as the Internet of Things (IoT) and artificial intelligence (AI), represents the culmination of four decades of technological advancement. This era has garnered sustained attention from both researchers and enterprises, prompting various organizations to endeavor to categorize the foundational tools and methodologies that will underpin the fourth industrial revolution. Notably, the Boston

Consulting Group (BCG) has identified nine key technologies as the cornerstone of Industry 4.0 (Hansen and Bøgh, 2021). Among these pivotal technologies are the IoT in industry and big data/analytics, which may encompass AI and machine learning (ML) techniques. While the terms "Smart Manufacturing" and "Smart Factory" are prevalent in the United States and Asia, respectively, "Industry 4.0" holds sway in Europe. Despite nuanced distinctions in their definitions, these terms are used interchangeably in this context due to their shared roots in technological progress within the business landscape. It is worth noting that while large corporations possess the requisite means and expertise to embrace and implement a tailored roadmap for transitioning to Industry 4.0, small and medium-sized enterprises (SMEs) often grapple with resource constraints in this regard.

The application of intelligent machines and the IoT may benefit conventional agriculture. Among the advantages are (a) enhanced data use from smart agricultural sensors, (b) management, and control of internal processes throughout the smart agriculture environment, including crop harvesting and storage management, (c) cost and waste reduction, (d) increased business efficiency through the automation of routine tasks, and (e) better product quality and volume (Alreshidi, 2019). Providing necessary information to farmers quickly is a significant challenge. As a result, AI holds a lot of promise to help conventional agriculture overcome its pressing problems.

AFFORDABLE CONNECTIVITY SOLUTIONS

Affordable and universally accessible infrastructure is required to provide equitable access to 5G and 6G technologies, regardless of geographic location or socioeconomic position. Accessible infrastructure catalyzes not only enhancing connectivity, but also stimulating innovation, fostering societal inclusion, and driving economic prosperity. Governments, telecom providers, and other stakeholders can bridge the digital divide and realize the revolutionary promise of sophisticated telecommunications technology in all areas by making infrastructure more affordable and accessible. The incorporation of the "affordable broadband" pillar has been seen as the fourth pillar in the development of 5G, which will ultimately lead to the establishment of an Enhanced 5G or 6G standard. (Yaacoub, Elias;Alouini 2024) The system primarily relies on three factors: (i) using unlicensed spectrum, including white space, (ii) prioritizing energy efficiency, and (iii) implementing SDN and NFV to decrease structure expenses. In rural regions, the construction of basic infrastructure may be initiated by using affordable pre-made equipment, which should be durable enough to withstand the unique weather conditions of the particular rural location. Additionally, open-source software should be used wherever feasible. A network management and monitoring platform was developed using a variety of open-source software applications to facilitate the cost-effective operation of a local wireless internet service provider within a rural area. The open-source tools used provide network monitoring, network administration, intrusion detection, and firewall features.

INCLUSIVITY IN TECHNOLOGY DESIGN

User-Centered Design Principles

The concept of user-centered design, originating from Donald Norman's research lab at the University of California San Diego, underscores the importance of addressing user needs and interests to ensure the usability of the design. This concept was further elucidated in his book, "The Design of Everyday Things," emphasizing the significance of making actions easily determinable, ensuring visibility, evaluating the current system state, and establishing natural mappings between intentions and required actions and effects, and visible information and the interpretation of the system state (Tellioğlu 2022). The designer's responsibility is to streamline the user's task and guarantee that the product can be utilized as intended with minimal effort. Norman recommends that products should be accompanied by a concise leaflet drawing on the user's worldly knowledge and easily digestible. This approach situates the user at the core of the design process (Norman, D., & Euchner, J. 2023).

Figure 9. User-centered Design Steps

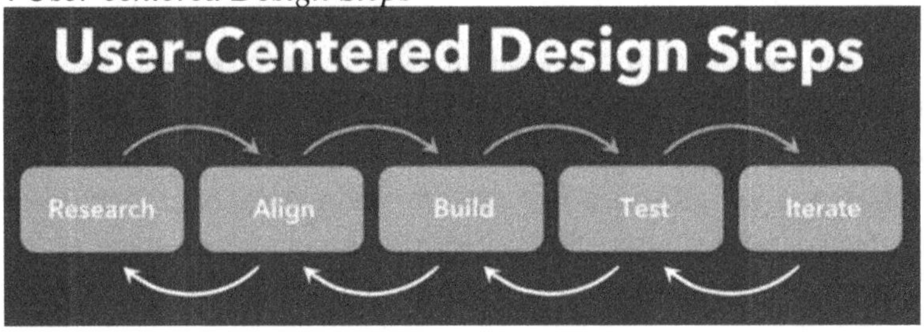

User-centered design places a strong emphasis on prioritizing the needs and perspectives of users in the development of information systems, ensuring a seamless integration into user workflows and an enhancement of user experiences. This approach is instrumental in helping organizations overcome challenges such as resistance during adoption, inefficiencies, and suboptimal outcomes (Crispin P. Noguerra, Jr. 2023). The study delves into the principles of user-centered design in technology development, with a focus on theoretical foundations, practical methodologies, and real-world applications. Its objective is to provide a complete understanding of how user-centered design can be effectively utilized to enhance information system development.

Inclusive Design Frameworks

The goal of the inclusive design philosophy is to provide goods that are helpful and available to the greatest number of people, with a focus on the elderly and disabled. This approach acknowledges that a completely inclusive product is an ideal rather than a practical outcome. The goal is to provide better products that limit the marginalization of less competent groups. Designers need information on the range of user abilities to assess their designs during the inclusive design process (Persad, Langdon, and Clarkson 2019). Guidelines and manuals are the majority of knowledge available on inclusive design, but additional visual forms, checklists, and written recommendations are needed. Quantitative data on the population's loss of functional capacity can also be beneficial for designers, company managers, and decision-makers. Therefore, user capability data is necessary to facilitate the assessment of design ideas at every stage of the design process, from requirement definition to prototype production and final product production.

Participatory design Practices

Participatory design, which has its roots in Scandinavia, involves incorporating users as co-designers in the process of product development. This approach was born out of the desire of labor unions to have more democratic control in work environments. To accommodate cultural variations, it is recommended to use prototyping techniques such as mock-ups or paper-based screen outlines. Additionally, other prototyping techniques include PICTIVE (Plastic Interface for Collaborative Technology Initiatives through Video Exploration) and CARD (Collaborative Analysis of Requirements and Design). PICTIVE primarily focuses on the detailed aspects of the system, while CARD is more concerned with task flow (Tellioğlu, 2022).

Accessible Interface Design

User Interface (UI) and User Experience (UX) Considerations User Experience (UX) is a crucial aspect of product development, with its elements ranging from the User Interface (UI) to the entire system. Employers often overlook the division of design into UI and UX, leading to a lack of productivity and efficiency (Sharma and Tiwari 2021). This is due to a lack of shared ideas and knowledge among designers, making problem-solving difficult. UI designers create interfaces that are relevant to user and system interactions, but UX involves factors like marketing, colors, speed, performance, visualization, and individuality. Similarly, UX requires competent problem-solving, including change recommendations, control processes, and implementation of solutions. The ability and focus of the designer are critical to the outcome of a problem-solving evaluation. However, one designer is not enough to handle both areas, and it is essential to have a team of skilled designers to effectively address and resolve issues.

COMPLIANCE WITH ACCESSIBILITY STANDARDS

Assistive Technologies and Services

Assistive Technology (AT) encompasses any item, equipment, or system used to enhance the functional capabilities of people with disabilities. The Technology-Related Assistance for Individuals with Disabilities Act originally passed in 1988 and reauthorized in 1994 and 1998, defines AT as any item, equipment, or system used to increase or improve functional capabilities (Zallio and Ohashi, 2022). An AT service refers to any direct assistance to individuals with disabilities in selecting, obtaining, or using an AT device. Technological advancements have significantly benefited individuals with disabilities by increasing their participation in society and reducing expenses associated with education, rehabilitation, training, employment, residential living, and independent living. The revised definition from the 1998 act provided a broader concept of AT, encompassing a wider range of applications and potential users.

Innovations in Assistive Tech

Assistive technology (AT) is a crucial tool for those with disabilities or aging into disability, helping them operate more easily. It bridges the “capability gap” that exists among one's goals and objectives and the constraints of his or her surroundings and special talents. In industrialized countries, technological advancement, human rights progress, and progressive public policy are transforming the historical lack of access to assistive devices (Buchanan and Layton 2019). There has been a recent focus on assistive technology's ability to satisfy sustainable development and universal healthcare has revealed numerous gaps in the system. Systems thinking, which identifies five important perspectives: people, products, personnel, policy, and provision, helps systematically assess assistive technology. As a result of supply chain complexity, consumer discontent, and evidence of technology desertion, assistive technology's potential has not yet been completely fulfilled. Matching people and technology is essential, but challenges arise when matching people, environment, and technology.

CHALLENGES IN BRIDGING THE DIGITAL DIVIDE

The phrase "digital divide" refers to three primary elements: (i) possessing the digital abilities required to make the most of online resources and services, (ii) using these skills in practice in terms of frequency and variety, and (iii) physical access to technology and connectivity, which involves the spread of internet access through policy implementation and the availability of affordable technologies. Financial and technical limitations are the primary barriers to achieving comprehensive access for everyone. The capital costs (CapEx) and operating expenses (OpEx) of building and maintaining telecommunication infrastructure are often considered too high by network providers. This is also true for electricity infrastructure. Additionally, the expensive and limited availability of spectrum resources further complicates the global connectivity challenge.

Addressing the digital divide is a crucial initial strategy for sustainable development, even though it may be difficult to measure the benefits of connecting a larger population. For example, online learning tools are expected to become increasingly important for acquiring knowledge and skills as the job market and economy evolve. Therefore, children who lack easy access to these tools as they grow up may struggle to succeed in their chosen fields. Similarly, the pandemic underscored the necessity of enhancing remote health services, particularly in rural regions with restricted access to contemporary medical facilities, and rendering remote work possibilities more attractive to employees in exurban areas. The deployment of 5G/6G networks should incorporate principles such as net neutrality to ensure equal access and prevent discrimination, data ownership and privacy to give users control over their data, robust security to protect user privacy, and ethical AI techniques to address bias and automation concerns.

REAL-TIME APPLICATION

Improving long-distance connectivity involves addressing spectrum availability. Licensed spectrum in sub-1 GHz bands is costly, and unique frequency reuse techniques may be necessary to enhance network coverage in remote areas. Implementing spectrum-sharing models, better co-existence strategies, and using unlicensed bands can reduce costs and attract new participants to build and operate local networks.

The Internet of Things (IoT) ascent has made global device connectivity more accessible, secure, scalable, and interoperable. However, IoT devices are vulnerable to cyber-attacks due to their numerous attack surfaces and the absence of established security standards. Cybersecurity research on IoT focuses on leveraging Artificial Intelligence (AI) to identify and prevent suspicious activities and potential intrusions. Industry 4.0, driven by modern inventions like IoT and AI, represents the culmination of decades of technological advancement. Compared to bigger firms, petite and medium-sized companies frequently lack the resources and knowledge necessary to make the transition to Industry 4.0. In agriculture, IoT and AI technologies offer opportunities for improved data utilization, internal process management, cost reduction, automation, and enhanced product quality and quantity. To ensure equitable access to 5G and 6G technologies, affordable and universally accessible infrastructure is essential. In addition to improving connection, accessible infrastructure promotes social inclusion, innovation, and economic prosperity. Governments, telecom providers, and other stakeholders can bridge the digital divide by making infrastructure more affordable and accessible.

CONCLUSION

In conclusion, the advancements in 5G and the potential for 6G technologies exist to build a more interconnected and effective community. However, the triumph of these improvements rests on our capability to traverse the digital divide. By prioritizing accessibility and inclusion, and addressing the various barriers that hinder universal access to ICT, we can maximize the potential of these technologies to establish a more balanced and thriving global community. The rapid evolution of ICT, marked by the emergence of 5G and the progress toward 6G, has the power to transform modern civilization by reshaping consumer behavior, improving business ecosystems, and raising societal standards through universal ICT access. Despite these advancements, the persistent digital gap remains a significant concern, with differences in access and usage based on wealth, organizational size, and regional development. The evolution from 4G to 5G and, eventually, to 6G introduces new architectural concepts that could potentially bridge this gap by reducing costs, enhancing network efficiency, and enabling specialized applications for underserved populations. Strategic initiatives should focus on promoting accessibility and diversity in technology design through deliberate legislative measures, targeted investments, and collaborative efforts across various sectors.

References

Aissaoui, N. (2022). The Digital Divide: A Literature Review and Some Directions for Future Research in Light of COVID-19. *Global Knowledge, Memory and Communication*, 71(8/9), 686–708. DOI: 10.1108/GKMC-06-2020-0075

Alreshidi, E. (2019). Smart sustainable agriculture (SSA) solution underpinned by internet of things (IoT) and artificial intelligence (AI). arXiv preprint arXiv:1906.03106.

Buchanan, R., & Layton, N. (2019). Innovation in Assistive Technology: Voice of the User. *Societies (Basel, Switzerland)*, 9(2), 1–10. DOI: 10.3390/soc9020048

Chaoub, A., Giordani, M., Lall, B., Bhatia, V., Kliks, A., Mendes, L., Rabie, K., Saarnisaari, H., Singhal, A., Zhang, N., Dixit, S., & Zorzi, M. (2022). 6G for Bridging the Digital Divide: Wireless Connectivity to Remote Areas. *IEEE Wireless Communications*, 29(1), 160–168. DOI: 10.1109/MWC.001.2100137

Chowdhury, M. Z., Shahjalal, M., Ahmed, S., & Jang, Y. M. (2020). 6G Wireless Communication Systems: Applications, Requirements, Technologies, Challenges, and Research Directions. *IEEE Open Journal of the Communications Society*, 1, 957–975. DOI: 10.1109/OJCOMS.2020.3010270

Clancy, J., Mullins, D., Deegan, B., Horgan, J., Ward, E., Denny, P., ... & Glavin, M. (2023). Feasibility study of V2X communications in initial 5G NR deployments. IEEE Access.

De', R., Pandey, N., & Pal, A. (2020). Impact of Digital Surge during Covid-19 Pandemic: A Viewpoint on Research and Practice. *International Journal of Information Management*, 55(June), 102171. DOI: 10.1016/j.ijinfomgt.2020.102171 PMID: 32836633

Dutton, W. H., & Reisdorf, B. C. (2019). Cultural Divides and Digital Inequalities: Attitudes Shaping Internet and Social Media Divides*. *Information Communication and Society*, 22(1), 18–38. DOI: 10.1080/1369118X.2017.1353640

Eissa, H. A., Bozed, K. A., & Younis, H. (2019). "Software Defined Networking." 19th International Conference on Sciences and Techniques of Automatic Control and Computer Engineering, STA 2019 (November): 620–25.

Farhadi Zavleh, A., & Bakhshi, H. (2021). Resource Allocation in Sparse Code Multiple Access-based Systems for Cloud-radio Access Network in 5G Networks. *Transactions on Emerging Telecommunications Technologies*, 32(1), e4153. DOI: 10.1002/ett.4153

Gantumur, Z. (2024). Secure Network Slicing in 5G.

Guterres, A. (2020). *Digital Divide 'a Matter of Life and Death'amid COVID-19 Crisis*. United Nations.

Hansen, E. B., & Bøgh, S. (2021). Artificial Intelligence and Internet of Things in Small and Medium-Sized Enterprises: A Survey. *Journal of Manufacturing Systems*, 58, 362–372. DOI: 10.1016/j.jmsy.2020.08.009

Hidalgo, A., Gabaly, S., Morales-Alonso, G., & Urueña, A. (2020). The Digital Divide in Light of Sustainable Development: An Approach through Advanced Machine Learning Techniques. *Technological Forecasting and Social Change*, 150, 119754. https://www.sciencedirect.com/science/article/pii/S0040162519310637. DOI: 10.1016/j.techfore.2019.119754

Huq, K. M. S., Busari, S. A., Rodriguez, J., Frascolla, V., Bazzi, W., & Sicker, D. C. (2019). Terahertz-Enabled Wireless System for beyond-5G Ultra-Fast Networks: A Brief Survey. *IEEE Network*, 33(4), 89–95. DOI: 10.1109/MNET.2019.1800430

Iivari, N., Sharma, S., & Ventä-Olkkonen, L. (2020). Digital Transformation of Everyday Life – How COVID-19 Pandemic Transformed the Basic Education of the Young Generation and Why Information Management Research Should Care? *International Journal of Information Management*, 55(June), 102183. DOI: 10.1016/j.ijinfomgt.2020.102183 PMID: 32836640

Imoize, A. L., Adedeji, O., Tandiya, N., & Shetty, S. (2021). 6G Enabled Smart Infrastructure for Sustainable Society: Opportunities, Challenges, and Research Roadmap. *Sensors (Basel)*, 21(5), 1709. DOI: 10.3390/s21051709 PMID: 33801302

Khatib, E. J., & Barco, R. (2021). Optimization of 5G Networks for Smart Logistics. *Energies*, 14(6), 1758. DOI: 10.3390/en14061758

Kuzlu, M., Fair, C., & Guler, O. (2021). "Role of Artificial Intelligence in the Internet of Things (IoT) Cybersecurity." Discover. *Internet of Things : Engineering Cyber Physical Human Systems*, 1(1), 7.

Mahmood, N. H., Böcker, S., Munari, A., Clazzer, F., Moerman, I., Mikhaylov, K., . . . Seppänen, P. (2020). White paper on critical and massive machine type communication towards 6G. arXiv preprint arXiv:2004.14146.

Mansouri, Y., & Ali Babar, M. (2021). A Review of Edge Computing: Features and Resource Virtualization. *Journal of Parallel and Distributed Computing*, 150, 155–183. DOI: 10.1016/j.jpdc.2020.12.015

Matinmikko-Blue, M., Aalto, S., Asghar, M. I., Berndt, H., Chen, Y., Dixit, S., . . . Ziegler, V. (2020). White paper on 6G drivers and the UN SDGs. arXiv preprint arXiv:2004.14695.

Matinmikko-Blue, M., Yrjölä, S., & Ahokangas, P. (2020, March). Spectrum management in the 6G era: The role of regulation and spectrum sharing. In 2020 2nd 6G Wireless Summit (6G SUMMIT) (pp. 1-5). IEEE.

Matracia, M., Rahman, A. U., Wang, R., Kishk, M. A., & Alouini, M. S. (2023). Bridging the digital divide. In *Fundamentals of 6G Communications and Networking* (pp. 113–139). Springer International Publishing.

Mihovska, A., & Prasad, R. (2020). "Overview of 5G New Radio and Carrier Aggregation: 5G and beyond Networks." In 2020 23rd International Symposium on Wireless Personal Multimedia Communications (WPMC), IEEE, 1–6.

Moon, N. W., Baker, P. M. A., & Goughnour, K. (2019). Designing Wearable Technologies for Users with Disabilities: Accessibility, Usability, and Connectivity Factors. *Journal of Rehabilitation and Assistive Technologies Engineering*, 6, 205566831986213. DOI: 10.1177/2055668319862137 PMID: 35186318

Network, I T U. (2019). "2030–A Blueprint of Technology, Applications and Market Drivers Towards the Year 2030 and Beyond."

Crispin P. Noguerra, Jr. (2023). "User-Centered Design Principles for Effective Information System Development." International Journal of Advanced Research in Science, Communication and Technology: 826–30.

Norman, D., & Euchner, J. (2023). Design for a Better World: A Conversation with Don Norman. *Research Technology Management*, 66(3), 11–18. DOI: 10.1080/08956308.2023.2183015

O'neill, J. L. (2021). "Accessibility for All Abilities: How Universal Design, Universal Design for Learning, and Inclusive Design Combat Inaccessibility and Ableism." Journal and Inclusive Design Combat Inaccessibility and Ableism. *Journal of Open Access to Law*, 9, 1–15.

Oinas-Kukkonen, H., Karppinen, P., & Kekkonen, M. (2021). 5G and 6G Broadband Cellular Network Technologies as Enablers of New Avenues for Behavioral Influence with Examples from Reduced Rural-Urban Digital Divide. *Urban Science (Basel, Switzerland)*, 5(3), 60. DOI: 10.3390/urbansci5030060

Pandi, S., & Albert, A. J. (2024). A Novel Enhanced Security Architecture for Sixth Generation (6G) Cellular Networks Using Authentication and Acknowledgement (AA) Approach. *Results in Engineering*, 21, 101669. DOI: 10.1016/j.rineng.2023.101669

Persad, U., Langdon, P., & John Clarkson, P. (2019). "A Framework for Analytical Inclusive Design Evaluation." Proceedings of ICED 2007, the 16th International Conference on Engineering Design: 1–12.

Pham, Q.-V., Fang, F., Ha, V. N., Piran, M. J., Le, M., Le, L. B., Hwang, W.-J., & Ding, Z. (2020). A Survey of Multi-Access Edge Computing in 5G and beyond: Fundamentals, Technology Integration, and State-of-the-Art. *IEEE Access : Practical Innovations, Open Solutions*, 8, 116974–17. DOI: 10.1109/ACCESS.2020.3001277

Prakasam, P., Shohel Sayeed, M., & Ajayan, J. (2021). Guest Editorials: P2P Computing for 5G, beyond 5G (B5G) Networks and Internet-of-Everything (IoE). *Peer-to-Peer Networking and Applications*, 14(1), 240–242. DOI: 10.1007/s12083-020-01001-5

Rajatheva, N., ... (2020). "White Paper on Broadband Connectivity in 6G." arXiv preprint arXiv:2004.14247.

Ramezanpour, K., Jagannath, J., & Jagannath, A. (2023). Security and Privacy Vulnerabilities of 5G/6G and WiFi 6: Survey and Research Directions from a Coexistence Perspective. *Computer Networks*, 221, 221. DOI: 10.1016/j.comnet.2022.109515

Rana, D. S., Dhondiyal, S. A., & Chamoli, S. K. (2019). Software Defined Networking (SDN) Challenges, Issues and Solution. *International Journal on Computer Science and Engineering*, 7(1), 884–889.

Rappaport, T. S., Xing, Y., Kanhere, O., Ju, S., Madanayake, A., Mandal, S., & Trichopoulos, G. C. (2019). Wireless communications and applications above 100 GHz: Opportunities and challenges for 6G and beyond. *IEEE Access : Practical Innovations, Open Solutions*, 7, 78729–78757.

Reddick, C. G., Enriquez, R., Harris, R. J., & Sharma, B. (2020). Determinants of broadband access and affordability: An analysis of a community survey on the digital divide. *Cities (London, England)*, 106, 102904.

Roth, R. (2017). User interface and user experience (UI/UX) design. Geographic Information Science & Technology Body of Knowledge, 2017(Q2).

Rupprecht, D., Kohls, K., Holz, T., & Pöpper, C. (2020). "Call Me Maybe: Eavesdropping Encrypted {LTE} Calls with {ReVoLTE}." In 29th USENIX Security Symposium (USENIX Security 20), 73–88.

Rush, K. L., Seaton, C. L., Corman, K., Hawe, N., Li, E. P. H., Dow-Fleisner, S. J., & Pesut, B. (2022). Virtual care prior to and during COVID-19: Cross-sectional survey of rural and urban adults. *JMIR Formative Research*, 6(8), e37059.

Saad, W., Bennis, M., & Chen, M. (2019). A Vision of 6G Wireless Systems: Applications, Trends, Technologies, and Open Research Problems. *IEEE Network*, 34(3), 134–142. DOI: 10.1109/MNET.001.1900287

Saarnisaari, H., ... (2020). "A 6G White Paper on Connectivity for Remote Areas." arXiv preprint arXiv:2004.14699.

Salameh, A. I., & El Tarhuni, M. (2022). From 5G to 6G—Challenges, Technologies, and Applications. *Future Internet*, 14(4), 117. DOI: 10.3390/fi14040117

Setthasuravich, P., & Kato, H. (2020). The Mediating Role of the Digital Divide in Outcomes of Short-Term Transportation Policy in Thailand. *Transport Policy*, 97(March), 161–171. DOI: 10.1016/j.tranpol.2020.07.008

Van Dijk, J. (2020). The Digital Divide. John Wiley & Sons. Docomo, N T T. 2020. "Inc. White Paper—5G Evolution and 6g; NTT Docomo." Inc.: Tokyo, Japan.

Chapter 4
Evaluating the Economic Impact of 5G-6G Implementations Across Diverse Industry Sectors

S. C. Vetrivel
https://orcid.org/0000-0003-3050-8211
Kongu Engineering College, India

P. Vidhyapriya
https://orcid.org/0000-0002-1421-8743
Kongu Engineering College, India

V. P. Arun
JKKN College of Engineering and Technology, India

ABSTRACT

The advent of 5G and the prospective rollout of 6G technologies represent transformative milestones in the telecommunications industry, promising to revolutionize connectivity and digital interaction across various sectors. This chapter aims to evaluate the economic impact of implementing 5G and 6G networks on diverse industry sectors, including healthcare, manufacturing, transportation, agriculture, and entertainment. By leveraging quantitative data analysis and qualitative case studies, the research explores how these advanced networks enhance operational efficiencies, drive innovation, and create new business opportunities. The chapter highlight significant productivity gains, cost reductions, and the emergence of new revenue streams, alongside challenges such as the need for substantial infrastructure investments and the potential for digital inequality.

DOI: 10.4018/979-8-3693-6725-4.ch004

1. INTRODUCTION

1.1 Overview of 5G and 6G Technologies

5G technology, the fifth generation of mobile networks, represents a significant leap forward from its predecessor, 4G. It is designed to provide higher speeds, lower latency, and greater connectivity, thereby enabling a wide array of new applications and services. With peak speeds potentially exceeding 10 Gbps, 5G is set to revolutionize industries by supporting high-definition video streaming, real-time gaming, and ultra-responsive smart devices. The technology also boasts a latency of as low as one millisecond, crucial for applications requiring immediate feedback, such as autonomous vehicles and remote surgeries. Moreover, 5G's enhanced connectivity, capable of supporting up to a million devices per square kilometer, will facilitate the growth of the Internet of Things (IoT), allowing for more efficient smart cities, healthcare systems, and industrial processes.As we look beyond 5G, the development of 6G technology is already underway, promising even more transformative capabilities. 6G is expected to offer speeds up to 100 times faster than 5G and latency reduced to microseconds, opening up possibilities for unprecedented advancements in various fields (Abdellatif et al., 2019). This includes the integration of advanced AI, augmented reality (AR), virtual reality (VR), and ubiquitous computing, which will collectively push the boundaries of what is technologically feasible. The ultra-reliable and low-latency communications (URLLC) of 6G will be critical for applications requiring extreme precision and reliability, such as advanced robotics, haptic communications, and immersive experiences that blend the physical and digital worlds seamlessly.The potential impact of these technologies is vast and far-reaching. In healthcare, for instance, 5G and 6G can enable remote patient monitoring and real-time health data analysis, improving patient outcomes and reducing healthcare costs (Andrews et al., 2014). In the automotive industry, these technologies will be integral to the development and deployment of autonomous vehicles, enhancing safety and efficiency on the roads. The manufacturing sector will benefit from smart factories where machines and systems are interconnected, allowing for more flexible and efficient production processes. Even in entertainment and media, the enhanced capabilities of 5G and 6G will facilitate the creation and delivery of high-quality, immersive content, transforming how consumers experience media. Furthermore, the rollout of 5G and the anticipated arrival of 6G are expected to drive significant economic growth (Ankarali et al., 2017)(Berardinelli et al., 2018). These technologies will create new business opportunities, spur innovation, and lead to the development of entirely new industries. They will also play a crucial role in addressing global challenges, such as climate change, by enabling smarter resource management and more efficient energy use.

1.2 Objectives and Scope of the Study

The primary objective of this study is to thoroughly evaluate the economic impact of implementing 5G and 6G technologies across various industry sectors. As these next-generation wireless technologies promise unprecedented advancements in speed, connectivity, and data processing capabilities, it is crucial to understand their potential economic implications. This study aims to provide a comprehensive analysis of how 5G and 6G implementations can drive economic growth, enhance operational efficiencies, and create new business opportunities across different sectors.The scope of this study encompasses a wide range of industries, including telecommunications, manufacturing, healthcare, transportation, energy, retail, financial services, and the public sector. By examining each of these sectors in detail, the study seeks to identify specific areas where 5G and 6G technologies can deliver significant economic benefits. For instance, in manufacturing, the focus will be on how these technologies enable smart factories and enhance supply chain management (Bochechka & Tikhvinskiy, 2014). In healthcare, the study will explore the potential for improved telemedicine services and advanced medical research. Additionally, the study aims to highlight the transformative potential of 5G and 6G in fostering innovation and driving digital transformation across industries. This includes examining the role of these technologies in enabling new business models, improving customer experiences, and facilitating the development of new products and services. The study will also consider the broader societal impacts, such as increased accessibility to high-speed internet and the potential for bridging the digital divide (Cambria, 2016).

2. THE EVOLUTION FROM 5G TO 6G

2.1 Key Differences and Advancements

The transition from 5G to 6G represents a significant leap in technological advancements, each generation building on the foundation of its predecessor while introducing new capabilities and enhancements (Chen et al., 2011). While 5G, the fifth generation of mobile network technology, brought substantial improvements in speed, latency, and connectivity over 4G, 6G aims to push these boundaries even further, creating an entirely new paradigm in wireless communication.

- **Speed and Latency Improvements:** One of the most notable differences between 5G and 6G is the expected increase in speed and reduction in latency. 5G networks already offer impressive data rates, with peak speeds of up to 10 Gbps and latencies as low as 1 millisecond. However, 6G is anticipated to provide data rates that are 10 to 100 times faster than 5G, potentially reaching speeds of 1 Tbps (terabits per second). This increase will facilitate instantaneous communication, significantly enhancing real-time applications such as augmented reality (AR), virtual reality (VR), and remote control of machinery and vehicles.

- **Frequency Spectrum Utilization**
- 5G networks primarily operate in the sub-6 GHz and millimeter-wave (mmWave) frequency bands. In contrast, 6G is expected to exploit higher frequency bands, possibly extending into the terahertz (THz) range (Cherkasova et al., 2021). Utilizing these higher frequencies will enable the transmission of larger volumes of data and support more devices per square kilometer. However, it also presents challenges such as increased signal attenuation and the need for advanced materials and technologies to manage signal propagation and penetration.
- **Network Capacity and Density:**Another critical advancement is the enhancement in network capacity and device density. 5G significantly increased the number of devices that can be connected simultaneously, accommodating the explosion of IoT devices. 6G aims to expand this capability even further, potentially supporting up to 10 million devices per square kilometer. This improvement will be crucial for the continued growth of IoT ecosystems, smart cities, and connected industries, where a vast number of sensors and devices need to communicate seamlessly.
- **Intelligent Network Management:** 6G networks will likely incorporate more sophisticated AI and machine learning techniques for intelligent network management. These technologies will enable dynamic allocation of resources, predictive maintenance, and real-time optimization of network performance ("Digital 2020: July Global Statshot," 2020). AI-driven management will also enhance security by detecting and mitigating threats more effectively, ensuring the integrity and reliability of the network.
- **Integration with Emerging Technologies:** While 5G has begun to integrate with emerging technologies such as AI, IoT, and edge computing, 6G is expected to deepen and expand these integrations. For example, 6G networks will likely have native support for AI, enabling real-time data processing and decision-making at the network edge. This will reduce the need for data to travel to centralized cloud servers, decreasing latency and improving the efficiency of applications that require immediate responses, such as autonomous vehicles and industrial automation.
- **Energy Efficiency and Sustainability:**Sustainability is becoming increasingly important in technology development, and 6G aims to address ener-

gy efficiency more comprehensively than 5G. Advanced materials, energy-efficient hardware, and intelligent power management techniques will be crucial in reducing the energy consumption of 6G networks. These improvements will help mitigate the environmental impact of massive data transmission and support the global push towards more sustainable technological solutions.

- **User-Centric Innovations:** 6G is expected to bring a more user-centric approach to connectivity. This includes enhanced quality of service (QoS) tailored to individual user needs, more immersive and personalized experiences through advanced AR and VR, and better support for diverse application requirements. By focusing on the end-user experience, 6G aims to create a more intuitive and seamless interaction with technology, making advanced services accessible and beneficial to a broader audience.

2.2 Technological Innovations

The transition from 5G to 6G represents a significant leap in technological advancements, promising to revolutionize the way we interact with digital infrastructure and reshape various industry sectors. 5G, with its enhanced mobile broadband, ultra-reliable low latency communications, and massive machine-type communications, has already set the stage for a more connected and efficient world. However, 6G aims to build on these foundations and introduce groundbreaking innovations that will further enhance connectivity, data transmission, and overall user experience. One of the key technological innovations in 6G is the expected increase in data transfer speeds, reaching up to 100 times faster than 5G. This dramatic improvement will be facilitated by the use of terahertz (THz) frequency bands, which offer significantly larger bandwidths compared to the millimeter-wave frequencies used in 5G. These higher frequencies will enable the transmission of vast amounts of data almost instantaneously, supporting applications that require real-time processing, such as augmented reality (AR), virtual reality (VR), and immersive gaming experiences.Another major innovation is the integration of advanced AI and machine learning algorithms within the 6G network architecture. These intelligent systems will enhance network management by optimizing resource allocation, predicting maintenance needs, and dynamically adjusting to varying traffic loads. AI-driven automation will also play a crucial role in ensuring seamless connectivity and minimizing latency, which is critical for applications like autonomous driving and remote surgery. Additionally, AI will facilitate more efficient spectrum sharing and interference management, ensuring optimal use of the available bandwidth.6G is also expected to bring significant advancements in network security and privacy (Fang et al., 2020) (Farris et al., 2018). As the number of connected devices continues to grow, the potential for cyber threats increases. To address this, 6G networks will incorporate advanced cryptographic techniques and decentralized security frameworks. These measures will not only protect data integrity and user privacy but also enable secure and trusted communication between devices in the Internet of Things (IoT) ecosystem. Furthermore, quantum communication technologies are being explored as a means to enhance security, offering virtually unbreakable encryption methods.Edge computing will be another cornerstone of 6G innovation, shifting data processing closer to the source of data generation. This will reduce latency and bandwidth usage, enabling faster and more efficient data analysis and decision-making processes. Edge computing will be particularly beneficial for applications that require real-time data processing, such as industrial automation, smart cities, and autonomous vehicles. By leveraging edge computing, 6G networks will support more resilient and responsive services, enhancing the overall user experience (Gangakhedkar et al., 2018).

2.3 Expected Deployment Timelines

The transition from 5G to 6G represents a significant leap in technology, with far-reaching implications for connectivity, speed, and overall network capabilities. Understanding the expected deployment timelines of these technologies is crucial for stakeholders across various sectors to prepare for and capitalize on the opportunities they present.

2.3.1 5G Deployment Timelines

The rollout of 5G technology began in earnest around 2019, with early adopters like South Korea, the United States, and China leading the charge. By 2020, many countries had initiated or were in the process of planning their 5G deployments, focusing on urban centers where the demand for high-speed connectivity was highest. The deployment of 5G is typically staggered in phases:

- **Initial Rollout (2019-2021)**: This phase involved the deployment of non-standalone (NSA) 5G networks, which rely on existing 4G infrastructure (Hoshina et al., 2021). The focus was on enhancing mobile broadband services and providing higher speeds in densely populated urban areas.
- **Expansion and Standalone (SA) Networks (2022-2024)**: As the technology matures, operators are increasingly shifting to standalone 5G networks that do not depend on 4G infrastructure. These networks offer lower latency and support advanced applications like IoT and autonomous vehicles. During this phase, coverage expands to suburban and rural areas.

- **Widespread Adoption (2025 onwards)**: By mid-decade, 5G is expected to achieve widespread adoption, with comprehensive coverage in both urban and rural areas. This phase will see a proliferation of 5G-enabled devices and the integration of 5G into various industry applications, from smart cities to industrial automation.

2.3.2 6G Deployment Timelines

While 5G is still in the process of global deployment, the vision for 6G is already taking shape. Research and development efforts for 6G began as early as 2020, with the aim of addressing the limitations of 5G and pushing the boundaries of wireless communication. The expected deployment timeline for 6G can be outlined as follows:

- **Research and Conceptualization (2020-2025)**: This initial phase focuses on fundamental research to define the technical standards and capabilities of 6G. Key areas of exploration include terahertz frequency bands, advanced antenna technologies, AI-driven network management, and the convergence of communication with sensing and computing.
- **Standardization and Early Trials (2025-2030)**: As research progresses, international standardization bodies like the International Telecommunication Union (ITU) and 3GPP will begin to establish 6G standards. Concurrently, early trials and pilot projects will be conducted to test and validate 6G technologies in real-world scenarios.
- **Pre-Commercial Deployment (2030-2032)**: By the early 2030s, 6G technology will enter the pre-commercial deployment phase. Leading telecom operators and technology companies will start implementing 6G networks in select areas, focusing on high-demand environments such as smart cities, industrial hubs, and advanced healthcare facilities.
- **Commercial Rollout (2032 onwards)**: The commercial rollout of 6G is anticipated to begin around 2032, gradually expanding to achieve global coverage. 6G will offer unprecedented data rates, ultra-low latency, and highly reliable connectivity, enabling a new era of applications like holographic communications, immersive virtual and augmented reality, and ubiquitous AI integration.

2.3.3 Challenges and Considerations

The deployment timelines for both 5G and 6G are influenced by several factors, including regulatory approvals, spectrum availability, infrastructure investments, and the readiness of the ecosystem. Governments and regulatory bodies play a crucial role in allocating the necessary spectrum and creating favorable policies to accelerate deployment (Huynh-The et al., 2020). Additionally, the significant capital expenditure required for infrastructure development necessitates collaboration between telecom operators, technology providers, and investors.

Moreover, the deployment of these advanced networks must address concerns related to cybersecurity, data privacy, and the environmental impact of increased electronic waste. Sustainable practices and robust security measures will be essential to ensure the responsible deployment of 5G and 6G technologies.

3. ECONOMIC IMPACT ON THE TELECOMMUNICATIONS SECTOR

3.1 Infrastructure Investments

The deployment of 5G and 6G technologies necessitates significant infrastructure investments, which play a crucial role in determining the success and economic impact of these next-generation networks. The transition from 4G to 5G involves upgrading existing network components and installing new infrastructure to support the increased data rates, lower latency, and higher device density characteristic of 5G. This includes the development of new base stations, small cells, and fiber optic networks to handle the vast amounts of data generated by 5G-enabled devices and applications. Additionally, the transition to 6G will likely require even more sophisticated infrastructure to support its anticipated capabilities, such as terahertz communication, advanced AI integration, and pervasive computing environments. One of the primary areas of investment is the expansion of small cell networks. Unlike traditional macro cell towers used in previous generations of mobile networks, 5G and future 6G networks rely heavily on small cells to provide dense coverage and maintain high-speed connectivity in urban environments (ITU-R M.2370-0, 2015). These small cells, often installed on streetlights, buildings, and other urban fixtures, require extensive planning and coordination with local authorities and property owners. The installation of small cells is not only a technical challenge but also a logistical and regulatory one, as it involves navigating complex zoning laws and public opposition.Moreover, fiber optic networks are essential for backhauling the vast amounts of data generated by 5G and 6G networks. Fiber optics provide the necessary bandwidth and low latency required to support high-speed wireless communication (Jiang et al., 2018). This necessitates significant investment in laying new fiber optic cables and upgrading existing ones, particularly in regions where current infrastructure is outdated or insufficient. Governments and private sector stakeholders must collaborate to fund these projects, as the cost of fiber deployment can be substantial, especially in rural or underserved areas where the return on investment may not be immediately apparent.Another critical component of 5G and 6G infrastructure is the implementation of edge computing. Edge computing involves processing data closer to the source of data generation rather than relying solely on centralized data centers. This reduces latency and allows for real-time data processing, which is essential for applications such as autonomous vehicles, smart cities, and industrial automation. Establishing edge computing infrastructure requires investment in local data centers, edge nodes, and related technologies to ensure seamless data processing and connectivity.The economic impact of these infrastructure investments is multifaceted. In the short term, the construction and deployment of 5G and 6G infrastructure generate jobs and stimulate economic activity in the telecommunications and construction sectors. In the long term, the enhanced connectivity and capabilities provided by these networks drive innovation across various industries, leading to the development of new products, services, and business models. For example, advanced manufacturing, telemedicine, augmented reality, and smart grid technologies all stand to benefit from the high-speed, low-latency connectivity of 5G and 6G networks.However, the scale of the required investments poses challenges, particularly for smaller telecommunications companies and emerging economies. Public-private partnerships and government incentives can play a vital role in facilitating these investments. Policymakers must also consider regulatory frameworks that encourage investment while ensuring fair competition and access. Balancing these factors is crucial to achieving the widespread deployment of 5G and 6G technologies and maximizing their economic benefits.

3.2 Revenue Models and Monetization

The advent of 5G and the anticipated rollout of 6G technologies represent a paradigm shift in the telecommunications industry, necessitating new revenue models and monetization strategies. Traditional revenue streams, primarily driven by voice and data services, are no longer sufficient to capitalize on the vast potential these next-generation networks offer. Consequently, telecom operators are exploring innovative business models to leverage the unique capabilities of 5G and 6G.One of the most promising revenue models is the adoption of network slicing. Network slicing allows operators to create multiple virtual networks within a single physical 5G or 6G infrastructure, each tailored to meet the specific requirements of different use cases (Juneja et al., 2021). This capability enables telecom companies to offer customized solutions to various industries, from high-bandwidth slices for media streaming to ultra-reliable low-latency slices for autonomous vehicles and critical IoT applications. By monetizing these specialized network slices, operators can tap into new revenue streams beyond traditional consumer markets.Another significant revenue opportunity lies in the expansion of the Internet of Things (IoT) ecosystem. 5G and 6G networks, with their enhanced capacity and connectivity, can support a massive number of IoT devices, facilitating advanced applications in smart cities, industrial automation, healthcare, and agriculture. Telecom operators can monetize this ecosystem by providing connectivity services, managing IoT platforms, and offering value-added services such as data analytics and edge computing. The ability to handle vast amounts of data generated by IoT devices and process it at the network edge will be critical in delivering real-time insights and driving efficiency gains across various sectors (Kimura et al., 2021) (Kota, 2011).The emergence of edge computing is another vital component of the new revenue models associated with 5G and 6G. Edge computing brings computation and data storage closer to the source of data generation, reducing latency and improving performance. Telecom operators can offer edge computing services to enterprises, enabling them to deploy applications that require low-latency processing and high reliability. This includes critical applications in areas like autonomous driving, augmented reality, and remote healthcare. By providing infrastructure and platforms for edge computing, telecom companies can create new revenue streams while enhancing their value proposition to enterprise customers.

3.3 Regulatory and Policy Implications

The deployment and expansion of 5G and the anticipated rollout of 6G technologies bring a myriad of regulatory and policy implications that are crucial for ensuring smooth implementation and maximizing economic benefits. Governments and regulatory bodies play a pivotal role in shaping the environment in which these advanced telecommunications networks operate, impacting everything from infrastructure development to consumer protection.

3.3.1 Spectrum Allocation and Management: One of the primary regulatory challenges associated with 5G and 6G is the allocation and management of spectrum. Spectrum is a finite resource, and its efficient allocation is critical for the performance and reach of 5G and 6G networks. Regulatory bodies must balance the needs of various stakeholders, including telecommunications companies, public services, and emerging tech firms. This involves auctioning spectrum licenses, re-farming existing bands, and ensuring that spectrum usage is optimized to prevent interference and promote fair competition. As we move towards 6G, which promises to use higher frequency bands, regulators will need to address new technical challenges and potential health concerns associated with these frequencies.

3.3.2 Infrastructure Deployment and Investment: The deployment of 5G infrastructure requires significant investment in new base stations, fiber optic cables, and other supporting technologies. Regulatory frameworks need to incentivize these investments while ensuring that they are economically viable for companies (Levy, 2005). Policies that streamline the permitting process for new infrastructure, provide financial incentives such as tax breaks or subsidies, and promote public-private partnerships can help accelerate the deployment of 5G networks. Moreover, as 6G approaches, the need for further investment in research and development will be crucial, necessitating forward-thinking policies that support innovation and collaboration between academia, industry, and government.

3.3.3 Data Privacy and Security: With the increased connectivity and data transmission capabilities of 5G and future 6G networks, ensuring data privacy and security becomes more complex and critical. Regulatory frameworks must evolve to address the heightened risks associated with more extensive data collection and real-time data processing. This includes implementing stringent data protection laws that mandate how data is collected, stored, and shared, as well as enhancing cybersecurity measures to protect against potential threats. Additionally, regulators need to enforce compliance with these laws and provide clear guidelines for handling breaches when they occur. The transition to 6G will likely introduce new security challenges, necessitating continuous updates to regulatory standards and practices.

3.3.4 Net Neutrality and Fair Competition: Ensuring that the benefits of 5G and 6G technologies are equitably distributed requires robust policies on net neutrality and competition. Net neutrality principles prevent service providers from discriminating against certain types of content or services, ensuring an open internet where innovation can thrive. Regulatory bodies must enforce these principles to prevent monopolistic behaviors and encourage competition among service providers (Li, Guo, Sun, Al-Rubaye, & Tsourdos, 2020). This is particularly important as new business models and services emerge with the advent of 5G and 6G, which could potentially lead to anti-competitive practices if not properly regulated.

3.3.5 Environmental and Health Regulations: The rollout of 5G and the development of 6G networks also raise environmental and health concerns. The increased number of base stations and antennas required for these networks can have environmental impacts, including energy consumption and electromagnetic emissions. Regulators must establish guidelines that minimize these impacts, promoting the use of energy-efficient technologies and ensuring that emissions remain within safe limits. Additionally, continuous research into the health effects of long-term exposure to higher frequency bands is essential to inform regulatory decisions and reassure the public.

4. TRANSFORMING INDUSTRY SECTORS WITH 5G-6G

4.1 Manufacturing and Industrial Automation

The implementation of 5G and future 6G technologies is poised to revolutionize the manufacturing sector, ushering in an era of smart factories and unparalleled industrial automation. At the core of this transformation lies the ability of 5G-6G networks to provide ultra-low latency, high reliability, and massive connectivity, which are essential for supporting advanced manufacturing processes.

4.1.1 Smart Factories and IoT Integration

Smart factories leverage IoT devices and sensors interconnected through 5G-6G networks to enable real-time monitoring and control of manufacturing operations. These networks facilitate seamless communication between machines, robots, and other equipment, optimizing production efficiency and minimizing downtime. For instance, predictive maintenance powered by AI algorithms can preemptively detect equipment failures, reducing costly disruptions and enhancing overall operational reliability. Moreover, real-time data analytics enable manufacturers to make data-driven decisions, improve resource utilization, and streamline supply chain management.

4.1.2 Supply Chain Optimization

5G-6G networks also play a crucial role in optimizing supply chain processes within manufacturing. Enhanced connectivity allows for real-time tracking of goods, enabling manufacturers to monitor inventory levels, predict demand fluctuations, and ensure timely deliveries. This level of transparency and efficiency not only reduces operational costs but also enhances customer satisfaction through faster order fulfillment and improved service reliability. Furthermore, by integrating blockchain technology with 5G-6G networks, manufacturers can secure and authenticate supply chain transactions, ensuring traceability and minimizing the risk of counterfeiting.

4.2 Transforming Healthcare and Medical Services

The integration of 5G and anticipated advancements with 6G technologies holds immense promise for revolutionizing healthcare and medical services. One of the primary benefits lies in enhancing telemedicine capabilities to deliver remote healthcare services with unprecedented efficiency and reliability. With 5G's low latency and high bandwidth, healthcare providers can conduct real-time consultations, monitor patients remotely, and even perform complex surgeries through augmented reality (AR) and virtual reality (VR) applications. This transformation not only improves access to healthcare for remote and underserved areas but also reduces healthcare costs and enhances patient outcomes through timely interventions and continuous monitoring.Moreover, the synergy between 5G and emerging technologies like artificial intelligence (AI) and big data analytics is set to further amplify these benefits. AI algorithms can analyze vast amounts of patient data in real-time, enabling personalized treatment plans and predictive diagnostics (Li, La Guardia, Liu, & Peng, 2019). This capability is crucial in early disease detection and management, optimizing healthcare resources, and improving overall patient care quality. Additionally, 5G-enabled medical devices and wearables can continuously monitor patient health metrics, providing proactive alerts to healthcare providers and empowering patients to actively manage their health.

However, alongside these transformative benefits, the implementation of 5G in healthcare faces significant challenges. These include ensuring data security and patient privacy, interoperability of healthcare systems, and regulatory compliance. Addressing these challenges requires robust cybersecurity measures, standardized protocols for data exchange, and collaboration among stakeholders to establish clear regulatory frameworks (Li, Liu, Zeng, & Xiong, 2021). By overcoming these hurdles, the healthcare industry stands to realize substantial economic savings and societal benefits, marking a paradigm shift towards more accessible, efficient, and patient-centered healthcare services.

4.2.1 Telemedicine and Remote Surgery

Telemedicine, enabled by advancements in 5G and anticipated enhancements with 6G, represents a transformative approach to healthcare delivery. It leverages high-speed, low-latency connectivity to facilitate remote consultations, diagnosis, and treatment, thereby overcoming geographical barriers and improving access to healthcare services. In telemedicine, patients can connect with healthcare providers through video conferencing, allowing for real-time communication and medical examination from virtually anywhere. This capability is particularly beneficial in rural or underserved areas where access to specialists may be limited. Moreover, telemedicine enhances patient convenience by reducing travel time and costs associated with in-person visits.Remote surgery, an advanced application of telemedicine, relies on ultra-reliable low-latency communication (URLLC) offered by 5G and future 6G networks. It enables surgeons to perform procedures on patients located in different geographical locations with precision and efficiency (Lovley, 2017). This is achieved through robotic systems controlled by surgeons remotely, supported by high-definition video feeds and haptic feedback mechanisms that simulate the sense of touch. Remote surgery has the potential to revolutionize healthcare by allowing experts to perform complex operations across long distances, collaborate with international colleagues, and provide immediate assistance in emergency situations.Both telemedicine and remote surgery promise to enhance healthcare outcomes by speeding up diagnosis, reducing treatment delays, and expanding access to specialized medical expertise. However, their widespread adoption depends on addressing technical challenges such as network reliability, cybersecurity concerns, and regulatory frameworks to ensure patient safety and data privacy. As these technologies continue to evolve alongside 5G and future 6G developments, they hold the promise of transforming the healthcare landscape, making quality medical care more accessible and efficient globally.

4.2.2 AI and Big Data in Healthcare

AI and big data are revolutionizing healthcare by enhancing diagnostic accuracy, improving patient outcomes, and optimizing operational efficiencies. In the realm of diagnostics, AI algorithms can analyze vast amounts of medical data, including imaging scans, pathology reports, and patient records, to detect patterns and anomalies that may be imperceptible to human eyes. For instance, AI-powered image recognition systems can assist radiologists in identifying early signs of diseases like cancer with higher precision and speed, thereby facilitating earlier interventions and improving survival rates.Moreover, AI enables personalized medicine by analyzing genetic data and patient histories to tailor treatment plans that are more effective and tailored to individual patient profiles (Mao et al., 2021). Machine learning algorithms can predict patient responses to specific treatments based on data from similar cases, helping clinicians make informed decisions that optimize therapeutic outcomes while minimizing side effects. This approach not only improves patient care but also contributes to the overall efficiency of healthcare delivery.In parallel, big data analytics plays a crucial role in healthcare by aggregating and analyzing vast datasets from diverse sources, such as electronic health records (EHRs), wearable devices, and population health data. By leveraging big data analytics, healthcare providers can identify trends, risk factors, and epidemiological patterns at population levels, enabling proactive healthcare management and disease prevention strategies.

For example, analyzing population health data can help public health officials detect and respond to disease outbreaks more effectively, leading to faster containment and reduced spread.However, the integration of AI and big data in healthcare is not without challenges. Issues such as data privacy and security, regulatory compliance, and ethical considerations surrounding the use of patient data must be carefully addressed to ensure trust and adherence to healthcare standards. Furthermore, there is a need for ongoing research and collaboration to refine AI algorithms, validate their effectiveness in clinical settings, and address biases that may affect diagnostic accuracy and treatment outcomes.Despite these challenges, the synergy between AI and big data holds immense promise for transforming healthcare into a more personalized, efficient, and accessible system. As technology continues to advance, continued investment in research, education, and infrastructure will be essential to fully harness the potential of AI and big data in improving healthcare outcomes and enhancing patient-centered care.

4.3 Transportation and Logistics

4.3.1 Autonomous Vehicles

The integration of 5G and future 6G technologies promises a transformative impact on transportation and logistics sectors, particularly through the advancement of autonomous vehicles (AVs). AVs rely on high-speed, low-latency connectivity provided by 5G and anticipated enhancements with 6G to operate safely and efficiently. These vehicles can communicate with surrounding infrastructure, other vehicles, and pedestrians in real-time, enabling seamless coordination and decision-making. Beyond enhancing safety through reduced human error, AVs offer potential benefits such as optimized traffic flow, reduced congestion, and lower emissions. In logistics, autonomous trucks and drones equipped with 5G-6G capabilities promise faster and more efficient delivery processes, with applications ranging from last-mile delivery solutions to warehouse management systems (Matinmikko-Blue et al., 2019). The economic implications are profound, as autonomous transportation not only reduces operational costs but also opens up new opportunities for innovative business models and services. However, challenges such as regulatory frameworks, public acceptance, and cybersecurity must be addressed to realize the full potential of AVs in transforming transportation and logistics industries.

4.3.2 Fleet Management and Optimization

Fleet management stands to undergo a revolutionary transformation with the implementation of 5G and eventual integration of 6G technologies. The advent of these advanced cellular networks promises to enhance the efficiency, safety, and sustainability of transportation and logistics operations worldwide. By providing ultra-low latency and high reliability, 5G and 6G will enable real-time monitoring and control of fleet vehicles, improving route optimization, vehicle maintenance, and fuel efficiency. Autonomous vehicles, powered by these networks, will navigate traffic more intelligently, reducing congestion and carbon emissions. Moreover, the integration of IoT sensors and AI-driven analytics will offer predictive maintenance capabilities, detecting potential issues before they cause disruptions (Mei et al., 2021). Fleet operators will benefit from enhanced visibility into their operations, enabling proactive decision-making and cost reductions. Overall, the evolution from 5G to 6G in fleet management promises a future where transportation networks are smarter, safer, and more sustainable, ushering in a new era of efficiency and innovation in logistics and transportation sectors globally.

4.4 Energy and Utilities

4.4.1 Smart Grids and Energy Management

Smart grids represent a paradigm shift in how electricity is generated, distributed, and consumed. By integrating advanced communication and control technologies, smart grids enable real-time monitoring, optimization, and management of energy flow across the grid. This capability enhances grid reliability, efficiency, and resilience, while also accommodating the growing adoption of renewable energy sources and electric vehicles. Smart grids leverage IoT sensors, data analytics, and machine learning algorithms to detect and respond to changes in energy demand and supply dynamically. This real-time responsiveness allows utilities to balance grid operations more effectively, reduce peak demand through demand response programs, and optimize energy distribution to minimize losses (Moldovan et al., 2017). Moreover, smart grids empower consumers with tools such as smart meters and energy management systems, enabling them to monitor their energy usage, adjust consumption patterns, and participate in energy-saving initiatives. As smart grid technologies continue to evolve, they are expected to play a crucial role in achieving energy sustainability goals by facilitating the integration of decentralized renewable energy sources and fostering a more efficient and resilient energy infrastructure.

4.4.2 Renewable Energy Integration

The integration of renewable energy sources, such as solar, wind, and hydroelectric power, into the traditional energy grid is central to achieving global sustainability targets and reducing carbon emissions. Renewable energy technologies offer clean and abundant sources of electricity, but their variable nature poses challenges for grid stability and reliability. Advanced AI algorithms and predictive analytics play a pivotal role in overcoming these challenges by forecasting renewable energy generation patterns, optimizing energy storage solutions, and managing grid operations in real-time ("NVIDIA Maxine," 2020). AI-driven predictive models analyze weather data, historical energy production, and demand patterns to anticipate fluctuations in renewable energy output and adjust grid operations accordingly.

4.5 Retail and Consumer Services

4.5.1 Enhanced Customer Experiences

In the realm of retail and consumer services, the advent of 5G and upcoming 6G technologies promises to revolutionize customer experiences like never before. With significantly higher speeds, lower latency, and greater capacity, 5G enables retailers to offer immersive, personalized experiences both in-store and online. Augmented Reality (AR) and Virtual Reality (VR) applications powered by 5G can allow customers to visualize products in real-time and in their own environments, enhancing their decision-making process and reducing returns. Interactive displays and smart mirrors can provide personalized recommendations based on customer preferences and past behaviors, creating a seamless shopping journey (Pan et al., 2019). Moreover, 5G facilitates ultra-fast mobile transactions, enabling retailers to offer frictionless payment experiences and expedited checkouts, thereby improving customer satisfaction and loyalty. As 6G emerges, with its potential for even faster speeds and more sophisticated capabilities, the possibilities for retail innovation expand further, promising to redefine how consumers interact with brands and make purchasing decisions.

4.5.2 Supply Chain and Inventory Management

Supply chain and inventory management in retail stand to benefit significantly from 5G and future 6G implementations, addressing longstanding challenges and inefficiencies. The high-speed, low-latency connectivity of 5G facilitates real-time tracking and monitoring of goods throughout the supply chain, from production facilities to distribution centers and retail outlets. This real-time visibility enables retailers to optimize inventory levels, reduce stockouts, and minimize wastage, thereby improving operational efficiency and cost-effectiveness. Furthermore, 5G-enabled IoT devices and sensors can provide granular data on product location, condition, and demand patterns, facilitating predictive analytics and demand forecasting (Parvez et al., 2018). This proactive approach helps retailers anticipate consumer demand fluctuations and adjust supply chain strategies accordingly, ensuring product availability and enhancing customer satisfaction. As 6G technologies evolve, with anticipated advancements in machine learning and AI integration, retailers can expect even greater automation and optimization capabilities across their supply chains, paving the way for more agile and responsive retail operations in the future.

5. IMPACT ON THE FINANCIAL SERVICES SECTOR

The integration of 5G and future 6G technologies is set to revolutionize the financial services sector, enhancing operational efficiency, security, and customer experiences across various domains.

5.1 Digital Banking and Payment Systems

Digital banking and payment systems stand to benefit significantly from 5G and 6G advancements. These technologies enable faster, more reliable data transmission, reducing latency and enhancing transaction speeds. This improvement is crucial for real-time payment processing, allowing financial institutions to offer instant payment solutions that rival traditional payment methods. Moreover, the increased bandwidth and connectivity of 5G-6G networks support seamless mobile banking experiences, facilitating access to banking services from anywhere at any time. Enhanced security protocols, such as biometric authentication and encrypted data transmission over robust networks, further bolster consumer trust in digital banking solutions. As financial institutions continue to innovate with AI-driven customer insights and personalized financial services, the convergence of 5G and 6G will accelerate the shift towards digital-first banking experiences, driving greater convenience and accessibility for customers globally.

5.2 Fraud Detection and Cybersecurity

In the realm of fraud detection and cybersecurity, 5G and 6G technologies present new opportunities and challenges. The speed and capacity of these networks enable real-time data analysis and proactive threat detection, crucial for identifying and mitigating fraudulent activities swiftly. Machine learning algorithms powered by high-speed data processing capabilities can analyze vast amounts of transaction data and user behavior patterns to detect anomalies indicative of fraud.Furthermore, secure communication protocols and advanced encryption techniques inherent in 5G-6G networks enhance data protection and privacy, safeguarding sensitive financial information from cyber threats (Rappaport, MacCartney, Sun, Yan, & Deng, 2017). However, the rapid proliferation of connected devices and IoT within financial ecosystems also introduces new attack surfaces, necessitating robust cybersecurity measures and continuous monitoring to prevent breaches. As financial institutions leverage AI-driven cybersecurity solutions and collaborative industry efforts to address emerging threats, the integration of 5G and 6G will play a pivotal role in fortifying digital defenses and maintaining trust in financial services.

5.3 Financial Inclusion and Accessibility

5G and 6G technologies have the potential to bridge the gap in financial inclusion by expanding access to underserved populations and remote regions. The high-speed connectivity and low latency of these networks enable the delivery of banking services and digital financial solutions to areas with limited infrastructure, empowering individuals and businesses to participate in the global economy. Mobile banking applications and digital wallets powered by 5G-6G networks provide convenient access to banking services, including savings accounts, loans, and insurance products, without the need for physical bank branches.

6. PUBLIC SECTOR AND SMART CITIES

The implementation of 5G and future 6G technologies holds immense promise for transforming urban environments into smart cities, where digital infrastructure integrates seamlessly with public services and governance. Infrastructure and urban planning benefit significantly from these advancements, enabling cities to optimize resource allocation, improve efficiency, and enhance quality of life for residents (Rappaport et al., 2019). With 5G's high-speed, low-latency capabilities, smart cities can deploy IoT sensors to monitor and manage critical infrastructure such as transportation networks, utilities, and waste management systems in real-time. This data-driven approach enables predictive maintenance, reduces operational costs, and supports sustainable urban development practices. Moreover, smart city initiatives leverage 5G to enhance connectivity across public spaces, improving accessibility to digital services and fostering community engagement (Saranovac & Glisic, 2020). As cities evolve towards 6G, advancements in AI and edge computing will further empower smart infrastructure to autonomously respond to dynamic urban challenges, ensuring cities remain resilient and adaptable in an increasingly interconnected world.

6.1 Infrastructure and Urban Planning

The integration of 5G and future 6G technologies into urban planning represents a paradigm shift towards more efficient, sustainable, and resilient cities. These technologies enable real-time data collection and analysis, facilitating informed decision-making in infrastructure development and management. Urban planners can leverage 5G's capabilities to monitor traffic patterns, optimize public transportation routes, and enhance energy efficiency through smart grid systems. Additionally, IoT sensors deployed across urban infrastructure enable predictive maintenance, reducing downtime and extending the lifespan of critical assets. 6G's advancements, including ultra-reliable and low-latency communications (URLLC), will further enhance these capabilities, enabling autonomous systems to respond to real-time demands and challenges. By integrating 5G and 6G technologies into urban planning strategies, cities can create more livable environments that prioritize sustainability, inclusivity, and economic growth.

6.2 Public Safety and Emergency Services

5G and future 6G technologies are poised to revolutionize public safety and emergency services by enabling faster response times, enhanced situational awareness, and more effective disaster management strategies. With 5G's high-speed connectivity and low latency, emergency responders can access real-time data streams, including live video feeds, location tracking, and sensor data, enabling them to make informed decisions swiftly and efficiently. Moreover, IoT devices and wearable technology powered by 5G facilitate seamless communication and coordination among first responders during crisis situations. In smart cities, 5G supports the deployment of advanced surveillance systems and predictive analytics tools that enhance threat detection and prevention efforts (Sato et al., 2019). Looking ahead to 6G, innovations such as holographic communication and augmented reality (AR) applications will further transform how emergency services operate, providing immersive situational awareness and remote assistance capabilities. By harnessing these technologies, cities can strengthen their resilience to emergencies and improve overall public safety outcomes.

6.3 Environmental Monitoring and Sustainability

The deployment of 5G and upcoming 6G technologies is instrumental in advancing environmental monitoring and sustainability initiatives within smart cities. These technologies enable the integration of IoT sensors and data analytics platforms to monitor air quality, water usage, waste management, and overall environmental health in real-time. By collecting and analyzing vast amounts of environmental data, cities can implement proactive measures to mitigate pollution, conserve natural resources, and promote sustainable practices. 5G's high bandwidth and low latency support the connectivity requirements of IoT devices spread across urban landscapes, facilitating seamless data transmission and analysis. Furthermore, 6G's advancements in energy efficiency and environmental monitoring capabilities will enable more precise modeling of climate change impacts and proactive adaptation strategies (Scully, Chen, & Yen, 2016). Smart cities leveraging 6G technologies can implement innovative solutions like smart grids, renewable energy integration, and green infrastructure development to achieve carbon neutrality goals and create healthier, more resilient urban environments for future generations.

7. CHALLENGES AND RISKS OF 5G-6G IMPLEMENTATIONS

7.1 Technical and Operational Challenges

Implementing 5G and preparing for 6G networks presents a range of technical and operational challenges that must be addressed for successful deployment. One significant challenge is the need for extensive infrastructure upgrades to support the higher frequencies and bandwidths that 5G and 6G technologies require. This includes deploying more cell towers, upgrading existing networks with fiber-optic cables, and ensuring seamless integration with existing telecommunications infrastructure. Additionally, the deployment of small cells and massive MIMO (Multiple Input Multiple Output) antennas introduces complexities in network planning and optimization to ensure adequate coverage and capacity.Moreover, managing the spectrum allocation for 5G and future 6G networks poses another technical challenge. Spectrum scarcity, especially in densely populated urban areas, can lead to interference and reduced network performance if not managed effectively. Coordination among telecommunications regulators, network operators, and technology providers is crucial to optimize spectrum use and mitigate interference issues.Operational challenges include the increased complexity of network management and orchestration due to the diverse range of devices and applications that 5G and 6G will support (Sharma & Shrivastava, 2021). These networks will enable a massive increase in connected devices, requiring robust network slicing capabilities to prioritize and allocate resources efficiently. Ensuring network reliability, low latency, and high availability across various use cases—from autonomous vehicles to industrial automation—adds further complexity to operational management.

7.2 Security and Privacy Concerns

Security and privacy remain paramount concerns in the deployment of 5G and future 6G networks, given the increased attack surface and potential vulnerabilities introduced by these technologies. One major concern is the security of IoT devices and endpoints connected to 5G networks. IoT devices often have limited security features and can be vulnerable to cyberattacks, potentially compromising network integrity and data confidentiality. Securing these devices through robust authentication, encryption, and continuous monitoring is critical to mitigating risks. (Sharma & Shrivastava, 2021)Additionally, the architecture of 5G networks introduces new security challenges, such as the virtualization of network functions and the use of software-defined networking (SDN) and network functions virtualization (NFV) technologies. These virtualized environments create potential entry points for cyber threats if not properly secured. Implementing strict access controls, network segmentation, and regular security audits are essential measures to safeguard against unauthorized access and data breaches.Privacy concerns arise from the extensive collection and utilization of user data in 5G and 6G networks. These technologies enable highly personalized services and applications based on real-time data insights, raising questions about user consent, data ownership, and compliance with data protection regulations. Implementing transparent data handling practices, providing users with control over their data, and conducting privacy impact assessments are crucial steps to address these concerns and build trust among consumers and stakeholders.

7.3 Economic and Social Disparities

The deployment of 5G and future 6G networks has the potential to exacerbate economic and social disparities if not managed inclusively and equitably. One concern is the digital divide, where rural and underserved communities may have limited access to high-speed 5G connectivity compared to urban areas. This disparity can widen the gap in access to educational resources, healthcare services, and economic opportunities, hindering socioeconomic development and digital inclusion efforts. Moreover, the cost of deploying 5G infrastructure—such as small cells, fiber networks, and spectrum licenses—can be prohibitive for some regions and operators, particularly in developing countries. Addressing these economic barriers requires innovative financing models, public-private partnerships, and regulatory incentives to facilitate equitable access to 5G technologies (Shen et al., 2020). Social disparities may also arise from the unequal distribution of benefits and opportunities enabled by 5G and 6G networks. Industries and communities with advanced technological capabilities and resources may leverage these technologies to achieve competitive advantages and economic growth, while others may struggle to keep pace. Bridging these disparities requires proactive policies and initiatives to ensure that all sectors of society can benefit from the transformative potential of advanced telecommunications technologies, fostering inclusive economic growth and social equity.

8. FUTURE TRENDS AND INNOVATIONS

8.1 AI and Machine Learning in 5G-6G Networks

The integration of Artificial Intelligence (AI) and Machine Learning (ML) in 5G and upcoming 6G networks represents a transformative shift in the telecommunications industry. This integration promises to enhance network performance, optimize resource utilization, and enable new, innovative applications. AI-driven network optimization is becoming more prevalent, with algorithms managing complex network configurations, predicting traffic patterns, and dynamically allocating resources to ensure optimal performance. This reduces latency and enhances user experience, particularly for applications requiring high-speed data transmission and low latency, such as augmented reality, virtual reality, and autonomous vehicles. Moreover, AI and ML facilitate predictive maintenance by analyzing network data to foresee potential issues and address them proactively, thereby minimizing downtime and improving network reliability.

In 5G networks, AI and ML are employed to manage the complexity and scale of the network. AI algorithms can process vast amounts of data generated by user devices, sensors, and network equipment to make real-time decisions. For instance, AI can optimize beamforming and handover processes, ensuring that users receive the best possible connection as they move. Machine learning models can also be trained to predict traffic congestion and automatically adjust network parameters to alleviate bottlenecks. This level of automation and intelligence significantly improves the efficiency and effectiveness of network operations.

As we look towards 6G networks, the role of AI and ML will become even more crucial. 6G aims to provide ubiquitous connectivity with data rates up to 100 times faster than 5G and latency as low as one microsecond (Shirvanimoghaddam et al., 2018) (Stojmenović & Lin, 2014). Achieving these goals will require highly intelligent and adaptive networks. AI and ML will enable advanced techniques such as network slicing, where multiple virtual networks can be created on a single physical infrastructure, each tailored to specific applications or user requirements. This will allow for more efficient use of network resources and better quality of service for diverse applications.

Edge computing will also play a significant role in 6G, and AI and ML will be essential in managing edge resources. By processing data closer to the source, edge computing reduces latency and bandwidth usage. AI can optimize the placement of workloads and manage the interaction between edge devices and the central network. Machine learning models can also be deployed at the edge to provide real-time analytics and decision-making capabilities, enabling applications like real-time video analytics, autonomous vehicle coordination, and smart city infrastructure management.

The integration of heterogeneous networks, including terrestrial and non-terrestrial networks such as satellites and high-altitude platforms, will be a hallmark of 6G. AI and ML will be pivotal in managing the seamless interaction between these diverse networks, ensuring continuous and reliable connectivity. By leveraging AI-driven analytics and automation, 6G networks will be able to adapt to changing conditions, user demands, and environmental factors, providing a resilient and robust communication infrastructure.

8.2 Integration with Emerging Technologies (IoT, Blockchain, etc.)

The convergence of 5G-6G networks with emerging technologies like IoT and blockchain promises to create synergies that amplify their individual impacts. IoT devices, empowered by high-speed, low-latency connections provided by 5G-6G, will proliferate across industries, enabling real-time data collection and analysis for smart applications. These applications range from smart cities with connected infrastructure to industrial IoT facilitating predictive maintenance and autonomous operations (Sun, Rappaport, Shafi, & MacCartney, 2016). Blockchain, with its decentralized and secure ledger technology, will enhance trust and transparency in transactions conducted over 5G-6G networks, particularly in sectors like finance, supply chain, and healthcare. Combining blockchain's immutable record-keeping with the speed and reliability of 5G-6G networks will facilitate faster and more secure transactions globally (Vetrivel, Sowmiya, & Sabareeshwari, 2024) (Wang et al., 2019). The integration of these technologies will foster innovation ecosystems, driving new business models and enhancing operational efficiencies across diverse industry sectors.

8.3 Predictions and Speculations for Post-6G Era

Looking beyond 6G, future telecommunications systems are expected to continue evolving towards even faster, more reliable, and pervasive connectivity. Post-6G technologies are speculated to transcend current limitations, possibly incorporating quantum computing, terahertz frequencies, and advanced nanotechnologies (Yadav, Dave, & Tiwari, 2019). These advancements could enable unprecedented data transfer rates, nearly imperceptible latency, and ultra-reliable communications supporting new applications such as telepathic communication and brain-computer interfaces. Moreover, post-6G networks may integrate with biotechnology, enabling advancements in healthcare, personalized medicine, and human augmentation. The era beyond 6G holds immense potential for reshaping communication infrastructures, societal interactions, and technological capabilities in ways that are currently difficult to fully comprehend, paving the way for a truly interconnected and intelligent future.

9. CASE STUDIES AND SUCCESS STORIES

9.1 Leading Countries and Pioneering Companies

Several countries have emerged as leaders in the deployment and utilization of 5G and are paving the way for future 6G technologies. South Korea, for example, has been at the forefront with widespread 5G coverage and significant investments in infrastructure. Companies like Samsung and LG have played pivotal roles in developing 5G-enabled devices and technologies, contributing to the country's leadership in telecommunications innovation. China has also made substantial strides, with major telecom providers like Huawei driving advancements in 5G infrastructure and applications (Yadav, Kwon, & Jang, 2020). The Chinese government's strategic investments in 5G deployment have supported initiatives ranging from smart cities to industrial automation, positioning China as a global leader in 5G adoption. In the United States, companies such as Verizon and AT&T have been instrumental in rolling out 5G networks across urban centers, focusing on enhancing broadband connectivity and enabling IoT applications. These leading countries and companies not only showcase the technical capabilities of 5G but also demonstrate the economic benefits of early adoption through improved productivity, innovation in digital services, and global competitiveness.

9.2 Successful Implementations Across Sectors

The successful implementation of 5G-6G technologies spans diverse industry sectors, each demonstrating unique benefits and transformative impacts. In manufacturing, for instance, companies like Bosch and Siemens have integrated 5G networks to enable real-time monitoring and control of production processes, optimizing efficiency and reducing operational costs. Healthcare sectors in countries like Japan have leveraged 5G for remote surgery and telemedicine, enhancing access to specialized medical services and improving patient outcomes. In transportation, autonomous vehicle trials in cities like Singapore and Helsinki have utilized 5G connectivity to enable safer and more efficient mobility solutions (Yan, Zheng, Zhang, & Xing, 2019). Retail sectors globally have embraced 5G to enhance customer experiences through augmented reality (AR) shopping experiences and personalized marketing strategies. These successful implementations underscore the versatility of 5G-6G technologies in driving innovation across industries, catalyzing economic growth, and addressing societal challenges.

9.3 Economic Impact of 5G-6G Implementations Across Diverse Industry Sectors

The advent of 5G and the upcoming 6G technology promises to revolutionize various industry sectors by providing unprecedented speed, reliability, and connectivity. This chapter delves into the economic impact of these technologies across diverse sectors and showcases success stories that highlight the transformative potential of 5G and 6G implementations.

9.3.1 Manufacturing Industry: Siemens' Smart Factories

Siemens has been at the forefront of integrating 5G technology into its manufacturing processes. By deploying 5G-enabled smart factories, Siemens has achieved significant improvements in efficiency and productivity. The high-speed connectivity and low latency of 5G networks have enabled real-time data analysis, predictive maintenance, and seamless communication between machines. This has resulted in reduced downtime, lower operational costs, and enhanced overall production capabilities. Siemens' success story exemplifies how 5G can drive the digital transformation of the manufacturing sector, leading to substantial economic gains.

9.3.2 Healthcare Industry: Remote Surgery and Telemedicine

The healthcare sector has experienced remarkable advancements through the implementation of 5G technology. One notable success story is the rise of remote surgery and telemedicine. For instance, China's People's Liberation Army General Hospital successfully performed a remote brain surgery using 5G technology. The ultra-low latency and high reliability of 5G networks allowed surgeons to operate with precision from a distance. Telemedicine has also seen significant growth, with patients in remote areas gaining access to high-quality medical consultations and treatments. These advancements have not only improved patient outcomes but also reduced healthcare costs and increased the efficiency of medical services.

9.3.3 Automotive Industry: Autonomous Vehicles by Tesla

Tesla's implementation of 5G technology in autonomous vehicles has set new standards in the automotive industry. The enhanced connectivity and real-time data processing capabilities of 5G have enabled Tesla to develop vehicles with advanced self-driving features. These vehicles can communicate with each other and with infrastructure in real time, ensuring safer and more efficient transportation. The economic impact of this technology is profound, as it reduces traffic congestion, lowers accident rates, and opens new opportunities for mobility services. Tesla's success underscores the potential of 5G to transform the automotive sector and contribute to economic growth.

9.3.4 Agriculture Industry: Precision Farming by John Deere

John Deere has leveraged 5G technology to revolutionize precision farming. By utilizing 5G-enabled sensors and drones, John Deere has developed advanced farming solutions that monitor crop health, soil conditions, and weather patterns in real time. This data-driven approach allows farmers to make informed decisions, optimize resource usage, and increase crop yields. The economic benefits are significant, as precision farming reduces waste, lowers costs, and enhances productivity. John Deere's success story highlights how 5G can drive innovation in agriculture and boost economic sustainability.

9.3.5 Entertainment Industry: Immersive Experiences by Disney

Disney has harnessed the power of 5G to create immersive entertainment experiences. Through the integration of 5G technology in its theme parks and media platforms, Disney has introduced augmented reality (AR) and virtual reality (VR) experiences that captivate audiences. These technologies offer visitors and viewers interactive and engaging content, enhancing their overall experience. The economic impact is evident in increased visitor numbers, higher customer satisfaction, and expanded revenue streams. Disney's innovative use of 5G showcases the potential for this technology to transform the entertainment industry and drive economic growth.

9.3.6 Retail Industry: Smart Stores by Walmart

Walmart has embraced 5G technology to develop smart stores that enhance the shopping experience. By implementing 5G-enabled devices and IoT solutions, Walmart has optimized inventory management, improved customer service, and streamlined checkout processes. Smart shelves, automated checkout systems, and personalized shopping recommendations are some of the innovations made possible by 5G. These advancements have led to increased operational efficiency, higher customer satisfaction, and greater sales. Walmart's success story demonstrates how 5G can revolutionize the retail sector and contribute to economic development.

9.4 Lessons Learned and Best Practices

Across various implementations of 5G-6G technologies, several key lessons and best practices have emerged to guide future deployments. Firstly, collaboration between stakeholders—including governments, telecom operators, technology providers, and industry stakeholders—is essential for successful deployment and adoption. Clear regulatory frameworks that support innovation while ensuring data privacy and security are crucial to fostering trust and accelerating deployments (Yan, Zhang, Cheng, & Zhang, 2017) (Zhang, Xiao, Wu, & Li, 2020). Moreover, scalability and interoperability of technologies are critical considerations to ensure seamless integration across diverse applications and sectors. Companies that have successfully implemented 5G-6G have prioritized ongoing education and upskilling of their workforce to harness the full potential of these technologies (Zhang, Li, Qian, & Shen, 2016). Lastly, continuous monitoring and evaluation of deployments are necessary to adapt strategies and mitigate risks effectively (Zhang, Yao, Shu, Liu, & Li, 2018). By applying these lessons learned and best practices, organizations can navigate the complexities of 5G-6G implementations and maximize their economic and societal impact.

10. CONCLUSION

In conclusion, the evaluation of the economic impact of 5G-6G implementations across diverse industry sectors reveals a transformative landscape poised for significant growth and innovation. As 5G networks continue to expand and 6G technologies emerge on the horizon, the potential benefits span from enhanced connectivity and operational efficiencies to revolutionary advancements in healthcare, manufacturing, transportation, and beyond. Leading countries and pioneering companies have demonstrated the feasibility and economic viability of these technologies through successful implementations that improve productivity, drive digital transformation, and stimulate job creation. However, realizing the full potential of 5G-6G requires addressing challenges such as regulatory alignment, cybersecurity concerns, and ensuring equitable access across communities. By embracing collaboration, investing in robust infrastructure, and fostering a conducive regulatory environment, stakeholders can harness the power of 5G-6G to propel economies forward, foster innovation, and improve quality of life globally. The journey towards realizing these benefits demands continued adaptation, strategic investment, and a commitment to leveraging technology responsibly to ensure a sustainable and inclusive future.

REFERENCES

Abdellatif, A. A., Mohamed, A., Chiasserini, C. F., Tlili, M., & Erbad, A. (2019). Edge computing for smart health: Context-aware approaches, opportunities, and challenges. *IEEE Network*, 33(3), 196–203. DOI: 10.1109/MNET.2019.1800083

Andrews, J. G., Buzzi, S., Choi, W., Hanly, S. V., Lozano, A., Soong, A. C., & Zhang, J. C. (2014). What will 5G be? *IEEE Journal on Selected Areas in Communications*, 32(6), 1065–1082. DOI: 10.1109/JSAC.2014.2328098

Ankarali, Z. E., Peköz, B., & Arslan, H. (2017). Flexible radio access beyond 5G: A future projection on waveform, numerology, and frame design principles. *IEEE Access : Practical Innovations, Open Solutions*, 5, 18295–18309. DOI: 10.1109/ACCESS.2017.2684783

Berardinelli, G., Mahmood, N. H., Rodriguez, I., & Mogensen, P. (2018). *Beyond 5G wireless IRT for Industry 4.0: Design principles and spectrum aspects. In 2018 IEEE Globecom Workshops (GC Wkshps)*. IEEE.

Bochechka, G., & Tikhvinskiy, V. (2014). *Proceedings of the 2014 ITU Kaleidoscope Academic Conference: Living in a Converged World-Impossible Without Standards?* Geneva, Switzerland: ITU.

Cambria, E. (2016). Affective computing and sentiment analysis. *IEEE Intelligent Systems*, 31(2), 102–107. DOI: 10.1109/MIS.2016.31

Chen, P., Wang, P., & Sun, J. (2011). Design and implement of the OFDM communication system. In *Proceedings of the 2011 IEEE International Workshop on Open-source Software for Scientific Computation* (pp. 59-63). Hangzhou, China: IEEE. DOI: 10.1109/OSSC.2011.6184695

Cherkasova, O. P., Serdyukov, D. S., Nemova, E. F., Ratushnyak, A. S., Kucheryavenko, A. S., Dolganova, I. N., Xu, G., Skorobogatiy, M., Reshetov, I. V., Timashev, P. S., Spektor, I. E., Zaytsev, K. I., & Tuchin, V. V. (2021). Cellular effects of terahertz waves. *Journal of Biomedical Optics*, 26(9), 090902. DOI: 10.1117/1.JBO.26.9.090902 PMID: 34595886

DIGITAL 2020: JULY GLOBAL STATSHOT. (2020). Retrieved March 11, 2022, from https://www.bonfire.com.au/wp-content/uploads/2021/08/Hootsuite-digital2021julyglobalstatshotreportv02-210721025325.pdf

Fang, L., Zhao, B., Li, Y., Liu, Z., Ge, C., & Meng, W. (2020). Countermeasure based on smart contracts and AI against DoS/DDoS attack in 5G circumstances. *IEEE Network*, 34(6), 54–61. DOI: 10.1109/MNET.021.1900614

Farris, I., Taleb, T., Khettab, Y., & Song, J. (2018). A survey on emerging SDN and NFV security mechanisms for IoT systems. *IEEE Communications Surveys and Tutorials*, 21(1), 812–837. DOI: 10.1109/COMST.2018.2862350

Gangakhedkar, S., Cao, H., Ali, A. R., Ganesan, K., Gharba, M., & Eichinger, J. (2018). Use cases, requirements and challenges of 5G communication for industrial automation. In 2018 IEEE International Conference on Communications Workshops (ICC Workshops) (pp. 1-6). Kansas City, MO, USA: IEEE.

Hoshina, H., Yamazaki, S., Tsubouchi, M., & Harata, M. (2021). Terahertz irradiation effects on the morphology and dynamics of actin biopolymer. *JPhys Photonics*, 3(3), 034015. DOI: 10.1088/2515-7647/ac0958

Huynh-The, T., Hua, C. H., Pham, Q. V., & Kim, D. S. (2020). MCNet: An efficient CNN architecture for robust automatic modulation classification. *IEEE Communications Letters*, 24(4), 811–815. DOI: 10.1109/LCOMM.2020.2968030

ITU-R M.2370-0. (2015). IMT Traffic Estimates for the Years 2020 to 2030. Geneva, Switzerland: ITU Publications. https://www.itu.int/dms_pub/itu-r/opb/rep/R-REP-M.2370-2015-PDF-E.pdf

Jiang, X., Shokri-Ghadikolaei, H., Fodor, G., Modiano, E., Pang, Z., Zorzi, M., & Fischione, C. (2018). Low-latency networking: Where latency lurks and how to tame it. *Proceedings of the IEEE*, 107(2), 280–306. DOI: 10.1109/JPROC.2018.2863960

Juneja, S., Gahlan, M., Dhiman, G., & Kautish, S. (2021). *Futuristic Cyber-Twin Architecture for 6G Technology to Support Internet of Everything*. Hindawi Scientific Programming. DOI: 10.1155/2021/9101782

Kimura, S., Aburakawa, Y., Watanabe, F., Torashima, S., Igarashi, S., Nakamura, T., & Yamaguchi, M. (2021). Holographic Video Communication System Realizing Virtual Image Projection and Frontal Image Capture. *ITE Transactions on Media Technology and Applications*, 9(1), 105–112. DOI: 10.3169/mta.9.105

Kota, S. L. (2011). Hybrid/integrated networking for ngn services. In 2011 2nd International Conference on Wireless Communication, Vehicular Technology, Information Theory and Aerospace &Electronic Systems Technology (Wireless VITAE) (pp. 1-6). Chennai, India: IEEE.

Kürner, T., & Priebe, S. (2014). Towards THz communications-status in research, standardization and regulation. *Journal of Infrared, Millimeter and Terahertz Waves*, 35(1), 53–62. DOI: 10.1007/s10762-013-0014-3

Levy, J. (2005). *The Great Pyramid of Giza: Measuring Length, Area, Volume, and Angles*. The Rosen Publishing Group, Inc.

Li, C., Guo, W., Sun, S. C., Al-Rubaye, S., & Tsourdos, A. (2020). Trustworthy in 6G-enabled mass autonomy: From concept to quality-of-trust key performance indicators. *IEEE Vehicular Technology Magazine*, 15(4), 112–121. DOI: 10.1109/MVT.2020.3017181

Li, H., La Guardia, M. J., Liu, H., Hale, R. C., Mainor, T. M., Harvey, E., Sheng, G., Fu, J., & Peng, P. (2019). Brominated and organophosphate flame retardants along a sediment transect encompassing the Guiyu, China e-waste recycling zone. *The Science of the Total Environment*, 646, 58–67. DOI: 10.1016/j.scitotenv.2018.07.276 PMID: 30048869

. Li, T., Liu, W., Zeng, Z., &Xiong, N. (2021). DRLR: A deep reinforcement learning based recruitment scheme for massive data collections in 6G-based IoT networks. IEEE Internet of Things Journal. Advance online publication.

Lovley, D. R. (2017). e-Biologics: Fabrication of sustainable electronics with "green" biological materials. *mBio*, 8(3), e00695–e17. DOI: 10.1128/mBio.00695-17 PMID: 28655820

Mao, B., Tang, F., Kawamoto, Y., & Kato, N. (2021). Optimizing Computation Offloading in Satellite-UAV-Served 6G IoT: A Deep Learning Approach. *IEEE Network*, 35(4), 102–108. DOI: 10.1109/MNET.011.2100097

Matinmikko-Blue, M., Latva-aho, M., Säily, J., & Mustonen, M. (2019). Industry 4.0 and 5G: Where do they meet? *Business Horizons*, 62, 47–56.

Mei, J., Wang, X., Zheng, K., Boudreau, G., Sediq, A. B., & Abou-Zeid, H. (2021). Intelligent radio access network slicing for service provisioning in 6G: A hierarchical deep reinforcement learning approach. *IEEE Transactions on Communications*, 69(9), 6063–6078. DOI: 10.1109/TCOMM.2021.3090423

Moldovan, A., Karunakaran, P., Akyildiz, I. F., & Gerstacker, W. H. (2017). Coverage and achievable rate analysis for indoor terahertz wireless networks. In *2017 IEEE International Conference on Communications (ICC)* (pp. 1-7). Paris, France: IEEE. DOI: 10.1109/ICC.2017.7996402

NVIDIA MAXINE. (2020). Retrieved March 11, 2022, from https://developer.nvidia.com/maxine

Pan, C., Ren, H., Wang, K., Elkashlan, M., Nallanathan, A., Wang, J., & Hanzo, L. (2019). Intelligent reflecting surface enhanced MIMO broadcasting for simultaneous wireless information and power transfer. *IEEE Journal on Selected Areas in Communications*, 38(8), 1719–1734. DOI: 10.1109/JSAC.2020.3000802

Parvez, I., Rahmati, A., Guvenc, I., Sarwat, A. I., & Dai, H. (2018). A survey on low latency towards 5G: RAN, core network and caching solutions. *IEEE Communications Surveys and Tutorials*, 20(4), 3098–3130. DOI: 10.1109/COMST.2018.2841349

Rappaport, T. S., MacCartney, G. R., Sun, S., Yan, H., & Deng, S. (2017). Small-scale, local area, and transitional millimeter wave propagation for 5G communications. *IEEE Transactions on Antennas and Propagation*, 65(12), 6474–6490. DOI: 10.1109/TAP.2017.2734159

Rappaport, T. S., Xing, Y., Kanhere, O., Ju, S., Madanayake, A., Mandal, S., Alkhateeb, A., & Trichopoulos, G. C. (2019). Wireless Communications and Applications Above 100 GHz: Opportunities and Challenges for 6G and Beyond. *IEEE Access : Practical Innovations, Open Solutions*, 7, 78729–78757. DOI: 10.1109/ACCESS.2019.2921522

Saranovac, L., & Glisic, S. (2020). 5G impact on e-commerce: Opportunities and challenges. *Journal of Systems and Software*, 163, 110555.

Sato, N., Hong, T., Hirose, K., & Yoshida, S. (2019). Terahertz wavefront sensing and real-time adaptation using a binary-phase spatial light modulator. *Journal of Infrared, Millimeter and Terahertz Waves*, 40, 823–834.

Scully, M. O., Chen, Y., & Yen, T.-H. (2016). Quantum optical tests of complementarity. *Proceedings of the National Academy of Sciences of the United States of America*, 113, 532–538.

Sharma, S., & Shrivastava, S. (2021). A review on artificial intelligence based edge computing in 6G wireless networks. *Wireless Personal Communications*. Advance online publication.

Shen, X., Lee, J.-S., Kim, K.-J., & Kwon, T. (2020). Artificial intelligence in edge computing: A review and research directions. *IEEE Access : Practical Innovations, Open Solutions*, 8, 47538–47556.

Shirvanimoghaddam, M., Nguyen, T. M., Chatzinotas, S., Ottersten, B., & Juntti, M. (2018). Terahertz band communications: Applications, research challenges, and standardization activities. *IEEE Journal on Selected Areas in Communications*, 36, 2524–2538.

Stojmenović, M., & Lin, X. (2014). Application of ambient backscatter for RF-powered cognitive radio networks. *IEEE Communications Magazine*, 52, 140–145.

Sun, S., Rappaport, T. S., Shafi, M., & MacCartney, G. R. (2016). MIMO techniques in millimeter wave wireless communications. *IEEE Transactions on Communications*, 64, 3201–3239.

Vetrivel, S. C., Sowmiya, K. C., & Sabareeshwari, V. (2024). Digital Twins: Revolutionizing Business in the Age of AI. In Ponnusamy, S., Assaf, M., Antari, J., Singh, S., & Kalyanaraman, S. (Eds.), *Harnessing AI and Digital Twin Technologies in Businesses* (pp. 111–131). IGI Global., DOI: 10.4018/979-8-3693-3234-4.ch009

Wang, H., Al-Abdullah, M., Mao, B., & Giupponi, L. (2019). Physical layer security in millimeter wave networks: Challenges and opportunities. *IEEE Communications Magazine*, 57, 112–118.

Wu, Q., Ding, Z., Lin, H., & Bhargava, V. K. (2016). Device-to-device communication in 5G cellular networks: Challenges, solutions, and future directions. *IEEE Access : Practical Innovations, Open Solutions*, 4, 6623–6639.

Yadav, R., Dave, M., & Tiwari, P. (2019). Secure communication using 5G and IoT: A survey. *Wireless Personal Communications*, 109, 1109–1124.

Yadav, S., Kwon, J., & Jang, Y.-M. (2020). Machine learning in 6G networks: Current status and future directions. *IEEE Open Journal of the Communications Society*, 1, 145–152.

Yan, J., Zheng, H., Zhang, Y., & Xing, C. (2019). THz-based ultra-high data rate wireless communication: Fundamental aspects, channel modeling, and applications. *IEEE Transactions on Terahertz Science and Technology*, 9, 542–563.

Yan, Z., Zhang, P., Cheng, Z., & Zhang, Y. (2017). Deep learning for image-based detection of unmanned aerial vehicles. *IEEE Transactions on Vehicular Technology*, 66, 8134–8145.

Zhang, H., Xiao, Y., Wu, D., & Li, Y. (2020). Data-driven autonomous security resource management for 5G and beyond mobile networks. *IEEE Transactions on Mobile Computing*, 19, 145–160.

Zhang, L., Li, Y., Qian, Y., & Shen, X. (2016). End-to-end security provisioning in mobile edge computing with NOMA. *IEEE Network*, 30, 46–51. DOI: 10.1109/MNET.2016.7474343

Zhang, Y., Yao, H., Shu, L., Liu, Z., & Li, D. (2018). Data-driven smart healthcare systems: A review. *IEEE Transactions on Industrial Informatics*, 14, 2230–2239.

Chapter 5

Transition From 5G to 6G Communication Technologies: Workforce Evolution and Skill Development Needs

Suryanarayana Alamuri
https://orcid.org/0009-0009-6875-3467
Osmania University, India

Ravi Aluvala
https://orcid.org/0000-0003-3862-5009
Mahatma Gandhi University, India

Ramesh Kumar Miryala
https://orcid.org/0000-0003-2282-1558
Mahatma Gandhi University, India

ABSTRACT

The technological transition from 5G to 6G is poised to revolutionize various sectors, creating unprecedented opportunities and challenges for the workforce. Today, industries that adopt 6G technologies promise enhanced connectivity, ultra-low latency, and advanced AI capabilities, pointing to the critical need to understand and address the evolving skill requirements. For successfully navigating this transition, employees must develop a diverse set of skills. In the 6G era, there is an increased demand for expertise in areas such as AI, Machine Learning, Edge Computing, & Quantum Computing. Professionals will need to master new networking protocols, Cybersecurity measures, and Data Analytics techniques to effectively manage

DOI: 10.4018/979-8-3693-6725-4.ch005

and secure 6G networks. As automation and AI become more integrated into the workplace, human-centric skills such as creativity, critical thinking, and emotional intelligence will become invaluable. This Chapter explores the workforce evolution necessitated by the shift from 5G to 6G and identifies the most essential skill sets required to thrive in this new era.

INTRODUCTION

Plenty of planned and progressive changes are likely to happen with the advent of 6G type of technology, which is very much facilitated by the highly proliferating Internet of Everything (IoE). Now, it is a known fact that the use of AI-powered accessories has huge potential resulting in totally altered and radical conditions in organizations. Once proper measures for effective implementation are put in place, these advancements will change the game in the fields of telecommunication, healthcare, manufacturing and entertainment to a large extent. Therefore, the workforce must evolve and transition into new roles and responsibilities. However, this calls for employees to understand these technologies at a deeper level. Organizations are mandated to take the initiative to educate them through education, online courses, and on-the-job training. Moreover, continuous workplace learning will be required by organizations to keep pace with these developments. To be able to effectively address the need of prevailing global market dynamics and foster policy environment that are conducive for rapid transition towards 6G: business organizations and educational institutions need to craft and adopt distinct strategies within three buckets viz., (i) Curriculum Design, (ii) Corporate Training Programs, and (iii) Policy and Regulatory Support.

In order to succeed, programs need to be updated according to the needs emerging from the adoption and application of new tools and methods before they come into fruition. In this context, academic–industry collaborations can help to keep curricula timely and in-line with the market needs. For this purpose, organizations must spend adequate time on comprehensive and complete training modules that offer employees all the technical and soft skills they need. This is definitely an area where partnerships with technology developers and training institutions can their focus efforts on. Governments and regulatory bodies need to enable innovation through the design of flexible frameworks that will support workforce retraining initiatives. Moreover, incentives to promote R&D in 6G technology can aid the progress. Lastly, the latest development signifies a key inflection point in technology evolution requiring a major change in workforce skills and capacity. This calls for understanding the specific skills before designing and developing an effective strategy for organizational efficacy. For this, organizations need to ensure that their workforce is ready to tap into what 6G has to offer as an imperative.

REVIEW OF RELEVANT LITERATURE

In this Section, the findings presented in the previous research studies are synthesized and supplemented with additional data excerpts from some of the original published research articles on 5G, its skill set requirements during roll-out and for transition to 6G, as well as implications for workforce development.

1. 6G TECHNOLOGICAL DEVELOPMENTS

As 6G era approaches, many breakthroughs in technology will take place demanding newer and ever changing skills. Varied and many important themes that have emerged from the literature consisted of:

— *AI and ML Integration:*

- Articles like "Artificial Intelligence and Machine Learning for 6G Networks" (Zhang *et al.,* **2023**) explain that AI/ML is a must-have when it comes to the orchestration and optimization of 6G networks.
- An example is a paper, "Quantum Communication in 6G: Opportunities and Challenges" (Chen & Wang, **2022**) which mentions that the training of quantum mechanics and cryptography should be needed.

Here are brief introductions to some of these works:

- Terahertz Communication: — "The 6G Wireless Terahertz Technology Pioneers the Ultimate High-Speed Data Transfer." (Huang *et al.*, 8)
- Leader of Science, Xuyan Chu and Mr. Kami Lohas: An Innovative Approach for Observing Thz Frequencies To Send Ultra-High-Speed Data Points.

2. SKILLSET REQUIREMENTS

A. Technical Skills

- *AI and Machine Learning:* Analysis carried out by Li *et al.* on "*Machine Learning for 6G: An Overview*" demonstrated the importance of AI/ML skills that will be mandatorily required in automating processes in next-generation networks.

- *Quantum Computing and Cryptography: "Quantum Technologies for Future Networks"* (Patel *et al.,* **2023)** describes the need for quantum computing and advanced cryptographic skill.
- *Terahertz Communication*: The theme of THz communication is covered in "Terahertz Waves for 6G Wireless Networks" (Xu & Yang, **2023**), which presents the importance of a critical future network infrastructure.
- *Advanced Network Architecture* "Network Slicing and Edge Computing for 6G" (Smith *et al.*, **2022)** explains the importance of network slicing, edge computing as well as cloud-native principles.
- *Cybersecurity: "Security Challenges in 6G Networks"* (Jones & Brown, **2023)** sets a cyber security focus and illustrates the increased vulnerabilities from 6G.
- *IoT and Sensor Networks:* The "*IoT Integration in 6G Networks"* (Kim *et al.,* **2023**) emphasize on the capacity needed to proficiently handling large-scale IoT ecosystems.

B. Soft Skills

- *Critical and Active Thinking:* It is critical for solving intricate technical challenges which have been highlighted as "Critical Thinking in Technical Professions" (Garcia, *et al.*)
- *Adaptability and Continuous Learning:*Lee & Kim **(2022)** discuss the *requirement for persistent secondary learning in "Adjusting* to Digital *Change in Telecom".*
- *Collaboration & Communication*: Requires teamwork as emphasized by Smith and Jones **(2013)** in their work on *Interdisciplinary Collaboration in Telecommunications.*

3. CURRENT SKILL GAPS

In this connection, many articles mention the current skill gap and some of them are forth following;

- Inadequate Ability in AI/ML: "*Bridging the AI Skills Gap in Telecommunications"* (Johnson *et al.,* **2022)** highlights there are not enough people with skills in using AI and ML.

- Discontent with Current Quantum Technology Knowledge: "Educating the Quantum Workforce" (Taylor & Green, **2023)** describes the scarcity of quantum technology education.
- Cybersecurity Readiness: 'Improving Cybersecurity Skills for 6G' (Davis *et al.*, **2022)** Emphasizes on Advanced Cybersecurity training to counter state-of-the-art threats from adversities.

4. THE DRIVERS FOR VOCATIONAL TRAINING ARE TWOFOLD

A. Training and Education

- *Revamped Curricula:* White *et al.,* on *Professional Development Programs: Post-*6G (Brown & Black, **2022**) – *Life-long learning opportunities* ·

B. Industry-Academia Collaboration

- An emphasis on partnership is evident from recent research outputs, such as the IC 4 G initiative; a paper entitled "Fostering Industry-Academia Collaboration for 6G" (Wilson *et al)*
- Research and Development programs: Miller, R., & Davis **(2022).** CODEN: GEIOSV6Title: *Joint R&D as a driver of innovation in the future* telecommunication landscape.

C. Policy and Strategy

- *Governmental Policies*: Clark & Adams, **2023** (Government Initiatives for 6G Skill Development) "*Analyzing the Ways in which government plays a part in skill and development.*"
- *"Corporate Strategies: Corporate Approaches to 6G Workforce Training"* (Hall & Brown, 2022) illustrates how companies can invest in employee upskilling.

5. CASE STUDIES

- *University Programs: "A Study on the University of Technology's 6G Curriculum"* (Nguyen & Smith, **2023**) provides an overview to how individual universities may fill the function gap of specific courses for 6G

- For corporate America, *"Nokia's 6G Training Program OP"* by Jacob Harris and Dan Clark **(2022),** a program meant to take employees on a tour through what 6G tech is and how it might change the future.

6. CHALLENGES AND OPPORTUNITIES

A. Challenges

- *Barriers to 6G Adoption*: Martin & White **(2023)** Report: This Report describes the resistance in adoption because new technology is unfamiliar.
- Resource Constraints—*"Addressing Resource Constraints in Skill Development"* (Garcia *et al.*, **2022**) highlights the limitations placed on educational resources as a result.
- Resistance to Change: Lee and Kim **(2022)** suggest working to instill a culture of lifelong learning can be proactively approached.

B. Opportunities

- *Innovation and Economic Growth: "Economic Impacts of 6G Innovation"* (Johnson, *et al.*, **2022**) showcases an essential means for sustainable economic growth. ·
- *More Global Connectivity*: *'Global Connectivity in the 6G Era'* (Jones & Brown, **2023**) discusses increased global collaboration
- *"Job Pathways: In 6G Emerging New Roles*" (Smith & Jones, **2022**) follows new jobs that 6G will drive as an innovation.

7. ADDITIONAL THEMES AND OTHER ALLIED MATTERS

- The seminal piece on the "The *Future of Employment: How Susceptible Are Jobs to Computerization"* Filed: March (to be published in Technological Forecasting and Social Change).
- There is one more Article "*Automation is creating the new embarrassment of labor riches*" **(2017)** by Carl Benedikt Frey, Michael A. Osborne.
- David Autor, Frank Levy, and Richard Murnane **(2003),** *"Skill Biased Technological Change and Employment: Evidence from OECD Countries,"* The Quarterly Journal of Economics.

- The Article by Mark Muro, Jacob Whiton, and Robert Maxim **(2019)** Brookings Institution: *Digitalization and the American Workforce*—examines the changing skill requirements within firms due to digital transformation.
- World Economic Forum White Paper entitled *"Technological Change and Skill Development in the Fourth Industrial Revolution"* prepared by Klaus Schwab and Nicholas Davis **(2018)** explores the required skills during Fourth Industrial Revolution and Its Effect on the Labor Market.
- *"The Skills that Matter: What Employers look for in Employees and How to Develop These Skills in the Workforce*" by James Heckman and Tim Kautz (Journal of Labor Economics, **2012**): This article provides a detailed analysis about the specific skills employers need discussed within the context technological transitions and workforce development strategies.
- The Report by McKinsey Global Institute, *Workforce Transitions in a Time of Automation* by Susan Lund, James Manyika, and Michael Spence **(2019)** takes on the workers' transitions that automation brings with itself and the skills they need to adapt.
- The American Economic Review with the article *"Human Capital in Age of Artificial Intelligence"* by Erik Brynjolfsson and Andrew McAfee (**2017**) details on how AI is transforming workforce and its implications for skill development and education.

The above literature review has specifically highlighted the need and necessity of expanding new skill sets to meet the 6G technological requirements. So, the focus now must be on bridging the skill gap, and for that the curriculum should change with time, learning must continue, as should collaborative initiatives among various stakeholders. These interventions alone can ensure a skilled and adaptable workforce in today's ever changing world.

IMPLICATIONS OF THE MONUMENTAL SHIFT FROM 5G TO 6G—A TECHNOLOGICAL LEAP

The quick evolution of mobile communication technology from 5G to 6G along with the tectonic shift in global migration is positioned to change the workforce landscape dramatically. The repercussions of this game-changing shift are going to be critical for organizations and economies that are bracing themselves for these changes, making it imperative for any workforce to be adaptive, resilient, and innovative. The 5G era is known for speed and low latency bringing the fastest speeds in history for mobile networks—allowing us to connect more devices, providing stronger bandwidth enabling the IoT (Internet of Things), enhanced vehicular automation, and even Smart Cities. 5G amplified this capability and now the sixth-generation (6G) network is promised to take this even further through ultra-low latency, terahertz frequencies, and AI-driven connectivity. Holographic communication, real-time VR (Virtual Reality) and AR (Augmented Reality), and bridging digital and physical environments are among the expected *use cases*. The main challenge will be the transition to 6G, which implies a major qualification requirement for AI and Machine Learning, Cybersecurity skills, Data Analytics and, of course, Quantum Computing. More works of routine and manual type will be automated so as to require a workforce able to control and capitalize on these advanced technologies. There will be new roles which would come into prominence in areas like innovation, interpreting data, and managing systems. People migrate in search of superior employment and living—attributable frequently to variations in economic development and prospects for work. Wars, global warming and disasters cause the displacement of people from one place to another leading to a workforce with various cultural backgrounds in the host country. Migration generates a wide and diverse human capital which promotes creativity and innovation. This is why companies need to embrace diversity and economic inclusion so that they will be able to take advantage of these benefits. Most migrants come with valuable skills and contribute to the local economy but they routinely face obstacles regarding language, credibility of degrees, and insertion into the local labor market. It is essential to address these gaps through training, certifications programs, and supportive policies.

THE TECHNOLOGICAL CONTEXT OF THE 5G TO 6G TRANSITION AND THE CONCOMITANT IMPACTS OF 6G ON WORKFORCE EVOLUTION, REQUIREMENTS, AND IMPERATIVES

The introduction of 5G technology promises to deliver an unprecedented level of speed, connectivity, and innovation, ultimately rebranding sectors such as healthcare, transportation, entertainment, etc., as *'breakthrough industries'*. Nevertheless, 6G takes these boundaries even further and profess to use the higher data rates, ultra-low latency, improved reliability, increased energy consumption. Holographic communications and ubiquitous connectivity are some of the newer applications which will be made possible thanks to 6G.

The improvements will be felt across multiple industries, from health care, to manufacturing, transportation and entertainment where new jobs would be created and the existing ones changed entirely. But the move from 5G to 6G is like getting five times better connectivity, speed, and other capabilities that translates into huge changes for all of us at work. Due to the properties and characteristics of 6G, researchers are starting to see that it will deliver extremely fast data speeds with low latency but also improved quality / reliability of connectivity, in addition to newer technologies incorporation as AI, ML, and XR. It's a given that machines will displace people from certain types of jobs. It is also likely to lead to new job categories as well as the re-definition of some roles. Since 6G Technology will be penetrating into different sectors, the expertise in this technique also requires specific skills.

As the workforce dynamic changes, tools, platforms, and methodologies are bound to become obsolete or require modification which is where continuous learning/ adaptability comes into play. From 5G to 6G migration, Lifelong Learning signals are crucial. Specifically, employers and educational institutions will need to work together to provide ongoing training—including both up-skilling employees and re-skilling workers presently with outdated skills. In addition to technology capability, flexible thinking and cross-border teamwork skills are becoming more important over technical capabilities. Policymakers and institutions have to develop policies that facilitate the workforce transition. They need to be investing in education, training programs, and social safety nets when jobs disappear. While it is absolutely essential to facilitate the emerging technologies, taking all parts of the labor force along the strong digital and physical infrastructure at the same time is yet another priority. And although we may expect a good portion or even most to return to the office sometime in the next year or two, the future of work is really going to be an amorphous blend between remote, hybrid, and traditional on-site business, driven as much by technology changes as it is by employer desire (for flexibility) and employee resistance (to commutes). Leaders need to be creating a culture that includes diversity, innovation, and supports employee wellness.

CURRICULUM DESIGN, WORKFORCE EVOLUTION, AND TRANSITIONING TO 6G

With the advent of the new 6G era, organizations should plan to make a curriculum design that will ensure they have *future-ready* employees. If done right, curriculum design can provide people of all ages with the skills and knowledge to leverage a constantly changing technological world. This is true as the role of the curriculum design is to make sure that educational programs are kept up-to-date with changing industry demands. This means and involves collaborating with the industry *i.e.,* partnering with experts and understanding current and future skill requirements and bring into one's curriculum. Curriculum changing has to be carried out on a periodical basis so that it remains updated with the current technologies and technologies evolving in the industry. One has to keep in mind the characteristic hallmarks of 6G in this regard. A highly multi-disciplinary education and learning becomes a pre-requisite for incorporating the technology across various domains of 6G and their integration. The curriculum design requires the inclusion of interdisciplinary courses in technical education with ethics, business studies, and social sciences to address complex problems that emerge within 6G capabilities and usages. Moreover, '*Project-Based Learning*' should allow students work on team-based projects mimicking real-world challenges and solutions stemming from inter-disciplinary findings.

With the rise of technology, which is growing more and more complex by the day, human-centric skills (literally all of what is written about them) need to be accentuated. One should focus on making sure that abilities are integrated by injecting human-centric skills into technical programs to provide a comprehensive education. First, there would absolutely be a provision for experiential learning, providing opportunities for students to obtain those skills by working as interns, on co-op or through collaborations with various projects. Secondly, the fast pace of technological change means learners need to adopt a lifelong learning mindset. Curricula must provide modular and flexible learning pathways that can be taken over the course of a career—for both upskilling and reskilling. Through the development of comprehensive online and continuing education programs, one can hope to open up and offer learning opportunities to a wider audience. Curriculum design should include several important areas of emerging technologies such as AI, ML, XR, and IoT with new generation tools. If the student-learners have to stay tuned with 6G, they need to get practical experience within all these areas as all of them relate to preparing the student for future workforce. In addition, the focus must be on modern pedagogical methods that include virtual labs, simulations, and '*gamification*' to enrich learning opportunities.

HOW CURRICULUM DESIGN SHAPES WORKFORCE EVOLUTION

In a well-designed curriculum, graduates leave with all the skills and cognitive abilities needed to thrive in our future 6G-enterprise world. Future-ready professionals are those who have a paradoxical combination of technical skills and critical thinking-led adaptability. Any curriculum design has to find and feel significant connections between education and industry. Only then, organizations would be able to promote a culture of innovation and generate economic growth. For example, Graduates who are equipped with skills to utilize the new 6G technologies can help spearhead the creation of products, services, and business models that have yet to be imagined. This can have an important part in making sure that 6G benefits are universal. Inclusive domestic policies have the potential to create new opportunities for underrepresented minorities and women in the workforce.

The rise of 6G technology brings with it the need for ethical and social considerations. Preparation of students to address these types of complex issues in the future will involve an elementary curriculum design capable of encompassing courses on ethics, privacy and social responsibility. Importantly, the curriculum must be co-created with a range of stakeholders such as industry experts from multiple disciplines and areas, educators, and policymakers to address the needs of all parties. They have to be tailored for offering continuous improvement opportunities and also for getting the feedback from students who undergo studies. This would go a long way to continuously update/refine the content, which is another key aspect. The learning process should be technological based, include the use of digital tools and online platforms, and new pedagogical practices. There is yet another urgent priority—providing more ongoing professional development for educators so they can stay current with the latest technologies and best practices.

MIGRATION FROM 5G TO 6G AND LIFELONG LEARNING IMPERATIVE

Given the rate at which technology is changing, it requires all the stakeholders to commit to a philosophy of "*always keep learning*". This phenomenon is even more pronounced during the shift from 5G to 6G networks. As we approach the crest of this technological wave, it is important to consider the consequences for individuals, organizations, and society. This has turned learning into a lifelong necessity as we need to be able to adjust, develop, and excel in the new digital maze.

Everyone has to continuously learn the skills of 5G TO 6G transition in rapid rate evolution. Everybody should be a lifelong learner in order to remain current on new technologies, tools, and techniques. This is even more important in areas like Network Engineering, Cybersecurity and Data Science where technological obsolescence can happen quickly. The available specialized skills continue to change with the advent of new technologies, as industries adapt. Professional development in the lifelong learner era helps keep professionals '*current*' and enables them to be more competitive. Such will be the case with mastery of 6G network landscapes, one has to watch out for and note AI implementation and comprehensive Cybersecurity measures. A culture of curiosity and innovation is nurtured by continuous learning. Participating in these latest discovery opportunities by engaging with what they offer and experimenting with their possible uses can further advance human innovation. It would also unlock new revenue streams for companies, create new products and services faster, better, cheaper, and help business operations run smoother and successful. 6G technology envisions solving several societal problems such as digital inclusion, smart cities, and sustainable development. It allows people—throughout their lives —to play an active role in the initiatives that come out of these changes, making sure that advancements are made for the betterment of society.

Now as this transformation may seem to be a shift in technology, it is also much bigger move being made about what skill and knowledge levels are demanded from the workforce. Lifelong learning is increasingly necessary, as the landscape changes faster than ever and requires individuals and organizations to adapt, innovate, and sustain. One needs to discuss the crucial need and importance of lifelong learning in 5G and 6G technologies. At the heart of the racing level of technological deterioration, 5G technology has significantly impacted connectivity by providing faster speeds, lower latency, scalability among connected devices and supporting leaps towards Internet-of-Things (IoT), smart cities, and autonomous cars. 6G is expected to come with improvements on all these fronts, plus terahertz frequencies, "beyond AI" connectivity besides real-time, virtual, and augmented reality components. It sounds great on paper—blazing fast transfer speeds, near-zero latency, and the final bridging of our physical and digital worlds. One has to also keep in mind the increasing demand for and the necessity of the Human Skills that are required for making this mandatory transition. The transition to 6G will also require new technical skills in the AI, ML, Cybersecurity, Quantum Computing, and Big Data Analytics.

Soft skills like adaptability, critical thinking, creativity, and cross-discipline collaboration are equally necessary. Only then would it make sense of technology in a complex multidimensional environment for the purpose of problem-solving and innovation. Technological change is moving so quickly that skills and knowledge of the last century are outdated in the current month. The one thing we must realize is that with the changes in technologies and best practices, if there is no lifelong learning, individuals cannot survive. That is why people are trying to keep up with new technologies and methodologies, learning every day about how they change. In the world of job market dynamics that includes career longevity and employability, some jobs might get outdated while others are created directly due to advancing technologies (5G and 6G). The answer in most cases will be lifelong learning that gives individuals the ability to easily move into new jobs and industries. Nonetheless, with every type of learning that is continuous in nature it can open up avenues for growth and expansion to keep the individual revitalized enough to be able to take on new challenges in life and even at work where they normally would get opportunities to move into leadership roles.

The need for going forward would compel the organizations to cultivate lifelong learning and remain competitive in order to innovate, pivot in response to market changes, and avoid obsolescence. This will increase employee satisfaction and retention, as team members will feel supported while growing in their careers along with the company.

STRATEGIES TO PROMOTE LIFELONG LEARNING

Given this context, it is also imperative for Educational Institutions to integrate emerging technologies and other interdisciplinary studies into current revised curricula for making students future workforce ready. Without partnership with the industry, they cannot provide the student with Industry experience that is very much needed to join the same later. Here, industry partners do offer internships, co-op programs, and real-world project experiences that have the potential to bridge academia-industry divide. They must create on-the-job training programmes to enable workers to learn and apply new skills in a practical business context. We need workshops, seminars, online courses, and certification programs to encourage and support a path of lifelong learning. They are suggested for acquiring and using of E-learning tools and online learning resources that can make learning opportunities available in a flexible, accessible, and personalized way through technology-enhanced learning. In this case, microlearning methods can break down information into small, bite-sized content bits. This would easily allow individuals to incorporate learning into their daily schedules in more digestible sections. Lifelong learning initiatives and training programs need government and policy support for the grants, tax incentives, etc. We need to come up with policy frameworks and formulae a self-sustaining set of policies that encourage lifelong learning. They do provide the support necessary for anyone, from any socio-economic background, to be able to access lifelong education and training.

ADVANTAGES IN 5G/6G ERA OF LIFELONG LEARNING

Lifelong learning in and with different fields will give an individual the creativity to innovate new methods by working under various domains. A smarter workforce, with access to the *best-in-class* knowledge and data can take better knowledge-based and well-informed decisions that can uplift overall organizational performance. It is a necessary preliminary for the organizations to have a workforce that is well-trained as it can shift to meet new demands. This is very much a *sine qua non* as it alone can lead to economic growth and stability, allowing for the advancements in modern industry that we have today. It also contributes to social inclusion by offering individuals from all backgrounds opportunities to up-skill themselves and participate more fully in a digital economy. It enables and empowers people by equipping them with the skills and sense of self that are needed to move through a dynamically changing world. With continuous and life-long learning, individuals never become or remain stagnant with their own curiosities while still being employable.

THE EMERGENCE OF 6G AND THE ROLE OF EDUCATION AND TRAINING IN PREPARING WORKERS WITH HUMAN-CENTRIC SKILLS

People have welcomed a bold new world of connectivity—the next phase in technology evolution from 5G to its successor 6G—and with it real-life implications for both the workforce and society. In the 6G era, possession and application of human-centric skills are more important than ever experienced or practiced anywhere and anytime before. Moreover, they will be crucial in the adaptation to new technologies, in the promotion of innovation, and also in making sure that 6G benefits everyone and always. They include a variety of skills that have at the heart the human part or dimension of technology and work, so people call them human-centric. These include:

- *Emotional Intelligence*(**EQ**)**:** EQ means awareness of your own as well as others' feelings and the ability to manage these emotions.
- *Critical Thinking and Problem Solving: the* ability to analyze complex problems and generate fresh perspectives and solutions.
- *Teamwork and Communication:* Working well in teams, including those across different contexts or backgrounds.
- *Adaptability and Learning Agility*: Acquiring and experiencing a willingness to learn new skills and adapt if required.
- *Ethical and Responsible Decision-making:* While making decisions, one has to think through and consider all the ethical implications and societal environments.

And, 6G Technologies—enabled with all the functionalities like Artificial Intelligence (AI) Integration, Extended Reality (XR), and ubiquitous connectivity—need a workforce that will bridge technology and humanities. Human-centric skills ensure that any technology developed is usable and responsive to the needs of people, while remaining ethically compliant with human values. 6G systems are too complex and require fresh thinking and new solutions. With soft skills that are geared towards the human experience, people can develop ways to creatively apply 6G to their potential. As 6G begins to take hold, the issue of ethics becomes increasingly important. A good grasp of ethical decision-making and social responsibility is also needed in order to produce technologies which maintain privacy, encourage inclusivity, and tackle wider societal problems. Building the next era of 6G technologies requires broad, interdisciplinary teams that have acquired cross-domain expertise. In today's multinational business environment, the collaboration and communication must work together as they form a horizontal function to coordinate efforts and share knowledge across a wide range of functional activities to achieve their common goals. Things move so fast in the tech world that what you know today could become irrelevant tomorrow. Possessing and demonstrating adaptability and learning agility ensures a constant evolution of empowered employees who can learn *what* they need to, *when* they need to, enabling them to move forward in the new world of work.

MIGRATING TO 6G: SOFT SKILLS AND WORKFORCE EVOLUTION

The future of 6G technology is just around the corner and it will bring some of the most radical changes in the way our society works. For the most part, technical skills are extremely valuable, but a business organization should also take into consideration how the way an employee relates to others through the so called 'soft skills'. Otherwise, they would woefully be missing something huge in terms of their effectiveness. Matters that are related to communications, interpersonal skills, problem-solving, and emotional intelligence are just as important as coding when confronting the rapid evolution of technology. In the age of 6G, a sea change for humankind is one that is driven by the migration to strengthen one's soft skills—personal attributes and interpersonal abilities that enhance an individual's interactions, job performance, and career prospects.

Key soft skills include:

- *Communication*—good written and verbal communication
- *Collaboration*—the ability to work with a team
- *Adaptability:* The ability to respond to changing circumstances.
- *Problem-Solving*: Creative and analytical thinking to solve problems; ·
- *Emotional Intelligence* (EI): Ability to perceive, understand and manage our own emotions and those of the others.

Within the Workforce Dynamics, every individual needs to rediscover the changing job roles. The more the technology merges with our everyday work, the greater the job roles will demand a mix of technical and soft skills! You will need to be able to intelligently interface both with technology and humanity. We now live in the era of 5G, and as we step into the future with 6G, workplaces will slowly transition to being fully connected, global, and virtual. At this point, the importance of soft skills (communication and collaboration) kicks in but also becomes an absolute imperative. This is Next Normal where we need to get better at working with distributed teams in different time zones and cultures referred to as '*remote work*'. Interdisciplinary teams that collaborate with professionals from different industries to innovate and tackle complex problems are also needed. Technology is changing too fast for us that we can no longer hope our workers would embrace that same sense of versatility—the ability to adjust on the fly and learn new skills, faster—like agility. *Reskilling* and *upskilling* are very much the need of the hour. They are a must for teams and professionals who need to acquire new skills as they are now required to complete their tasks quickly, often involving completely new technologies and methodologies. Given the nature of a tech industry, which is constantly in a flux, they need to champion change as something that is new and positive must be managed effectively. This coincides with innovation that is built based on creative thinking and problem-solving together, where soft skills are essential when creating ideas and concepts to utilize 6G technologies.

- ***Collaborative Problem-solving:***

Soft skills include collaborative problem-solving also where we solve challenges together to find new ways of being and doing. If technology is becoming embedded in human activities, it is important to address the emotional and ethical stakes. It ensures emotional intelligence, which is very much needed in dealing with underlying human relations and team dynamics—the "soft" side of becoming a true professional. This means to show and demonstrate to the people the way ethical decision-making is done so that they make responsible decisions and weigh the 6G technology`s societal impact. For being 6G-enabled world, the emphasis has to be on two aspects of soft skills—providing pleasant user experiences and satisfying the customer-centric needs. Effective communication with customers is all about really getting to know them and present offers personalized offers that are tailor-made according to their needs. Generally, engineering teams are the sole reason any products and services become intuitive enough for people to actually 'use'. Hence, user-experience design requires a very close collaboration with technical teams so that they keep going-forward.

HOW TO PROMOTE LIFELONG LEARNING / HUMAN-CENTRIC AND SOFT SKILL DEVELOPMENT

In the upcoming 6G era, the educational institutions in future should have to update their curricula on a regular basis to tick with the latest technological trends as well as the required skills. Interdisciplinary studies that combine technology with ethics, law, and business should be encouraged to deal with the multiplicity of challenges presented by 6G. Moreover, support for lifelong learning initiatives in grants and tax benefits will also serve as incentives for continuous education. This will involve a more robust marketing emphasizing the value of lifelong learning, mass advertising campaigns and working with industry partners. The corporate sector must also look at dishing out more employee training programs which zero in on the upcoming technologies and interdependencies like those of 6G. And one way to approach this is by creating a culture of continuous learning where all personnel are encouraged to update their skills with available mentors, workshops, or any self-learning portals.

In the context of self-directed learning, always, every person needs to take more special care for continuing education that necessitate in pursuing online courses, training events, and self-reflection. They have to chase Professional Certifications on the latest technologies and domains related to 6G and get assessed on practical skills through them.

- ***Curriculum Integration:***

This is indispensable because, at the conclusion of the day, human-Centric skills need to be developed alongside building capacities in soft skills among the student-learners. Therefore, **STEM** education and technical programs should include human-centric skills in the syllabus. On another hand, soft skills training must be blended with the technical and professional education programs. By offering more experiential learning (e.g., internships, co-op programs, project inspiration), students get an opportunity to see how these skills are relevant in real world applications. Organizations in the corporate sector have to eventually come about with full-fledged training programs that focus on internationalizing humans with by fusing technical and human involved skills. They also should create guiding, mentoring, and coaching systems that teach employees Emotional Intelligence, (EI), communication, and leadership. All of this needs to be buttressed by government with the formulation of policies that facilitate the incorporation of human-centric skills into educational and professional development initiatives.

This requires government funding and incentives (*in order to promote similar skill development initiatives across all the sectors*). Participation in professional networks and projects may help to increase capacity for communication, teamwork, and leadership. Implementation of e-Blockchain network in u-Medicus must undergo rigorous pilot testing and further professional development from 5G to 6G technologies. Along with the technical setup, the transition from 5G to 6G technologies for people and society will require a combination of human-centric skills as well. The need to have these skills will be essential as we move into an age of 6G with all the complexities and new possibilities. For this, all our professions need to develop for our current workforce in this transformative era. When the focus is on being a business design technologist, there would be demands for competencies such as the Adaptability and Agility to change. One has to be fast in adapting to new technologies and capabilities. Moreover, with rapid changes in technology, they have to be aligned with a very strong commitment for continuous learning.

- ***Critical thinking and Problem-solving:***

Analytical skills, as mentioned above, consist of the ability to analyze complex problems, understand the underlying factors behind them and/ or making informed decisions based on data and situational awareness when needed and coming up with new solutions. Creative thinking is required to bring about innovative ideas and solutions as well as to take advantage of novel technologies in the field of creativity and innovation. A mindset of innovation can drive experimentation and the genesis of new products, services or processes. In the same way, it is also requisite to collaborate and work laterally with other congressman across interdisciplinary researches. This includes requiring great communication skills to share ideas, impart information, and organization efforts. The inclusion of *'Emotional Intelligence'*—which includes self-awareness—understanding one's own emotions, strengths, and weaknesses; 'Empathy" (being able to understand share the feeling of others) would in together ensure a progressively congruent work atmosphere.

'Ethical and Responsible Leadership' (no brainer): If ethical Judgment is about providing for the right decisions, weighing them in terms of the ethical implications and societal contributions, responsible leadership is about leading with integrity, transparency with a special focus on sustainability as well as social responsibilities.

STRATEGIES FOR PROFESSIONAL DEVELOPMENT

1. Education for Life and Structure Education:

- *Continuing Education:* Training programs are up to date in order to keep employees knowledgeable about the most recent technologies and methodologies.
- *Micro-learning Modules:* these are most often short, focused learning modules that can be easily woven into an employee's daily work activities.

2. Cross-Disciplinary Learning:

- *Interdisciplinary Courses*: Programs designed to target both technical aspects and soft skills, allow students to understand how technologies interact with human-centric elements for well-rounded perspectives.
- *Project-Based Learning:* Synergy based projects that require interdisciplinary collaboration.

3. Mentorship and Coaching:

- *Mentorship Programs*—Pair less experienced employees with more seasoned professionals to provide guidance, support, and knowledge.
- *Coaching Sessions*—Coaching to improve specific soft skills (e.g., leadership, communication, emotional intelligence)

4. Cooperative Working Spaces

- *Innovation Labs*—Spaces to Innovate—Innovative projects and new ideas through collaboration of employees.
- *Cross-Functional Teams:* Advancing the creation of teams made up of a variety of backgrounds and skill sets tackling different problems to spark creativity.

5. Usage of Technology in an Ethical and Responsible Manner

- *Ethics Training*—Education programs able to remind employees about the ethics connection of their work and ensuring they apply technology effectively.
- Diversity and Inclusion Initiatives—establish a culture that promotes inclusion in the workplace, scrap on different perspectives and experiences.

6. Better Communication Streams:

- *Feedback Mechanisms:* Develop processes to get feedback routinely and ensure a proper channel exists for open communication.

- *Knowledge Sharing Platforms*—facilitate the sharing of knowledge, best practices and innovative ideas across the organization.

THE CHALLENGES AND OPPORTUNITIES IN THIS MOVE TO A 6G WORLD FOR THE WORKFORCE

There exist several and varied challenges and opportunities in the current move to a 6G World for the modern workforce. Some of them are briefly discussed hereunder:

- ***Challenges***

A mere migration from 5G to 6G will definitely signal one of the quantum leaps the likes of which we can scarcely imagine across sectors, economies and labor pools. This advancement of technologies poses some challenges as well as opportunities for human beings. These dynamics may be important for organizations and individuals to make the most of this changing landscape.

Rapid technological change renders current skills obsolete and forces people to learn new things on the go. At the same time, demand will soar for specialist skills in areas such as Artificial Intelligence (AI), Machine Learning, Quantum Computing and Advanced Data Analytics—a shift that could leave those without these competencies trailing further still. It is evident that in the future of 6G technologies, jobs are likely to be at risk going by rising automation and AI but it would mostly involve replacement of routine and manual jobs from the markets. However, workers in impacted roles will have to reskill or upskill to move into new positions—a process likely to be time and resource intensive. In turn, not all employees may have access to the latest technologies and training—a situation that could widen existing inequalities even further. It would also result in the building of a digital divide within our workforce. Infrastructure and resources for capitalizing on the 6G technologies will likely be less available in remote and underdeveloped regions. As many connected 6G devices and systems will become widespread so will be the avenues for cyber threats increased, hence providing a stimulus to further leverage Cybersecurity measures and awareness. Management of a large quantity of personal and confidential data will cause worry about privacy this is where stringent data protection methods should be followed.

- ***Opportunities***

When 6G technologies make an appearance in the market, it will subsequently generate and come up with new jobs that never existed before such as Network Management, Cybersecurity, Data Science, and AI Development. Greater collaboration and access to new technology will stimulate innovation, which has a domino effect in spawning start-ups and entrepreneurial ventures. With the rise of 6G technology, productivity will be bettered through efficient and effective tools and processes across different industry verticals. More cohesive connectivity will enable remote and flexi-enterprise working benefits, the option for work-life balance, and access to global talent pool. A need for constant automation technology upgrades will increase the growth of continuous learning programs broadening the skill and flexibility in the workforce as a whole. As employees learn and hone their abilities, they will likely find themselves being pulled into more specialized roles which are challenging and fulfilling. Improved connectivity will make it possible for teams collaborating globally, combining individuals from all over the world with no need to establish a physical address. To foster inclusive work environments, 6G Technologies can offer tools and platforms that accommodate various aspirations and desires.

APPROACHES ON HOW TO TACKLE AND CAPITALIZE THE 6G CHALLENGES

Of late, bodies of organizations and governments alike should commit to investments in up-skilling and re-skilling programs for the workforce to learn new roles and technologies. To shape curricula tailored to the needs of the 6G era, Partnerships and collaborations with Educational Institutions are indeed become desirable and even inevitable. It means that they have to be able to keep the most recent technologies and infrastructure while also closing the digital divide to support all employees. It demands the development of a compatible regulatory framework that facilitates innovation and balances with Cybersecurity, Privacy, and Ethical issues. So the need in the organizations is to build a culture of continuous learning—education should be ongoing and never-ending, companies should promote employees for higher studies or professional degrees on regular intervals. Another approach would be to deliver learning in different forms such as online courses, micro-learning modules, and on-the-job training so that it works for everyone. They will also have to deploy advanced Cybersecurity solutions to mitigate the new risks and vulnerabilities emerging from 6G technologies and also keep in place elaborate data protection mechanisms that can safeguard their critical information from interception or breach. Innovation hubs—labs for experimenting with new ideas and technologies—have to be created where the workforce gets a nurturing environment to operate. There has to be a provision for giving incentives to the employees for sharing their innovative ideas and solutions while fostering a culture of creativity and experimentation become the need of the hour.

SUMMARY THOUGHTS

5G to 6G—the transition and migration represents a technological leap that will revolutionize our lives. It calls for a holistic approach that includes lifelong learning to navigate this transition. This is necessary to ensure they are future-ready, capable of driving innovation and part of a more connected and integrated society. Likewise, life-long learning is a prerequisite that ensures all individuals are empowered with required tools and insights to traverse 6G frontiers—supported by policies, infrastructural underpinnings to foster inclusive leadership by design. Adapting to these changes means building a workforce that is truly agile, innovative, and inclusive enough to excel in the future. Thus, there is no other way for this to happen except the way of everlasting and ongoing learning. The program provides people with the competences and insight needed to adjust to new technologies, raises employability levels, promotes innovation, and is linked with organizational and societal growth. The time of lifetime learning can be the only way to have a workforce that is resilient, dynamic, and capable of anticipating what the future may need. Like all kinds of Gs that have gone before, the journey from 5G to 6G technologies will necessitate substantial innovation, investment, and coordination.

Successful capabilities development and deployment will require partnerships and collaboration across the many stakeholders, including governments, industries, academia, and standardization bodies. The migration of this particular technology in hands also demands us to have forward looking curriculum designing techniques. From aligning industry needs with education, to encouraging learning across disciplines to supporting human centric skills and a importance of lifelong learning *albeit* introduction of latest technologies, we can develop a workforce for the 'Digital Age.' Curriculum design is as critical for our personal careers and job prospects as it is in creating the new high-paying professions of the 6G era that will make our economy and society more prosperous. Failures and lessons learnt from 5G Technologies as well as the appearance of 6G Technology Generation also highlight the imperative necessity in work force, for human-centric knowledge. These include, among others, the ability to relate cutting-edge technologies and human requirements—a role that combines science with what researcher Else calls being an artisan; as well as inventiveness (or innovation), principles in designing technology responsibly (production) and effectiveness in collaborating. Prioritizing the cultivation of human-centric skills could equip the labor force to optimize the capabilities of 6G in ways that promote sustainable prosperity and a fairer, more inclusive society. A more common workforce theme as we head into 6G will be on the need for lifelong learning and human-centric skills that enable us to create adaptable, creative teams to meet the opportunities and challenges of the digital age.

With the evolving and ongoing move to 6G technology, it is increasingly important for soft skills to be developed among workers. The future lies in the ability to communicate, collaborate, adapt, solve problems, and manage emotions effectively as technology continues to change. The need to develop soft-skills take the spotlight again as this serves to ensure a workforce that is proficient not only at the technical and process levels but also one that is able to navigate complexities of a 6G-enbled world. Taking a two-fold approach to workforce development can spur innovation, drive productivity, and make sure we see the full fruits of our ongoing labor with 6G technology—in our professional lives and otherwise as well.

REFERENCES

Aspin, D. N., & Chapman, J. D. (2007). Lifelong Learning: Concepts and Contexts. *International Journal of Lifelong Education.*

Battarbee, K., & Koskinen, I. (2005). The Importance of Empathy in Design: Perspectives from the Field. *Design Issues.*

Bessen, J. E. (2019). Emerging Technical Skills for the Future of Work: Trends and Implications. *The Journal of Economic Perspectives.*

Brown, S., & Black, P. (2022). Professional Development in the Age of 6G. *Journal of Continuing Education*, 28(4), 150–165.

Brown, T. (2008). Design Thinking and Human-centric Innovation. *Harvard Business Review.*

Carnevale, A. P., & Desrochers, D. M. **(2001),** *"Technical Skills and Competencies in the Modern Workplace",* Workforce Education Forum.

Chen, L., & Wang, H. (2022). Quantum Communication in 6G: Opportunities and Challenges. *IEEE Communications Magazine*, 60(8), 45–51. DOI: 10.1109/MCOM.2022.9755279

Clark, A., & Adams, P. (2023). Government Initiatives for 6G Skill Development. *Public Policy and Technology Journal*, 21(4), 145–160.

Davis, M., Clark, J., & Lewis, R. (2022). Enhancing Cybersecurity Skills for 6G. *Cybersecurity Education Journal*, 13(5), 210–225.

Field, J. (2006). Lifelong Learning Policies and Strategies in Europe: Trends and Issues. *Comparative Education.*

Garcia, L., Brown, M., & Nguyen, T. (2022). Addressing Resource Constraints in Skill Development. *Resource Management Journal*, 29(4), 180–195.

Garcia, L., Nguyen, T., & Brown, M. (2022). Critical Thinking in Technical Professions. *Journal of Professional Development*, 20(3), 88–99.

Goleman, D. (1998). Emotional Intelligence and its Impact on Leadership Effectiveness. *Harvard Business Review.*

Hager, P., & Halliday, J. (2009). Learning Throughout Life: An *Interdisciplinary Perspective. Studies in Continuing Education.*

Hall, T., & Brown, J. (2022). Corporate Approaches to 6G Workforce Training. *Corporate Training Journal*, 25(3), 110–125.

Harris, P., & Clark, R. (2022). Nokia's 6G Training Program. *Journal of CorporateTraining Initiatives*, 17(4), 140–155.

Huang, M., Xu, Z., & Liu, Y. (2023). Terahertz Technology for 6G Wireless Communications. *Wireless Networks Journal*, 38(2), 98–112.

Johnson, P., Smith, J., & Davis, M. (2022). Economic Impacts of 6G Innovation. *Journal of Economic Development and Innovation*, 26(3), 220–235.

Johnson, P., Taylor, A., & White, G. (2022). Bridging the AI Skills Gap in Telecommunications. *Journal of AI Research and Development*, 18(2), 200–215.

Jones, T., & Brown, R. **(2023).** Global Connectivity in the 6G Era. Journal ofGlobal Communications, 14(2), 80-95.

Jones, T., & Brown, R. (2023). Security Challenges in 6G Networks. *Cybersecurity and Privacy Journal*, 29(7), 90–105.

Kim, J., Park, S., & Lee, H. (2023). IoT Integration in 6G Networks. *Sensors and Actuators Journal*, 45(9), 321–334.

Lee, H., & Kim, S. (2022). Adapting to Technological Change in Telecom. *Technology and Innovation Journal*, 33(4), 150–165.

Lee, H., & Kim, S. (2022). Overcoming Workforce Resistance to Technological Change. *Journal of Change Management*, 34(2), 100–115.

Li, S., Yang, T., & Chen, Q. (2022). Machine Learning for 6G: An Overview. *Future Internet Journal*, 14(5), 200–215.

Martin, J., & White, S. (2023). Barriers to 6G Adoption. *Telecommunications Barriers Journal*, 11(2), 77–89.

Miller, J., & Davis, L. (2022). R&D for Future Telecommunications. *Research and Development Journal*, 32(6), 90–105.

Mitchell, G. W., Skinner, L. B., & White, B. J. (2010). Essential soft skills for success in the twenty-first century workforce as perceived by business educators. *Delta Pi Epsilon Journal*, 52(1).

Nguyen, T., & Smith, K. (2023). Case Study: University of Technology's 6G Curriculum. *Journal of Higher Education in Technology*, 22(1), 55–70.

Norman, D. A. (2002). Human-Centered Systems: Information, Interactivity, and Intelligence. *Interacting with Computers.*

Patel, R., Singh, P., & Green, D. (2023). Quantum Technologies for Future Networks. *Journal of Quantum Computing*, 12(1), 67–79.

Robles, M. M. (2012). Soft skills and job performance: An empirical study. *Business Communication Quarterly.*

Smith, A., & Jones, M. (2022). Emerging Job Roles in 6G. *Future Jobs Journal*, 21(5), 175–190.

Smith, A., Jones, M., & Lee, K. (2022). Network Slicing and Edge Computing for 6G. *Communications Systems Review*, 52(4), 233–248.

Smith, J., & Jones, L. (2022). Interdisciplinary Collaboration in Telecommunications. *Collaboration and Technology Journal*, 15(2), 101–115.

Taylor, R., & Green, S. (2023). Educating the Quantum Workforce. *Journal of Quantum Education*, 8(1), 45–58.

Von Hippel, E. (1994). The Role of Technical Skills in Innovation: A Perspective from the Field of Engineering. *Research Policy.*

White, G., Black, A., & Thompson, L. (2023). Updating Academic Curricula for 6G. *Educational Technology Review*, 19(2), 133–149.

Wilson, K., Miller, J., & Davis, T. (2023). Fostering Industry-Academia Collaboration for 6G. *Collaboration in Education Journal*, 14(3), 122–138.

Xu, W., & Yang, S. (2023). Terahertz Waves for 6G Wireless Networks. *International Journal of Wireless Communications*, 27(3), 156–170.

Zhang, X., Li, Y., & Wang, J. (2023). Artificial Intelligence and Machine Learning for 6G Networks. *Journal of Telecommunications Technology*, 45(3), 123–134.

Chapter 6
5G to 6G:
Empowering the Workforce Skills for Next-Gen Connectivity

A. S. Anurag
https://orcid.org/0009-0005-5790-9084
Central University of Kerala, India

A. Swathi
https://orcid.org/0009-0004-6736-8480
Central University of Kerala, India

Rohith Ramesh
https://orcid.org/0009-0009-6957-0152
Central University of Kerala, India

M. Johnpaul
https://orcid.org/0000-0002-1133-0588
Central University of Kerala, India

ABSTRACT

The global launch of 6G communication is anticipated to occur in the near future. This high-speed communication technology will propel the growth of disruptive technologies that are currently underperforming due to our limited communication capabilities. With the establishment of 6G, advancements in AI, the Internet of Things, Machine Learning, and Edge Computing will reach new heights. These changes in the technological landscape will redefine workforce skills requirements. This chapter aims to examine the technological transformations resulting from the shift from 5G to 6G, particularly in the context of empowering workforce skills. Utilizing existing literature, the authors identified several workforce changes and proposed strategies to adapt to this evolving landscape. As communication speeds

DOI: 10.4018/979-8-3693-6725-4.ch006

and data transfer rates increase, society will transition completely towards a smart and digital era. This paper will assist researchers, policymakers, and management personnel in preparing for the changes that 6G technology will bring.

INTRODUCTION

The history of cellular networks began in 1947 when Bell Lab's engineers W. Rae Young and Douglas H. Ring designed a cellular phone network. Their eight-page document introduced vital concepts such as cell towers and frequency reuse, which are still central to modern networks (Madrigal, 2011). A typical cellular network architecture is composed of components such as a Base Station (BS), Cells, a Mobile Telephone Switching Office (MTSO) or a Mobile Switching Center (MSC), and a Central Office (CO). A typical cellular network architecture is illustrated in Figure 1.

Figure 1. A typical Cellular Network Architecture (Author)

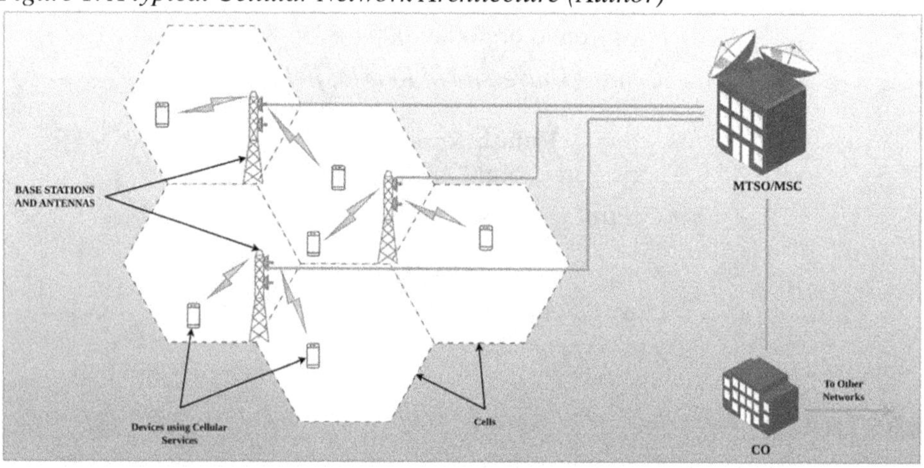

Cells are a crucial component of cellular networks. Cells are the geographical area covered by a BS. In the typical illustrations of a cellular network, the single cell is depicted as a regular hexagon. Multiple cells are interconnected to form a cellular network. The BS in a mobile network includes two primary components: the Base Station Transceiver (BST) and the Base Station Controller (BSC). The BST transmits and receives signals over the radio interface to enable mobile device communication with the network. At the same time, the BSC manages and coordinates multiple BSTs, connecting them to a single MSC to ensure efficient call and data transmission within the network. (Walters & Kritzinger, 2000). This part of the cellular network, which includes cells, BS, BST, and BSC, is collectively known as the Radio Access Network (RAN). RAN is the heart of the wireless communication system. RAN uses radio connectivity technology to connect the user device to the core network and manages the radio resources in the network (Singh et al., 2020).

Multiple BSCs are connected to a single MSC, and several of these MSCs are interconnected to form a meshed network. Riis et al. (2004) explain that MSC handles many cells at once and stores the customer information in the cells in the Visitor Location Register (VLR) database. However, the VLR's storage capacity is limited, and thus, the number of customers should be restricted. BTS and BSC help MSC to limit the number of customers to serve (Riis et al., 2004). The CO in the network acts as a local exchange; it connects calls between users of different local loops, manages call routing, call waiting, and caller ID provision, and links to other COs in the broader network.

The fundamental principle of cellular technology has remained essentially unchanged, but its technology and capabilities have advanced significantly over the years. With the advent of new technologies and more sophisticated hardware, the capacity of cellular network technology has grown exponentially. We have transitioned through several generations of cellular technology over the years. Currently, in the Fifth Generation (5G), we are already developing the Sixth Generation (6G) cellular technology, and we anticipate it will launch worldwide by 2030 (Meah, 2024). The 6G technology is expected to surpass 5G in all the capabilities. The 6G will be 1000 times faster than 5G, with top data transmission rates potentially reaching up to 1 Tbps (Meena et al., 2022).

This significant transformation will not only revolutionize current technologies but also give rise to new ones. We will witness the integration of technologies such as Artificial Intelligence (AI), Machine Learning (ML), blockchain, Digital Twins (DTs), and the Internet of Things (IoT). Consequently, global workforce skill expectations will evolve. It is crucial to understand these emerging technologies and their impact on future workforce skills. This chapter is dedicated to exploring the technological differences between 5G and 6G, projecting future changes in the workplace, and evaluating the skills required to address upcoming challenges.

The chapter is organized into multiple sections for comprehensive understanding and readability. The introduction provides an overview of cellular technology, followed by a detailed examination of the evolution of cellular networks with emphasis on 5G and 6G technology. The subsequent section addresses workforce skill development in light of technological advancements. The chapter then discusses anticipated business transformations and new opportunities arising from 6G technology. Additionally, it includes case studies to exemplify the arguments presented. The chapter concludes with a discussion of security challenges and recommendations. The main objective is to help readers understand 6G, anticipate the changes it will bring to the workforce, and enable stakeholders and decision-makers to prepare their strategies ahead of its launch.

EVOLUTION OF CELLULAR NETWORKS

Cellular networks form the foundation of modern digital technology. The performance and reliability of these technologies depend heavily on the speed and connectivity provided by cellular networks. The First Generation (1G) cellular network was an analog, voice-only system utilizing circuit-switched technology, operating at a peak frequency of 900 MHz and employing the Frequency Division Multiple Access (FDMA) protocol (Hodara & Skaljo, 2021). Due to the analog nature of the 1G network, it was inherently insecure. The conversations could be listened to with the help of nothing but a receiver equipped with an all-band radio (Bhandari et al., 2017).

The Second Generation (2G) network was introduced in 1991. It was the first digital technology in all generations of cellular networks, with a data rate of 64 kbps, and was based on the Global System for Mobile (GSM) Communications standard (Hodara & Skaljo, 2021). While 1G only has voice communication capability, 2G introduced communication through text through the Short Message Service (SMS) (Shukla et al., 2013). During this time, the popularity of 2G communication among the general public has drastically improved and forced network providers to increase the capacity of their communication systems to accommodate the increasing number of users. The popularity of 2G was further enhanced by integrating GSM with Value Added Services (VAS), including Voice Mail Systems (VMS) and SMS Centers. Later, the combination of GSM and General Packet Radio Service (GPRS) facilitated the transmission of packet data over the air interface, thereby enabling wireless access to the Internet (Shukla et al., 2013).

The increase in data traffic due to wireless internet access necessitated an improved system. To manage this high rate, carriers developed an enhanced version of 2G, known as 2G+ technology, which is based on High-Speed Circuit Switched Data (HSCSD), GPRS, and Enhanced Data Rates for Global Evolution (EDGE) technologies (Shukla et al., 2013).

The Third Generation (3G) network, capable of handling much higher data traffic, was launched to replace 2G+ services. Code Division Multiple Access (CDMA) and Time Division Multiple Access (TDMA) or Wideband CDMA (W-CDMA) access protocols were used with the CDMA 2000 1xEV-DV and the Universal Mobile Telecommunications Service (UMTS) W-CDMA mobile systems, respectively, with circuit and packet switching (Hodara & Skaljo, 2021). Compared to 2G, 3G offers several groundbreaking technological applications such as Video on Demand

(VOD), Global Positioning System (GPS), Location-Based Services (LBS), Mobile TV (M-TV), and video conferencing (Ezhilarasan & Dinakaran, 2017).

In 2009, Long-Term Evolution (LTE) technology was developed under the 3rd Generation Partnership Project (3GPP) to meet the requirements of the Fourth Generation (4G) cellular network. The initial version of LTE was not able to meet the standards set by the International Telecommunications Union-Radio (ITU-R). It was later modified to create a newer version called the LTE-Advanced (LTE-A), and thus, the 4G cellular network was established (Hodara & Skaljo, 2021). The security, access methodology, data transfer rate, and transmission terminology have enhanced significantly in 4G compared to 3G (Ezhilarasan & Dinakaran, 2017). Orthogonal Frequency Division Multiple Access (OFDMA) access protocols were used with packet-switching capable of achieving data rate between 100-300Mbps. High Definition (HD) M-TV, video conferencing, 3D television, Mobile Web access, IP telephony, and gaming services are some of the applications of the 4G cellular technology (Mihret & Haile, 2021).

The 5G network, launched in 2019, boasts several defining features, including enhanced mobile broadband, massive machine-type communications, ultra-reliable low-latency communications, cloud computing, and Software-Defined Networking (SDN) (Salameh & El Tarhuni, 2022). With a maximum downlink throughput of up to 20 Gbps, 5G technology supports the 4th Generation World Wide Wireless Web (4G WWWW), which is based on Internet Protocol version 6 (IPv6) (Dangi et al., 2021). Some of the applications of 5G technology, as discussed (Ezhilarasan & Dinakaran, 2017), are Virtual vision and Navigation, which enable users to access a database in street view and support a broad range of user equipment while meeting diverse data demands. Tele-geo-processing capabilities facilitate LBS upon request. Virtualized homes are realized through Cloud Radio Access Networks (C-RAN), offering seamless access to M-TV and other services with low-cost equipment. Smart cities, industrial support, educational opportunities, crisis management, and healthcare applications have also improved as a result of 5G technology.

The sudden growth in disruptive technologies demands a more improved version of the cellular network than the 5G. Hence, academia and industries have already discussed the standard requirement for the 6G network. 6G networks will revolutionize mobile networks by integrating AI and ML, seamlessly merging the physical, digital, and biological worlds, thereby enabling new use cases and applications beyond the capabilities of 5G (Chataut et al., 2024). According to (Jiang et al., 2021a), the 6G system is poised to revolutionize communication by transitioning from traditional electronic radio networks to advanced radio-optical systems. It will evolve into a connected, intelligent platform, harnessing the power of AI to enhance mobile networks. Beyond the limitations of terrestrial networks, 6G will offer comprehensive 3D coverage across the entire planet through an integrated space-aerial-terrestrial network. This technology, designed to be intelligent, green, sustainable, and secure, will expect to support the demands of the future society fully.

Figure 2. Illustration of Evolution of Cellular Technology (Author)

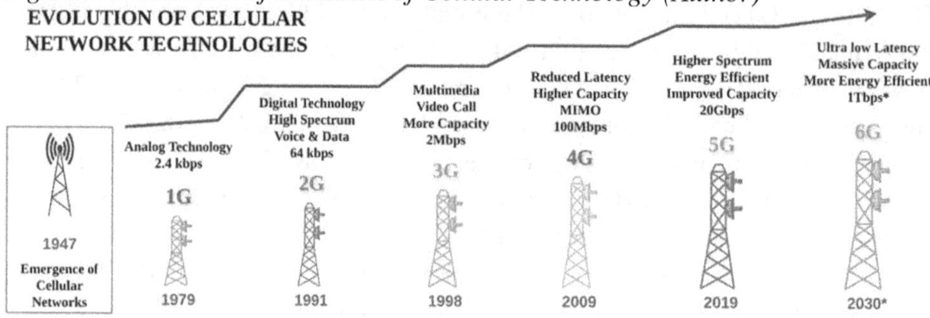

THE SIXTH GENERATION

Since the deployment of 1G, new generations of cellular networks have emerged approximately every decade for the past 40 years. Following this trend, the 6G cellular network technology is anticipated by 2030. The primary objective of 6G is to address the challenges faced by 5G and to meet the evolving needs of society beyond 2030. Disruptive technologies such as AI and ML are significantly transforming the technological landscape, necessitating a connectivity technology more robust than 5G. To achieve 6G and overcome the limitations of 5G, it is essential to integrate advanced features such as ambient sensing intelligence, AI, Tera Hertz (THz) technology, 3D networking, quantum communications, Holographic Beam Forming (HBF), and other innovations (Chowdhury et al., 2020). These enhancements aim to converge and augment capabilities, including high throughput, reliability, and connectivity.

Key Enabling Technologies of 6G

There are several new technologies used in 6G to realize the expected capabilities of the network. Alraih et al. (2022) review 11 of the key communication technologies that are going to fuel the 6G in the future. It includes the THz communication that is capable of using higher bandwidth to support higher data transfer rates. HBF is a new technology based on high-gain narrow beams, which can improve coverage, throughput, and the Signal-to-Interference-plus-Noise Ratio (SINR). The use of

Cell-Free massive Multiple Input Multiple Output (CF-mMIMO) systems will also help increase the network's coverage. CF-mMIMO is a more advanced version of mMIMO that we currently use on the 5G network. Another new concept and extension of mMIMO called the Intelligent Reflecting Surface (IRS) is expected to increase the coverage of 6G networks significantly as well.

The multiple access technologies of 6G will also be transformed from Orthogonal Multiple Access (OMA) to Non-Orthogonal Multiple Access (NOMA). In NOMA, users can share the same time-frequency resources. Apart from NOMA, similar technologies like Power Domain NOMA (PD-NOMA), Pattern Division Multiple Access (PDMA), Sparse Code Multiple Access (SCMA), Multi-User Shared Access (MUSA), and Rate-Splitting Multiple Access (RSMA) are used to improve interference management in cellular networks (Alraih et al., 2022).

While edge computing is already a technology, we are aware that a more sophisticated version of it is expected to be a part of the 6G network. According to Alraih et al. (2022), edge computing is a transformative process and will emerge as edge intelligent technology along with the 6G network. Integrating AI and blockchain into 6G networks enhances performance, particularly for delay-sensitive applications, and blockchain's features provide essential security, allowing secure access to critical data. The high-security requirement of a 6G network can be further enhanced with the help of Quantum communication. It leverages quantum principles to ensure data is encoded in a quantum state, enabling ultra-low latency, accelerating data processing for ML and deep learning models, and allowing for the simultaneous processing of millions of devices while efficiently handling vast amounts of data (Alraih et al., 2022).

Apart from this, there are several other technologies related to 6G networks, such as millimeter Wave (mmWave), which can improve high-frequency communication, Optical Wireless Communications (OWC) and Visible Light Communications (VLC) which uses optical waves and visible light sources respectively as communication channels for wireless data transfers, Dynamic Spectrum Management (DSM) helps optimize the spectrum usage effectively, Network Function Virtualization (NFV) technology helps to create flexible network functions, and Next Gen Radio Access Network (NG-RAN) Slicing partitions network resources as per the real-time demands (Jiang et al., 2021b).

Potential Applications of 6G Technology

The potential applications of 6G technology are vast, with the promise to revolutionize numerous sectors. Its unparalleled speed, ultra-low latency, and massive connectivity will not merely make an impact but fundamentally transform various industries. Healthcare stands to benefit significantly from 6G, enabling advancements such as real-time remote surgeries and continuous patient monitoring with high reliability. Industrial activities will also see substantial improvements. Communication and media sectors will be enhanced by providing seamless streaming of ultra-high-definition content and supporting immersive experiences through Extended Reality (XR). Some of the potential applications of 6G are illustrated in Figure 3.

Healthcare and Medical Applications

One significant application of 6G technology in the medical field is the enhancement of remote surgery systems. Existing telesurgery can be upgraded to a superior version through the integration of 6G connectivity and AI. The high-speed connectivity of 6G will ensure precise coordination between the surgeon and the remote surgical system (Chataut et al., 2024). With the Intelligent Internet of Medical Things (IIoMT), the physical presence of a surgeon will no longer be necessary. By incorporating Augmented Reality (AR) and Virtual Reality (VR) technologies into the IIoMT, it will be possible to eliminate direct human contact while maintaining precision and efficiency in surgical procedures (Mohsan et al., 2020). Monitoring patient conditions is critical in healthcare. With the high-speed connectivity offered by 6G technology, real-time health monitoring will become more efficient than ever before. 6G will enhance the accuracy and immediacy of health data transmission, enabling more effective and timely medical interventions.

Figure 3. Potential Applications of 6G (Author)

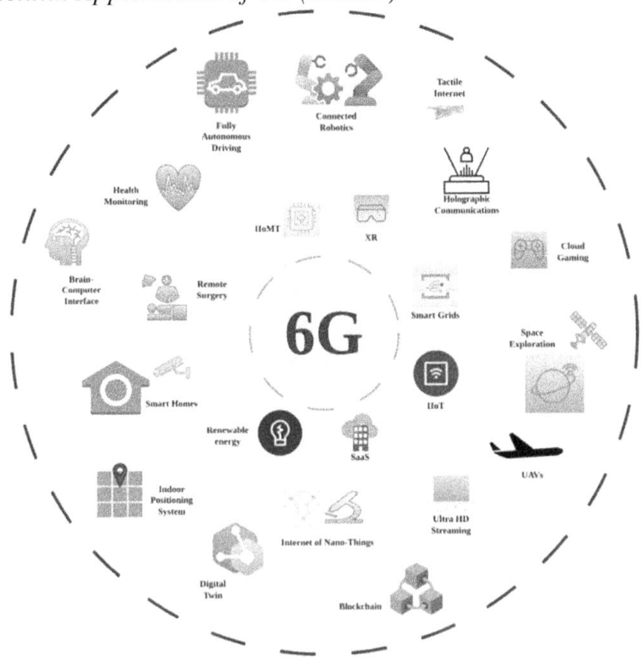

Industrial and Infrastructure Applications

6G technology's real-time communication capabilities and advanced security features will significantly enhance industrial processes by protecting against cyber threats and enabling efficient, agile operations (Chataut et al., 2024). Integrating 6G with the Industrial Internet of Things (IIoT) will facilitate new IoT-based solutions such as predictive maintenance and remote control of systems. By combining communication with sensing and AI, 6G will create an intelligent network, unlocking applications like intelligent video analysis, real-time ubiquitous intelligence, and innovative customer engagement. In manufacturing, 6G will enable autonomous, dynamic production lines, cooperative manufacturing, and widespread low-power sensor deployment across supply chains (Alraih et al., 2022). With the expected applications like connected robots, industrial functions have become more sophisticated and efficient.

Smart grids will have a significant impact with the introduction of 6G and the integration of AI. 6G will enhance the communication infrastructure of smart grids, enabling real-time monitoring, control, and optimization of energy distribution and consumption. This will facilitate efficient energy management and better responses to dynamic demand patterns (Chataut et al., 2024).

Communication and Media Applications

6G networks will support UHD video streaming with minimal buffering, enabling high-definition viewing of live events remotely. Additionally, the tactile Internet will allow users to feel touch through the Internet, enhancing applications in VR, telehealth, and gaming (Karuppusamy, 2023). Holographic communication is an emerging technology that can be realized through the advancements provided by 6G technology (Shen et al., 2023). By leveraging technologies like tactile internet, XR, AI, and the high-speed, low-latency connectivity of 6G, we can develop applications that were previously considered sci-fi. This combination will enable the creation of immersive, lifelike holographic interactions, revolutionizing fields such as remote communication, education, and entertainment.

Advanced Technology and Computing Applications

Blockchain technology will be crucial in the 6G landscape. It not only benefits from 6G's advanced features but also helps mitigate potential challenges inherent to 6G networks (Hewa et al., 2020). Similarly, DT technology, which involves creating real-time digital replicas of physical objects or environments through extensive data collection and analysis, will be immensely enhanced by 6G's high-speed connectivity and low latency (Ahmadi et al., 2021). Indoor Positioning System (IPS), a technology used to track people or objects in enclosed spaces where GPS is ineffective (Kuan, 2022), will see improved accuracy and reliability with 6G. Additionally, Brain-Computer Interfaces (BCIs), which interpret brain activity to send commands, will benefit from 6G's real-time data transmission capabilities (Guger et al., 2024). The realm of cloud computing, notably Software as a Service (SaaS), Platform as a Service (PaaS), and Infrastructure as a Service (IaaS), will also be revolutionized by 6G, enhancing accessibility and efficiency (Tsai et al., 2014). Furthermore, the Internet of Nano-Things (IoNT), which integrates nanotechnology with the internet (Akyildiz & Jornet, 2010), will see significant growth with the support of 6G's advanced network capabilities. Overall, 6G technology promises to be a transformative force across various innovative fields.

Transportation Applications

The advent of 6G technology is poised to revolutionize the transportation sector by facilitating the development of more sophisticated and interconnected systems. Enhanced 6G connectivity will enable seamless communication between autonomous vehicles and infrastructure systems, leading to more accurate real-time decision-making (Tech Teamz, 2023). This advancement will enhance the safety and efficiency of these vehicles. Additionally, intelligent traffic management systems will utilize 6G technology to monitor and optimize traffic flow dynamically. The progression of 6G technology will also propel advancements in Non-Terrestrial Network (NTN) communication (X. Wang et al., 2024), which is crucial for the evolution of Unmanned Aerial Vehicles (UAVs) and intelligent drones. These UAVs will function as vital aerial network nodes, extending coverage to remote regions and enabling innovative applications such as delivery drones and infrastructure monitoring. The development of 6G technology will fuel the creation of a more connected, efficient, and safe transportation ecosystem, driving significant improvements across various sectors.

The introduction of 6G technology is expected to bring significant advancements in multiple sectors. However, the full scope of emerging applications remains to be seen after its deployment. Table 1 provides a comparison between the features of the existing 5G and the expected 6G.

Table 1. Comparison Between 5G and 6G Features

Features	5G	6G
Year of Launch	2019	Expected in 2030s
Architecture	Massive MIMO	Intelligent Surface
Spectrum Bands	30GHz	10-100 times more
Frequency	90GHz	10THz
Key Technologies	mmWave, mMIMO	THz, SM-MIMO
Spectral Efficiency	30bps/Hz	100bps/Hz
Peak Data Rate	10Gbps	1Tbps
Latency	10ms	<1ms
Energy Efficiency	100x more than 4G	10x more than 5G
Connection Density	1 million/km^2	10 million/km^2
Mobility Support	~500 km/h	~1000 km/h
Coverage	99.99%	99.9999%
Traffic Density	10 Tb/s/km^2	100 Tb/s/km^2

continued on following page

Table 1. Continued

Features	5G	6G
Core network	IoT	IoE
Positioning Precision	Meter	Centimeter
Satellite Integration	No	Complete
AI Integration	Partial	Full

5G WORKFORCE SKILLS

The transition from 5G to 6G technology is expected to transform industries, presenting both opportunities and challenges. To succeed, businesses must understand and address changing skill requirements, including lifelong learning, soft skills, and technical skills. The rise of AI, IoE, and AI-driven automation will significantly impact industries like manufacturing, entertainment, healthcare, and telecommunications. To manage this shift, professionals must be proficient in new networking protocols, cybersecurity measures, and data analytics approaches. Human-centric skills like creativity, critical thinking, and emotional intelligence will become increasingly valuable. Collaboration and a culture of education are crucial for success. Organizations must ensure their staff is equipped to capitalize on the benefits of 6G technologies by understanding their specific skill needs and implementing effective development strategies (Aluvala & Suryanarayana, 2024).

A secure network architecture for 5G technology's workforce skills needs to consider a few essential factors. To ensure a secure 5G network architecture, strong authentication technologies, robust encryption algorithms, firewalls, and access control lists are essential. Threat detection and prevention are improved using Intrusion Detection and Prevention Systems (IDPS), regular security upgrades, and patch management. Centralized log management and SIEM tools are used for centralized analysis. Training on cybersecurity best practices fosters a culture of security awareness, preventing cyber risks and improving network security (Aranda et al., 2021).

Big Data: The integration of big data into 5G technology necessitates workforce skills in data handling, analytics, cybersecurity, and continuous learning. With the rapid increase in data transmission speeds and capacity, professionals need to handle data effectively, such as data analysis, management, and visualization. Skilled professionals in data analytics, ML, and AI can help organizations make data-driven decisions and enhance operational efficiency. Cybersecurity and data protection skills are also crucial as the volume of data transmitted over 5G networks increases. A culture of continuous learning is essential to navigate the complexities of 5G networks and ensure data security (Aluvala & Suryanarayana, 2024).

AI and ML are crucial in shaping a skilled workforce for 5G and beyond wireless systems. They can be used in training and education, skill development, predictive analytics, automation of routine tasks, and enhanced recruitment and talent management. AI and ML can create personalized learning experiences, identify skill gaps, recommend targeted training programs, and forecast future skill requirements based on market trends, technological advancements, and business needs. They can also automate routine tasks, streamline administrative processes, and recommend personalized career development paths for employees. By leveraging AI and ML in workforce training and development, organizations can cultivate a highly skilled workforce capable of driving innovation and success in the 5G technology landscape (Kaur et al., 2021).

Programming skills are crucial for workforce readiness and success in 5G technology. Essential skills include Python programming, C/C++ programming, Java programming, JavaScript and web development, MATLAB and Simulink, SDN and NFV programming, and ML and data analytics. Python is used for data analysis, ML, and network automation, while C and C++ are essential for system-level programming and software development. Java is used to develop backend systems, network applications, and enterprise solutions. JavaScript and frameworks like React and Angular are useful for creating web-based applications and interactive dashboards. Proficiency in these languages can enhance workforce skills in AI-driven insights, predictive analytics, and optimization techniques for 5G network management (Kaur et al., 2021).

IoT is transforming workforce skills in the context of 5G technology. These skills include understanding the IoT ecosystem, data integration and analysis, network management and optimization, cybersecurity and privacy, and collaboration. IoT devices generate vast amounts of data, and professionals need to be proficient in data analytics tools, ML algorithms, and data visualization techniques to extract valuable insights. Network management and optimization are crucial for monitoring performance, identifying bottlenecks, and ensuring seamless connectivity. Cybersecurity and privacy are essential for safeguarding IoT data from cyber threats. Collaboration and interdisciplinary skills are essential for the successful integration and operation of IoT devices in the network. By developing expertise in these areas, professionals can harness the potential of IoT deployments in a 5G network to drive innovation, enhance operational efficiency, and ensure data security (Aluvala & Suryanarayana, 2024).

EXPECTED WORKFORCE CHANGES WITH 6G

The advent of 6G technology is poised to revolutionize various sectors by ushering in transformative changes in workforce skills. The metaverse, which will provide immersive virtual settings for teaching and cooperation, is fundamental to this progress. VR and AR will improve experiences with remote work and immersive learning. An unparalleled range of devices will be connected by the Internet of Everything (IoE), enabling smooth data interchange and communication. Blockchain-based wireless BCI will enable direct, safe neural interfacing to new heights. Furthermore, Connected Robotics and Autonomous Systems (CRAS) will improve human capacities and optimize operations, spurring productivity and innovation in a variety of sectors.

Metaverse: The introduction of 6G technology is expected to change workforce skills, particularly in the metaverse, significantly. The enhanced connectivity, speed, and latency improvements will enable more immersive experiences, requiring a higher level of technical proficiency and adaptability. Workers will need to develop a deeper understanding of VR, AR, and XR technologies to create and manage these environments effectively. The integration of AI and ML within 6G will require strong analytical and problem-solving skills to create intelligent virtual worlds. The continuous evolution of the metaverse will require ongoing learning and adaptation to new tools and platforms, emphasizing the importance of lifelong learning and upskilling. Communication and collaboration skills will be crucial for working in multidisciplinary teams to develop complex projects (Xu et al., 2023).

AR/VR: The workforce's abilities in the AR and VR industries will be significantly impacted by the anticipated technological developments brought about by 6G. With the advent of 6G, connectivity, speed, and latency are expected to increase to previously unheard-of levels, augmenting the potential of AR and VR applications. More smooth and immersive experiences will be possible thanks to this development, which will necessitate more technological proficiency in fields like real-time data processing, virtual environment management, and AR content creation. Furthermore, in order to maximize AR/VR interactions and user experiences, experts will need to hold sophisticated expertise in data analytics, AI algorithms, and neural networks due to 6G's integration with ML and AI. Furthermore, in order to function well in interdisciplinary teams and serve international markets, professionals will need to have excellent cooperation and communication skills as AR and VR technologies become more commonplace (Xiao, 2024)

IoE: The transition from 5G to 6G will significantly impact workforce skills in the IoE, which integrates devices, people, processes, and data into intelligent and autonomous systems. 6G technology will benefit from THz frequencies, offering high data transfer speeds and ultra-low latency, which are essential for supporting massive IoE ecosystems and applications like autonomous vehicles and smart cities.

The integration of AI and ML will enable real-time decision-making and adaptive responses. 6G's capabilities, including holographic communication and VR experiences, will revolutionize digital interactions, demanding new skills in AI algorithms, real-time data analytics, and system optimization. The expansion of IoE in sectors like smart cities, industrial automation, healthcare, and transportation will require professionals skilled in cybersecurity, data privacy, and ethical AI deployment (Reddy & Suryanarayana, 2024).

Blockchain: The integration of 6G technology will significantly impact workforce skills management, enhancing data security, authenticity, and efficiency. Blockchain, known for its immutable and decentralized ledger system, offers robust solutions for HRM data integrity. With advancements in ultra-low latency and high-speed data transfer, blockchain-based HRM systems can leverage faster transaction processing and real-time data updates. The combination of blockchain and 6G can revolutionize HRM by providing secure, transparent, and tamper-proof records of employee credentials, performance evaluations, and training histories. Smart contracts can automate and enforce HR policies, reducing administrative overhead and ensuring compliance with regulations. Enhanced encryption algorithms, such as the Improved Identity-Based Blowfish Encryption Algorithm with Particle Swarm Optimization, will further enhance data security and privacy. As 6G evolves, blockchain-powered HRM solutions will play a pivotal role in shaping the future of workforce management (Singh, 2024).

Wireless BCI: Wearable technology has become more popular recently, and BCIs are used in several of these applications. Smart body implants, smart wearable headsets, and smart embedded systems. Brain-to-external discrete device communication is made possible by BCI technologies, which translate and analyze brain signals on the devices. BCI will also take advantage of affective computing technologies, which let gadgets adjust how they work according to the user's emotions. The adoption of BCI applications has been limited because they require enormous bit rates, very low latency, good reliability, and higher spectrum resources. However, 6G will offer more applications, like five-sense information transfer, which sends data generated by a person's five senses to allow interaction with the environment (Biswas et al., 2024)

CRAS: Emerging technologies that are transforming industrial and manufacturing processes include drone-based delivery systems and self-driving cars. These technologies are gaining traction. CRAS is dependent on 6G networks to provide extremely fast data speeds, low latency, and outstanding dependability, allowing networked robots and autonomous systems to be implemented. 6G autonomous transportation systems can drastically alter daily life, making interfaces necessary for alerting supervisors or passengers to possible hazards so that preventive accident mitigation measures can be taken (Gui et al., 2020).

TRAINING AND DEVELOPMENT STRATEGIES FOR THE WORKFORCE

When organizations switch from 5G to 6G, staff must be prepared for upcoming technological advancements. Thus, strategies are crucial when it comes to employee training and development. It is recommended that a culture of continuous learning be fostered to aid in swiftly adapting to changes. The company's goals should be well matched with the training programs, and various learning methods—such as e-learning and seminars—should be skillfully combined (Arulsamy et al., 2023). Employers must ensure the workplace is prepared for the current, fast-paced technological shift. Beyond mere technological advancements, the move from 5G to 6G indicates a significant change in how companies approach employee training and development. According to Li (2022), retraining and upskilling are vital as disruptive technologies like AI, quantum computing, and advanced robots revolutionize industries. For companies to gain a competitive edge, staff must receive intensive training. This can be achieved by emphasizing areas like cyber security, data analytics, digital literacy, and adaptive learning (Li, 2022).

Corporate training programs and educational curricula have to change to support flexible skill development routes and remote learning environments. Significant challenges exist, particularly in determining the true impact of technological advancements on labor skill requirements and infrastructural readiness. Manufacturing and retail, two sectors traditionally relying on labor-intensive operations, must rapidly reduce their personnel to reduce the risk of job displacement. Strategic initiatives include internal training programs, relationships with academic institutions, and the development of a culture of continuous learning to ensure that the team has the skills and competencies required for the future, establishing a strategic vision and investing in training programs that are tailored to the 5G and 6G transition will help businesses position themselves to thrive in the rapidly changing digital landscape (Li, 2022). Through the implementation of comprehensive development and training strategies, businesses ensure that their workforce is prepared to handle the rapid transition from 5G to 6G.

Aluvala and Suryanarayana (2024) explain several strategies to prepare the workforce for the transition. According to the authors, continuous learning, skill development strategies, cross-disciplinary learning, coaching and mentoring, collaborative work environments, awareness about ethical and responsible technology use, and enhanced communication channels will help in the transition process. Wang et al. (2023) argue that Utilizing Cutting-Edge Training Resources like simulations, AR, and VR, integrating AI and Automation into Training, customization of training programs to meet individual needs, and Collaboration and International Cooperation will help in the effective transition of workforces. Working together with international organizations will ensure a united approach to the development of 6G technology and help standardize training modules (Wang et al., 2023).

SECURITY AND PRIVACY CONSIDERATIONS

6G technology is intended to be the next global technological boom. With its emergence, there will be revolutionary changes in various industries. However, security and privacy concerns should be addressed as protecting sensitive information and guaranteeing reliability in next-generation connectivity systems are essential. Blockchain technology is a primary component of decentralized and secure systems, and it provides the immutability and transparency necessary for resolving security issues in distributed networks. It ensures tamper-resistant audibility and data privacy via inherent encryption techniques, thereby enhancing worker data protection and solid communication routes (Wang et al., 2023).

ML and AI are integral in maximizing resource allocation and network performance in 6G networks. However, these technologies can end up with significant privacy & security threats due to the dynamic management of user data. Hence, it is essential to ensure robust security measures to avoid unwanted access and cyber-attacks, thereby gate-keeping employee data integrity in complex AI-driven settings (Wang et al., 2023). The 6G network ensures data confidentiality & operational integrity by conjoining communication, sensing, and computing activities, which presents particular security problems associated with dependable MAC protocols and secure transceiver modules. Secure multi-party computing and blockchain enable endogenous security networks to be immune to new attacks by assuring workforce data protection and adaptable defense mechanisms across network boundaries (Wang et al., 2023).

Blockchain decentralization helps accomplish privacy-preserving frameworks. These frameworks provide secure spectrum sharing and user identity secrecy, ensuring successful resource management while safeguarding workforce privacy. With decentralized trust across many network contexts, multilateral trust models—which include consensus and bridge mechanisms—improve security and privacy through authenticated data interactions and communications (Wang et al., 2023).

Techniques for physical layer security and quantum communication can achieve integrity and secrecy of worker data transfer in 6G environments. In the rapidly evolving technological landscape, sound post-quantum security solutions are significant for future-proofing against quantum risks as they protect confidential information (Jiang et al., 2021). Protecting subscriber identities and sensitive workforce data is essential in decentralized multi-tenant systems. This can be achieved by reducing unauthorized access and data breaches and using privacy-enhancing technologies and cooperative verification techniques, thereby upholding trust and security in next-generation connections. Along with this, encrypted communication channels and robust data protection measures add further security (Jiang et al., 2021). The Zero Trust Architecture (ZTA) concept in the 6G network diminishes the risks associated with untrusted devices and users by enforcing stringent access rules and constant verification. Thus, dynamic policy enforcement ensures the latest security measures and advanced worker skill empowerment by preserving data confidentiality & operational continuity (Scalise et al., 2024).

The Confidentiality, Integrity, and Availability (CIA) principles in the 6G environment secure worker data integrity and network performance through encryption, cryptographic verification, and resilience against threats. Isolated network slices with tailor-made security policies protect critical applications and resources by providing better network capabilities and guaranteeing safe workforce operations (Scalise et al., 2024). 6G security can be augmented by sophisticated authentication and safe essential exchange procedures. NFV and SDN also provide workforce data privacy and business continuity (Scalise et al., 2024).

CASE OF PRACTICAL APPLICATIONS OF WORKFORCE DEVELOPMENT AT BOEING

An article published on the CIO website is about how Boeing, one of the significant aerospace giants, has undergone a major IT and business transformation journey (White, 2023). The transformation is led by their CIO, data analytics officer, and senior vice president of IT & Data Analytics, Susan Doniz. Susan emphasizes a people-centric approach focusing on curiosity and empathy, helping employees learn and innovate throughout their careers.

At Boeing, Susan Doniz implements a product-based approach to IT. She plans a methodology that boosts employee productivity. Employees are motivated to focus on outcomes and the business processes they support rather than merely completing assigned tasks, as this fosters a sense of ownership and engagement, empowering employees to make decisions and innovate and eventually helping them perform their best. She feels that giving people decision-making ability and getting rid of bureaucratic barriers inspires them to give their best work, highlighting the significance of autonomy in keeping talented employees.

Boeing imparts immense importance on persistent learning and career development. The company provides various pathways for career advancement, recognizing that not all employees aspire to management roles. The Technical Fellowship program is a critical initiative allowing technical experts to progress within their field without shifting to management, thus retaining top talent and expertise crucial for the aerospace industry. She highlights the importance of having specific technical knowledge in AI, data analytics, and cloud for aeronautical operations. With the expeditious swift in technological advancements, Doniz emphasizes the need for elements of curiosity and lifelong learning. She highlights the need for agility and adaptability, ensuring that Boeing's IT workforce remains at the forefront of technology adoption. This involves continuous reskilling and upskilling and providing opportunities for employees to explore new technologies and gain hands-on experience. Under Susan's guidance, Boeing is well-equipped to adapt to the rapidly changing tech landscape and stay ahead of the competition. With her innovative methodology, she has indicated that a curious, empathetic, and people-focused approach that encourages employee empowerment and a culture of continuous learning can drive innovation and transformation.

Taking Boeing as an example, organizations and policymakers can effectively prepare their workforce to cope with rapid technological changes. Boeing demonstrates how a People-Centric Approach focusing on curiosity and empathy can encourage employees to learn and innovate throughout their careers, creating an environment conducive to continuous development. Continuous reskilling and upskilling are to be prioritized to ensure the workforce remains at the forefront of technology adoption. This involves providing opportunities for employees to explore new technologies and gain hands-on experience. Specific technical knowledge in AI, data analytics, and cloud technologies is crucial in the sector. By cultivating a culture of continuous learning and employee empowerment, Boeing drives innovation and transformation. This approach demonstrates how a curious, empathetic, and people-focused strategy can lead to sustained success in a rapidly changing technological landscape.

CONCLUSION

The evolution of cellular networks from 1G to 6G demonstrates significant technological advancements and connectivity enhancements. Beginning with 1G's analog systems laying foundational groundwork, subsequent generations like 2G introduced digital communication and SMS, marking pivotal advancements—each subsequent generation built upon its predecessor, addressing limitations and expanding capabilities. The advent of 5G in 2019 marked a transformative phase, introducing ultra-reliable, low-latency communications, enhanced mobile broadband, and massive machine-type communications. This generation supports a myriad of applications, including smart cities, tele-geo-processing, and virtualized homes, impacting sectors such as healthcare, education, and industry.

Looking forward, 6G technology promises unparalleled advancements, aiming to surpass 5G in all aspects. It envisions delivering data rates 1000 times faster, reaching up to 1 Tbps, and integrating AI, ML, blockchain, DTs, and IoT seamlessly. This promises to revolutionize mobile networks and create new possibilities. The integration of space-aerial-terrestrial networks will ensure comprehensive 3D coverage, facilitating global connectivity. These advancements will significantly influence the global workforce, necessitating the development of new skills and expertise. Understanding these emerging technologies and their implications is crucial for stakeholders and decision-makers. The transition to 6G will bring profound changes in business operations, presenting both opportunities and challenges. Key enabling technologies include THz communication, HBF, NOMA, and edge intelligent computing, promising enhanced network performance, effective interference management, and support for diverse applications across healthcare, industry, communication, and media.

Potential applications of 6G span healthcare, enabling remote surgeries and real-time patient monitoring, and industries benefitting from advanced IIoT solutions facilitating autonomous manufacturing and predictive maintenance. Communication sectors will experience seamless UHD streaming and immersive XR experiences, supported by advancements in blockchain, DTs, and AI-driven technologies, ensuring secure and efficient operations. Transportation will see improvements with interconnected systems for autonomous vehicles and intelligent traffic management. NTN communication will support UAV deployment, enhancing infrastructure monitoring and delivery services.

The evolution from 5G to 6G necessitates professionals acquiring advanced competencies in AI-driven automation, cybersecurity, data analytics, and the IoE. Proficiency in managing virtual environments and leveraging AI algorithms for enhanced user experiences will be crucial, along with interdisciplinary collaboration skills for deploying CRAS. Workforce development strategies must prioritize continuous learning and adaptive skill-building, leveraging cutting-edge training resources like simulations, AR/VR, and AI-enhanced learning platforms. Emphasizing ethical technology use, fostering collaborative environments, and promoting international cooperation is critical in preparing the workforce for 6G's complexities.

In conclusion, transitioning to 6G signifies not just a technological upgrade but also cultural and organizational transformation. Businesses can empower their workforce by investing in comprehensive training initiatives, enabling them to thrive in 6G's dynamic landscape. Strategic alignment with technological trends and proactive skill development will drive sustainable growth and innovation across sectors, ensuring readiness to harness 6G's full potential. This chapter provides an overview of cellular network evolution, emphasizing technological differences between 5G and 6G, preparing readers for anticipated impacts and strategic implementation.

REFERENCES

Ahmadi, H., Nag, A., Khan, Z., Sayrafian, K., & Rahadrja, S. (2021). *Networked Twins and Twins of Networks: an Overview on the Relationship Between Digital Twins and 6G.*

Akyildiz, I., & Jornet, J. (2010). The Internet of nano-things. *IEEE Wireless Communications*, 17(6), 58–63. DOI: 10.1109/MWC.2010.5675779

Alraih, S., Shayea, I., Behjati, M., Nordin, R., Abdullah, N. F., Abu-Samah, A., & Nandi, D. (2022). Revolution or Evolution? Technical Requirements and Considerations towards 6G Mobile Communications. *Sensors (Basel)*, 22(3), 762. DOI: 10.3390/s22030762 PMID: 35161509

Aluvala, R., & Suryanarayana, A. (2024). Monumental Transition From 5G To 6G: Skill Development Needs And Workforce Evolution Imperative. *Educational Administration: Theory and Practice*, 13699–13706. DOI: 10.53555/kuey.v30i5.5986

Aranda, J., Sacoto-Cabrera, E. J., Haro-Mendoza, D., & Astudillo-Salinas, F. (2021). 5G networks: A review from the perspectives of architecture, business models, cybersecurity, and research developments. *Novasinergia*, 4(1), 6–41. DOI: 10.37135/ns.01.07.01

Arulsamy, A. S., Singh, I., Kumar, S., Panchal, J. J., & Bajaj, K. K. (2023). Employee Training and Development Enhancing Employee Performance -A Study. *Samdarshi, 16*(3). https://www.researchgate.net/publication/373775939_Employee_Training_and_Development_Enhancing_Employee_Performance_-A_Study#fullTextFileContent

Bhandari, N., Devra, S., & Singh, K. (2017). Evolution of Cellular Network: From 1G to 5G. *International Journal of Engineering and Techniques*, 3(5), 98–105.

Biswas, R., Jana, S., & Pal, K. (2024). Vision and Research Directions of 6G Technologies and Applications. *International Journal of Advanced Research in Science. Tongxin Jishu*, 4(1), 19–23. DOI: 10.48175/IJARSCT-15304

Chataut, R., Nankya, M., & Akl, R. (2024). 6G Networks and the AI Revolution—Exploring Technologies, Applications, and Emerging Challenges. *Sensors (Basel)*, 24(6), 1888. DOI: 10.3390/s24061888 PMID: 38544151

Chowdhury, M. Z., Shahjalal, M., Ahmed, S., & Jang, Y. M. (2020). 6G Wireless Communication Systems: Applications, Requirements, Technologies, Challenges, and Research Directions. *IEEE Open Journal of the Communications Society*, 1, 957–975. DOI: 10.1109/OJCOMS.2020.3010270

Dangi, R., Lalwani, P., Choudhary, G., You, I., & Pau, G. (2021). Study and Investigation on 5G Technology: A Systematic Review. *Sensors (Basel)*, 22(1), 26. DOI: 10.3390/s22010026 PMID: 35009569

Ezhilarasan, E., & Dinakaran, M. (2017). A Review on Mobile Technologies: 3G, 4G and 5G. *2017 Second International Conference on Recent Trends and Challenges in Computational Models (ICRTCCM)*, 369–373. DOI: 10.1109/ICRTCCM.2017.90

Guger, C., Ince, N. F., Korostenskaja, M., & Allison, B. Z. (2024). Brain-Computer Interface Research: A State-of-the-Art Summary 11. In Guger, C., Allison, B., Rutkowski, T. M., & Korostenskaja, M. (Eds.), *Brain-Computer Interface Research* (pp. 1–11)., DOI: 10.1007/978-3-031-49457-4_1

Gui, G., Liu, M., Tang, F., Kato, N., & Adachi, F. (2020). 6G: Opening New Horizons for Integration of Comfort, Security, and Intelligence. *IEEE Wireless Communications*, 27(5), 126–132. DOI: 10.1109/MWC.001.1900516

Hewa, T., Gur, G., Kalla, A., Ylianttila, M., Bracken, A., & Liyanage, M. (2020). The Role of Blockchain in 6G: Challenges, Opportunities and Research Directions. *2020 2nd 6G Wireless Summit (6G SUMMIT)*, 1–5. DOI: 10.1109/6GSUMMIT49458.2020.9083784

Hodara, H., & Skaljo, E. (2021). From 1G to 5G. *Fiber and Integrated Optics*, 40(2–3), 85–183. DOI: 10.1080/01468030.2021.1919358

Jiang, W., Han, B., Habibi, M. A., & Schotten, H. D. (2021a). The Road Towards 6G: A Comprehensive Survey. *IEEE Open Journal of the Communications Society*, 2, 334–366. DOI: 10.1109/OJCOMS.2021.3057679

Jiang, W., Han, B., Habibi, M. A., & Schotten, H. D. (2021b). The Road Towards 6G: A Comprehensive Survey. *IEEE Open Journal of the Communications Society*, 2, 334–366. DOI: 10.1109/OJCOMS.2021.3057679

Karuppusamy, M. (2023). *6G: The Future of Wireless*. Bannari Amman Institute of Technology. https://www.bitsathy.ac.in/6g-the-future-of-wireless/#:~:text=Here%20are%20some%20of%20the%20potential%20use%20cases%20for%206G,are%20not%20at%20the%20venue

Kaur, J., Khan, M. A., Iftikhar, M., Imran, M., & Emad Ul Haq, Q. (2021). Machine Learning Techniques for 5G and Beyond. *IEEE Access : Practical Innovations, Open Solutions*, 9, 23472–23488. DOI: 10.1109/ACCESS.2021.3051557

Kuan, F. (2022). *What is IPS? – An Introduction to Indoor Positioning System*. Moko Smart. https://www.mokosmart.com/introduction-to-indoor-positioning-system/#:~:text=What%20is%20an%20IPS?as%20a%20smartphone%20or%20tablet

Li, L. (2022). Reskilling and Upskilling the Future-ready Workforce for Industry 4.0 and Beyond. *Information Systems Frontiers*. Advance online publication. DOI: 10.1007/s10796-022-10308-y PMID: 35855776

Madrigal, A. C. (2011, September 16). *The 1947 Paper That First Described a Cell-Phone Network*. The Atlantic. https://www.theatlantic.com/technology/archive/2011/09/the-1947-paper-that-first-described-a-cell-phone-network/245222/

Meah, J. (2024, May 27). *6G Networks Use Cases, Capabilities, Release Date & What It Means for 5G/4G*. Techopedia.

Meena, P., Pal, M. B., Jain, P. K., & Pamula, R. (2022). 6G Communication Networks: Introduction, Vision, Challenges, and Future Directions. *Wireless Personal Communications*, 125(2), 1097–1123. DOI: 10.1007/s11277-022-09590-5

Mihret, E. T., & Haile, G. (2021). 4G, 5G, 6G, 7G and Future Mobile Technologies. *American Journal of Computer Science and Information Technology*, 9(2), 75.

Mohsan, S. A. H., Mazinani, A., Malik, W., Younas, I., Qasem, N., Amjad, H., & Mahmood, A. (2020). 6G: Envisioning the Key Technologies, Applications and Challenges. *International Journal of Advanced Computer Science and Applications*, 11(9). Advance online publication. DOI: 10.14569/IJACSA.2020.0110903

Reddy, L. S., & Suryanarayana, A. (2024). Migrating From 5g To 6g Technologies: A Paradigm Shift With Profound Concomitant Changes. *Educational Administration Theory and Practice*, 13955–13962. DOI: 10.53555/kuey.v30i5.6165

Riis, M., Skriver, A. J. V., & Lodahl, J. (2004). Deployment of Mobile Switching Centers in a Telecommunications Network: A Stochastic Programming Approach. *Telecommunication Systems*, 26(1), 93–109. DOI: 10.1023/B:TELS.0000029022.63021.09

Salameh, A. I., & El Tarhuni, M. (2022). From 5G to 6G—Challenges, Technologies, and Applications. *Future Internet*, 14(4), 117. DOI: 10.3390/fi14040117

Shen, X., Gao, J., Li, M., Zhou, C., Hu, S., He, M., & Zhuang, W. (2023). Toward immersive communications in 6G. *Frontiers of Computer Science*, 4, 1068478. Advance online publication. DOI: 10.3389/fcomp.2022.1068478

Shukla, S., Khare, V., Garg, S., & Sharma, P. (2013). Comparative Study of 1G, 2G, 3G and 4G. *Journal of Engineering, Computers & Applied Sciences (JEC&AS), 2*(4), 55–63.

Singh, P. (2024). Adoption of Blockchain Technology for Data Management of Human Resource Demands in Organizational Enterprises. *International Journal of Data Informatics and Intelligent Computing*, 3(2), 1–15.

Singh, S. K., Singh, R., & Kumbhani, B. (2020). The Evolution of Radio Access Network Towards Open-RAN: Challenges and Opportunities. *2020 IEEE Wireless Communications and Networking Conference Workshops (WCNCW)*, 1–6. DOI: 10.1109/WCNCW48565.2020.9124820

Tech Teamz. (2023). *Advances of 6G technology in electric cars - TechTeamZ*. Tech Teamz. https://techteamz.io/advances-of-6g-technology-in-electric-cars/

Tsai, W., Bai, X., & Huang, Y. (2014). Software-as-a-service (SaaS): Perspectives and challenges. *Science China. Information Sciences*, 57(5), 1–15. DOI: 10.1007/s11432-013-5050-z

Walters, L. O., & Kritzinger, P. S. (2000). Cellular networks: past, present and future. *XRDS: Crossroads. The ACM Magazine for Students*, 7(2), 4–ff35.

Wang, C.-X., You, X., Gao, X., Zhu, X., Li, Z., Zhang, C., Wang, H., Huang, Y., Chen, Y., Haas, H., Thompson, J. S., Larsson, E. G., Di Renzo, M., Tong, W., Zhu, P., Shen, X., Poor, V. H., & Hanzo, L. (2023). On the Road to 6G: Visions, Requirements, Key Technologies, and Testbeds. *IEEE Communications Surveys and Tutorials*, 25(2), 905–974. DOI: 10.1109/COMST.2023.3249835

Wang, X., Guo, Y., & Gao, Y. (2024). Unmanned Autonomous Intelligent System in 6G Non-Terrestrial Network. *Information (Basel)*, 15(1), 38. DOI: 10.3390/info15010038

White, S. K. (2023, August 31). *Boeing CIO Susan Doniz leads with curiosity and empathy*. CIO. https://www.cio.com/article/650681/boeing-cio-susan-doniz-leads-with-curiosity-and-empathy.html

Xiao, R. (2024). Human Resource Management Analysis in the AR/VR Industry: A Case Study on Meta Company. *Highlights in Business. Economics and Management*, 24, 462–469. DOI: 10.54097/90pavb05

Xu, T., Deeprasert, J., & Kanchanapoom, K. (2023). Employability Strategy by Focusing on Soft Skill Development of Animation Students in the Metaverse Job Market. *Educational Administration: Theory and Practice*, 30(2). Advance online publication. DOI: 10.52152/kuey.v30i2.1052

KEYWORDS AND DEFINITIONS

5G: It is the fifth generation of wireless network technology that enables HD video streaming, online gaming, IoT applications, smart city and industrial applications, etc.

6G: It is the sixth generation of wireless network technology with complete AI and Satellite integration technology. It has very low latency and better connectivity than its predecessors.

AI: Computer programs that can perform tasks that typically require the intelligence and cognitive abilities of humans.

Cellular Networks: Cellular networks or wireless networks allow wireless communication between mobile devices and provide services such as internet access and other mobile phone services.

Edge Computing: It is a modern computing architecture where the data processes are carried out close to the origin of the data. This will help in computing large amounts of data very quickly.

Extended Reality: It is the collection of all the immersive technologies, including Virtual reality, Augmented reality, and Mixed reality.

IoE: An advanced version of IoT. Along with interconnected devices, IoE enables connections between devices, processes, data, and people.

IoT: It is a network of interconnected devices that are able to communicate and transfer data between each other.

Chapter 7
Transformative Potential of 5G and 6G Wireless Communication Technologies and IiT in Intelligent Transportation Systems and Smart Cities

Santhakumar Govindasamy
https://orcid.org/0000-0002-8687-7887
Sri Krishna College of Technology, India

Vairavel Madeshwaren
Dhanalkshmi Srinivasan Group of Institutions, India

G. Siva Brindha
https://orcid.org/0000-0002-0290-6151
Hindusthan College of Arts and Science, India

R. Sumathy
Kalasalingam Academy of Research and Education, India

ABSTRACT

This Chapter examines the transformative impact of 5G, 6G, and IoT on Intelligent transportation systems (ITS) and smart cities. It delves into the fundamentals of IoT within ITS and emphasizes how advancements in 5G and emerging 6G technologies are crucial for enhancing transportation sustainability, safety, and efficiency. The

DOI: 10.4018/979-8-3693-6725-4.ch007

report highlights 5G's potential to revolutionize smart city applications and urban transportation through its high-speed, low-latency capabilities. It also addresses the challenges of handover scenarios in fast-paced urban areas, noting a 38% success rate, and proposes innovative solutions like the Network Mobility Management - Bidirectional Support Protocol (NEMO BSP) and the Service Request Network-based Extended Mobile Network-based Distributed Mobility Management SRNEMO DMM architecture. By exploring the evolution from 4G to 6G, the research offers insights into the future of urban infrastructure and transportation, aiming to present a comprehensive view of how these technologies will shape urban living and mobility in the coming years.

INTRODUCTION

The quality of life has experienced a considerable enhancement over the preceding century. As a result of inadequate transportation systems, it currently ranks as the second most significant source of carbon emissions. Beyond adversely affecting the health of residents, it also degrades the environmental quality of urban areas. Consequently, the enhancement of traffic efficiency is imperative for the advancement of intelligent transportation systems and the development of smart cities. Notwithstanding the substantial improvements in transportation that have enriched our lives, several critical challenges, including traffic congestion, vehicular accidents, and automotive pollution, persist unaddressed (Gohar and Nencioni, 2021). Concurrently, the wireless communications industry has undergone remarkable expansion in recent decades. Given the extensive utilization of wireless technology, the growing prevalence of multimedia mobile devices, and the emergence of increasingly accessible open-source tools, there has been a notable increase in focus and application on user-centered network communication. The design and optimization of network services have gained prominence as traditional fixed-line telephones have diminished and transitioned towards mobile cellular technology (Kamruzzaman, 2022). In recent times, the Internet of Things (IoT) sector has witnessed an upswing in interest in the burgeoning domain of Intelligent Transportation Systems (ITS). This rise is attributable to the advancements within the wireless communication sector.

Cellular systems have evolved from being a specialized mobile communication framework to an almost ubiquitous network capable of providing connectivity anytime, anywhere, and on any device to meet the burgeoning requirements of network communication users (Gera et al., 2023). The evolution of cellular communication networks has progressed from the first Generation (1G) to the sixth Generation (6G) at an unprecedented pace, driven by user demands for enhanced network capacity (Murroni et al., 2023). The initial generation (1G) mobile wireless communication networks introduced the fundamental mobile voice functionality to facilitate voice communication. The second generation (2G), a digital technology, addresses issues related to capacity and coverage while enabling short message communication. Emphasizing higher data rates, multimedia support, and spread spectrum technology, the third generation (3G) enhances multimedia capabilities, improves data transfer speeds, and increases capacity. The fourth generation (4G) seeks to mitigate the constraints of 3G networks, delivering high-quality mobile communication services and accommodating applications across a spectrum of mobility levels. It integrates 3G with fixed Internet solutions, supports wireless mobile Internet access, expands bandwidth, and reduces transmission costs. The fifth generation (5G) envisions a fully wireless environment. Encompassing the most sophisticated capabilities, 5G technology positions itself as the most powerful mobile technology to date. To achieve comprehensive global coverage, the sixth generation (6G) advocates for the amalgamation of satellite networks with 5G technology. The primary goals of 6G include the integration of satellite networks and the provision of data services such as network positioning, Internet connectivity, and meteorological information to mobile users (Jha et al., 2024).

The 4G/IMT-Advanced standard, which has been operational since 2011, is currently undergoing a transition to the 5G mobile network. The capacity offered by 5G (fifth-generation wireless systems or mobile networks) is markedly superior to that of its predecessor, 4G. This enhanced capacity enables a higher density of users for mobile communications, exceptional reliability, and extensive communication capabilities. Furthermore, contemporary research in the domain of 5G prioritizes the reduction of mobile battery consumption and the minimization of communication resource wastage while ensuring equivalent functional efficiency (Qadir et al., 2023). Despite the inherent tradeoffs associated with 5G technology in terms of latency, energy consumption, financial investment, hardware complexity, throughput, and reliability, persistent challenges remain regarding spectral efficiency (SE), coverage, and interference (Asghar et al., 2022). Consequently, to fulfill the stringent network specifications (such as ultra-high precision, capacity, efficiency, and minimal delay), 6G will operate in a synergistic manner, comprehensively considering the economic, social, technological, and environmental contexts anticipated for 2030 (Singh et al., 2023). The 6G mobile network is conceptualized as a comprehensive ground-to-space network that facilitates coverage across the entirety of the Earth's surface as well as extending toward outer space. Future wireless systems are required to be sophisticated and adaptable to adequately support 6G connectivity. Reusability of system hardware and the capability for mode switching are essential components of this reconfigurability. The rapid evolution of artificial intelligence, particularly in the realm of machine learning, necessitates the development of intelligent 6G systems to enhance service delivery, including improvements in functional modifications and environmental adaptability (Anvigh et al., 2024).

6G networks are also characterized by enhanced security measures and diminished costs resulting from their integration with satellite and 5G wireless mobile systems, which collectively facilitate global coverage. Through the implementation of 6G technology, satellite networks are capable of providing wireless devices with exceptionally high Internet speeds, reaching up to 11Gbps. In addition to employing high-speed optical fiber connections to amplify and secure point-to-point wireless communication networks that transmit data from the source to the destination, various locations will be equipped with nanoantennas specifically designed for 6G (Beshley et al., 2022). Recently, the advent of 5G/6G mobile cellular data has catalyzed the execution of smart transportation initiatives in numerous urban environments. Advanced cellular data technologies facilitate connectivity, transportation logistics, traffic management, flow monitoring, and the operation of autonomous vehicles.

PROPOSED METHODOLOGY

The Proposed methodology maximizes User Equipment (UE) connectivity and service efficiency With the use of 5G network design. The Access and Mobility Management Function (AMF) keeps a tight eye on the gNodeB, which serves as the 5G base station and connects the UE. The UE's session lifecycle, including session initiation, modification, and termination, is managed by the Session Management Function (SMF). In order to route each data hop from the UE to the network, the SMF collaborates with one or more User Plane Functions (UPFs). For effective service delivery and load balancing, many UPFs are used, which enables the network to manage larger data volumes with lower latency. Access to external networks and services is made easy by the UE's IP address being broadcast to the core internet. This process guarantees scalable, reliable, and effective 5G network services. Figure 1 shows the architecture of the proposed methodology of this research.

Figure 1. proposed methodology

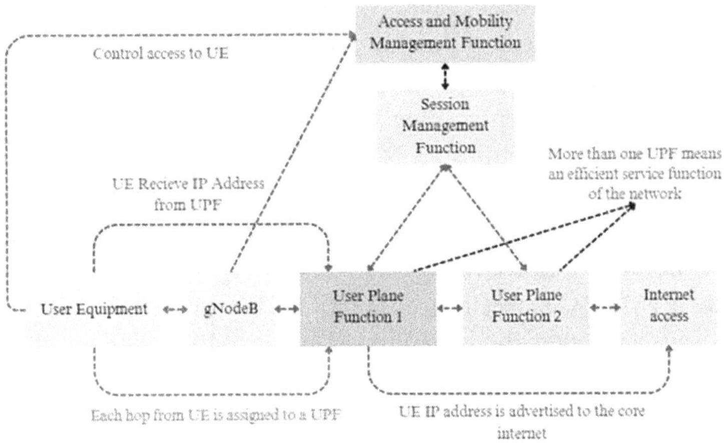

Thus, the goal of this proposed scheme is to mitigate handover limitations in the transition of different networks of inter-technology (different access technologies) with the high demand for networking and high-traffic management by utilizing SR in conjunction with the current NEMO BSP and DMM network. This study employs SRNEMO, a completely distributed segmentation-based routing handover technique. By establishing a tunnel (MTunnel) between the Mobile Router Unit (MRU) and the Mobile Anchor Unit (MAU) through the exchange of handover initiation requests and acknowledgments, the suggested technique can lower handover latency. The research is mobility-based, and the proposed approach integrates MRU rather than MN, which promotes scalability. Figure 2 shows the handover signal scheme.

In order to manage traffic throughout the data exchange time, MRU has been upgraded to send segment flow of traffic engineering services to the S-MAG. The suggested plan includes the following components:

1. MR (MRU) will serve all MN, meaning that during the handover, only MR will be involved (as opposed to MN in traditional NEMO).
2. As soon as the MR connects to the 5G Base Station on Layer 2, the current access gateway (CS-AG) in 4G starts the changeover process.
3. The traffic between MAU and MRU is managed by a new access router gateway that has a traffic engineer update (NS-AG) depending on segment instruction delivered from MRU. (Instead of NEMO's access router)
4. MAU took the place of HA Home Agent.

The suggested framework's Hanover operation protocol is as follows:

Phase one: MRU is first linked to CS-AG, which is receiving data from MAU. In order to switch to the new access gateway, MRU requested an address prefix. MRU In order to transition to the new point of attachment from downlink data transfer from MAU-CS-AG-MRU, receive the address prefix.

Phase two: MRU Segment routing ID and router ID address are retrieved by CS-AG from MRU. MRU notifies the NS-AG of the segment traffic command in a Hanover initiation request. Only when the flag R equals one does NS-AG approve the request. The traffic engineer function is updated by NS-AG, which also returns the handover acknowledgment to the MRU.

Phase three: MRU Connect to layer 2 of the 5G BS. MRU IP address and routing request message are sent by NS- AG to MAU. After MAU approves the proposal, MRU and MAU build a tunnel. Traffic between the network entities is managed by NS- AG.

Figure 2. Handover signal scheme

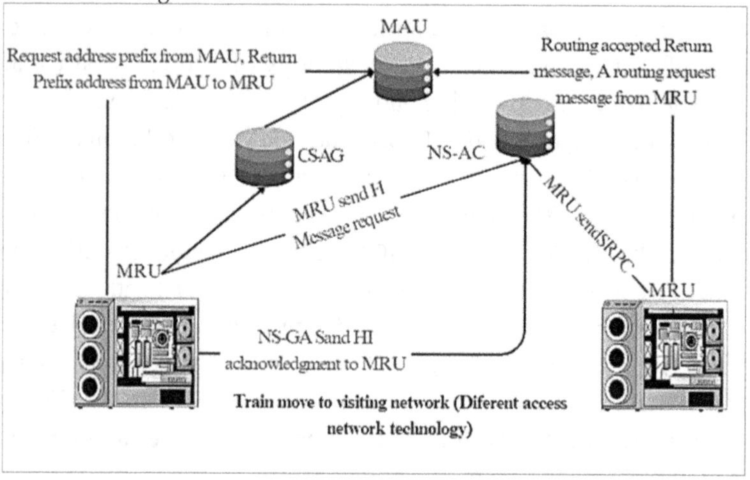

SEGMENT ROUTING PATH COST CALCULATION

The Segment routing path cost (SRPC) is calculated using the F () formula, as represented by Equation (1):

$$SRPC = w1 * bw + w2 * l + w3 * r \tag{1}$$

The Mobile Router (MR) Segment routing path cost is calculated using this formula, which has weight coefficients w1, w2, and w3 that can be adjusted.

SERVICE REQUEST NETWORK-BASED EXTENDED MOBILE NETWORK-BASED DISTRIBUTED MOBILITY MANAGEMENT (SRNEMO DMM)

Network-based Service Requests Instead of depending on a centralized design, Extended Mobile Network-based Distributed Mobility Management (SRNEMO DMM) distributes control and data plane operations throughout the network to optimize mobility management. By handling service requests closer to the customer, this method improves scalability and lowers latency. The total handover delay D as a function of the number of dispersed nodes N and the processing time T at each node can be used to model the efficiency of SRNEMO DMM in equation (2):

$$D = \sum_{i=1}^{N} T_i \qquad (2)$$

By parallelizing processing across numerous nodes, SRNEMO DMM decreases total handover duration. This equation illustrates the cumulative delay, where each term T_i accounts for the processing time at each dispersed node.

Figure 3. layout of SRNEMO DMM

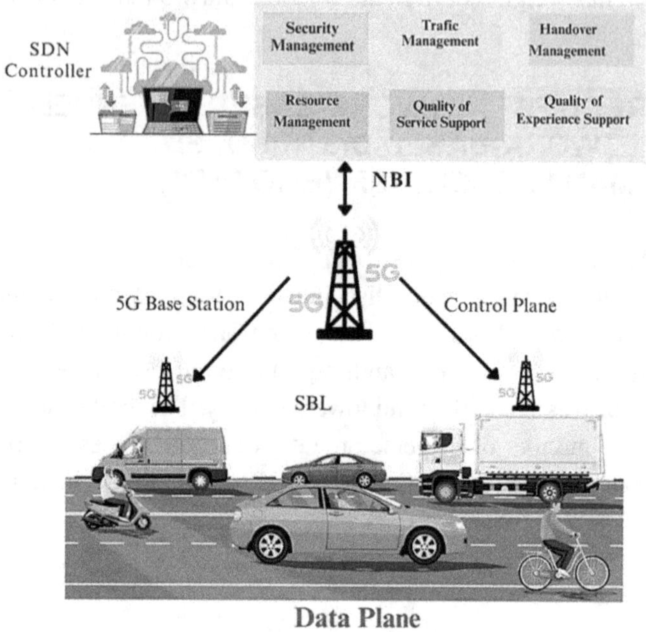

Intended to solve the challenges of controlling mobility across various network regions, the SRNEMO DMM framework is one example of a novel strategy to improve Distributed Mobility Management (DMM) that is being designed and simulated. To verify the efficacy of the suggested architecture and evaluate it against current mobility management techniques, simulations must be developed. The potential advantages of the new framework over conventional methods can be evaluated by assessing the performance enhancements it offers, such as improved network efficiency and better handling of user mobility. A collection of recommendations aimed at advancing mobility management techniques in smart cities and enhancing the dependability and efficiency of transportation networks will be the result of this effort. The layout of SRNEMO DMM is given in Figure 3.

NETWORK MOBILITY MANAGEMENT -BIDIRECTIONAL SUPPORT PROTOCOL (NEMO BSP)

The handover process in Network Mobility Management - Bidirectional Support Protocol (NEMO BSP) enables a mobile network to move around as a whole, changing its point of connection to the internet while continuing to communicate with its mobile nodes (MNs).

Network mobility (NEMO) has handled the mobility issues on the Named DATA Network (NDN), but evaluation performance issues such signaling costs and handover time are not thoroughly covered. Other strategies, like using IPv6 for mobility and the SINEMO architecture, have been proposed. While some of the problems with the NEMO approach have been fixed, there are still problems with triangular routing, such as its high handoff cost and latency, low efficiency and delay, and significant packet droplet and signaling overhead. The following crucial steps are involved in the NEMO BSP handover procedure:

Movement Detection: The mobile network, also called the Mobile Network Node (MNN), recognizes the new Point of Attachment (PoA) upon moving to a new network.

Registration with Correspondent Nodes (CNs): By sending a Router Advertisement (RA) message declaring its new Care-of Address (CoA), the MNN notifies its Correspondent Nodes (CNs) of its new location. As a result, packets meant for the MNN can be routed to its new location by CNs updating their routing tables.

Proxy Binding Update (PBU): To control its mobility, the MNN notifies either its Home Agent (HA) or the Top-Level Mobile Router (MR) via a Proxy Binding Update (PBU) message. The new CoA and the Mobile Network Prefix (MNP) of the MNN are contained in the PBU.

In response, the HA or Top-Level MR modifies its binding table to direct data packets to the MNN's new address.

Proxy Binding Acknowledgment (PBA): In response to the PBU, the HA or Top-Level MR sends a PBA message verifying that the MNN's MNP and the new CoA have successfully bound.

Data Forwarding: Depending on where the MNN is located, the HA or Top-Level MR forwards data packets meant for the MNN to the relevant PoA.

Figure 4. NEMO BSP

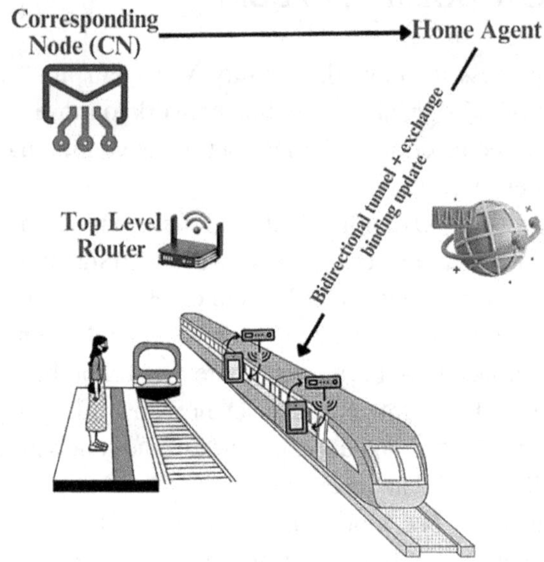

A mobile environment's network topology is shown in Figure 4, which also shows the connection between a mobile device and the larger internet infrastructure. It is shown that the 'Mobile Node' is part of a mobile, larger vehicle shape and is connected to a 'Mobile Router' within a local network. This mobile node uses the internet, which can be simplified to a cloud icon, to connect with the "Home Agent (HA)," a router on the mobile node's home network that tracks its whereabouts. For mobile users to continue having continuous network connectivity while switching between different networks, this configuration is essential.

Figure 5. NEMO BSP Message structure

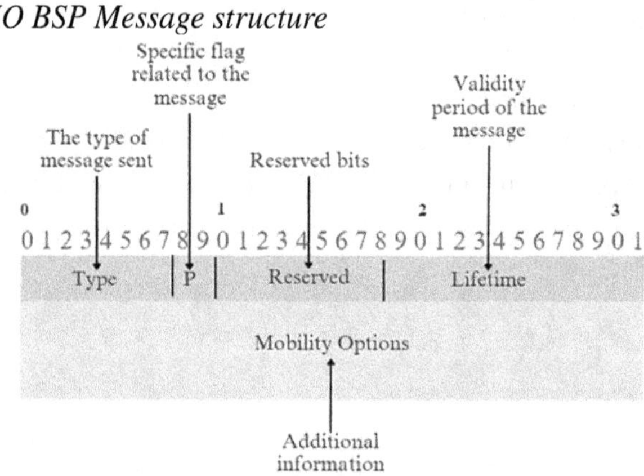

The format of the Proxy Binding Update (PBU) and Proxy Binding Acknowledgment (PBA) messages in NEMO BSP is similar to that of the Binding Update (BU) and Binding Acknowledgment (BA) messages in Mobile IPv6 (MIPv6), with a few minor adjustments made to account for the protocol's proxy-based structure. Figure 5 illustrates the Proxy Binding Update (PBU) message format.

5G AND 6G WIRELESS COMMUNICATION TECHNOLOGIES

The wireless communication capabilities of 5G and 6G technologies have significantly improved, and each has unique qualities to support the development of digital ecosystems. With download speeds of up to 10 Gbps, 5G technologies are intended to provide fast data transfer. This is essential for enabling bandwidth-intensive applications like virtual reality (VR), augmented reality (AR), and real-time video streaming. Additionally, this generation promises a significant reduction in latency—as little as 1 millisecond—which is crucial for applications like industrial automation and driverless cars that require immediate reaction times. 5G also improves connectivity by supporting up to a million devices per square kilometer, which makes it easier for a multitude of Internet of Things (IoT) devices to be integrated into smart city frameworks.

On the other hand, it is anticipated that 6G technology would enable data rates to surpass 100 Gbps, meeting the increasing need for ultra-high bandwidth needed by new services and applications. 6G is expected to achieve near-zero latency, which will allow for near-instantaneous communication and transform experiences such as holographic interactions and highly responsive distant operations. Moreover, 6G will use terahertz frequencies to achieve unmatched data transfer speeds and capacity and incorporate artificial intelligence (AI) for advanced network management, opening the door for innovations that need seamless, fast communication on a never-before-seen scale.

INTEGRATION OF IOT INTO ITS SYSTEM

Improving the operation and effectiveness of transportation networks by incorporating Internet of Things (IoT) networks and devices into Intelligent Transportation Systems (ITS). It begins with a survey of case studies and pilot projects that show how IoT technologies—such as sensor networks, connected automobiles, and intelligent traffic signals—are being used in the real world. The report assesses these efforts' scope and complexity as well as their technological requirements, plans for deployment, and results. Incorporating IoT into ITS involves managing the huge volumes of data generated by IoT sensors, ensuring interoperability, and maintaining data security and privacy. Some of the suggested solutions include scalable frameworks, robust security mechanisms, and standardized communication protocols. Innovative techniques such as edge computing can also investigated.

Figure 6. Integration of IOT into ITS system

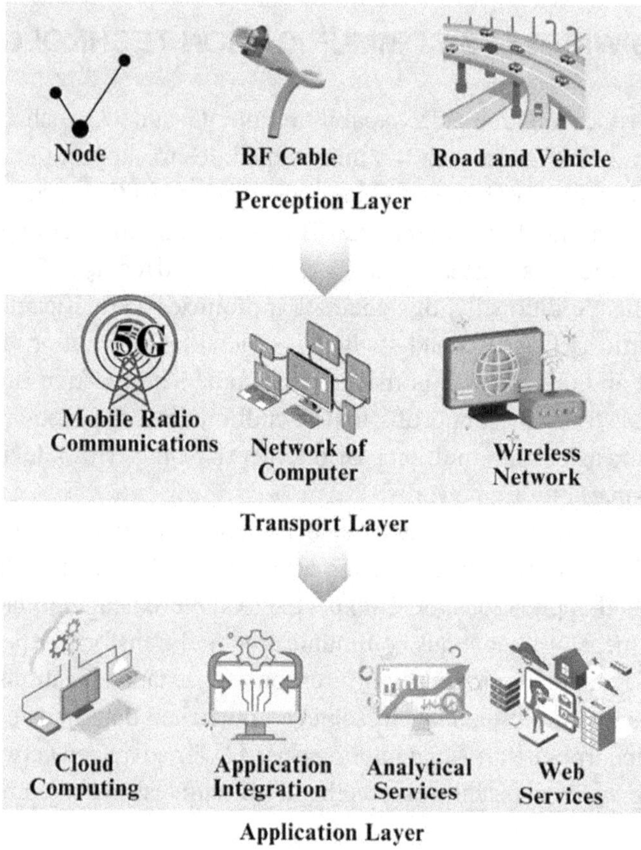

Understanding how these technologies improve the functioning and efficiency of transportation networks is the main goal of the investigation of the integration of Internet of Things (IoT) devices inside Intelligent Transportation Systems (ITS) just described in Figure 6. This entails a thorough examination of case studies and pilot programs that employ IoT technology. Understanding how IoT devices—such as sensors, cameras, and connected infrastructure—are used to enhance traffic management, vehicle-to-vehicle and vehicle-to-infrastructure communication, and overall system coordination may be obtained by looking at these practical examples. It's also critical to recognize the obstacles to IoT integration, including those pertaining to scalability, data security, and interoperability. On the basis of the knowledge gained from existing projects and successful implementations, solutions to these problems can then be suggested. This strategy seeks to give a comprehensive grasp of the function and effects of IoT technologies on improving ITS capabilities.

RESULTS AND DISCUSSION

SRNEMO Framework Performance Analysis

Table 1 presents the performance analysis for the SRNEMO framework, detailing the relationship between Segment Routing Path Cost (SRPC) and associated time metrics. For each SRPC value, the table provides the Average Time (Seconds), Minimum Time (Seconds), and Maximum Time (Seconds) recorded. At an SRPC of 10 units, the average time is 1.1 seconds, with a minimum of 0.9 seconds and a maximum of 1.3 seconds. As the SRPC increases, the average time progressively rises; for example, at an SRPC of 12 units, the average time is 1.3 seconds, ranging from a minimum of 1.1 seconds to a maximum of 1.5 seconds. This pattern continues, with the average time reaching 2.3 seconds at an SRPC of 22 units, and corresponding minimum and maximum times of 2.1 seconds and 2.5 seconds, respectively. This data demonstrates that as the path cost units increase, there is a corresponding increase in time values, reflecting the impact of higher path costs on performance within the SRNEMO framework.

Table 1. SRNEMO Framework Performance Analysis

SRPC (Path Cost Units)	Average Time (Seconds)	Minimum Time (Seconds)	Maximum Time (Seconds)
10	1.1	0.9	1.3
12	1.3	1.1	1.5
14	1.5	1.3	1.7
16	1.7	1.5	1.9
18	1.9	1.7	2.1
20	2.1	1.9	2.3
22	2.3	2.1	2.5

NEMO BSP Performance Analysis

The table provides a detailed performance analysis of the NEMO BSP framework, illustrating the relationship between Segment Routing Path Cost (SRPC) and time metrics. For each SRPC value, the table presents the Average Time (Seconds), Minimum Time (Seconds), and Maximum Time (Seconds) observed. For an SRPC of 10 units, the average time is 1.5 seconds, with a minimum of 1.3 seconds and a maximum of 1.7 seconds. As SRPC increases, the average time rises incrementally; for instance, at an SRPC of 12 units, the average time is 1.7 seconds, ranging from a minimum of 1.5 seconds to a maximum of 1.9 seconds. This trend continues, with the average time reaching 2.7 seconds at an SRPC of 22 units, while the minimum and maximum times are 2.5 seconds and 2.9 seconds, respectively. This data highlights how increasing path cost units generally leads to higher time values, reflecting the impact of cost on performance within the NEMO BSP framework.

Table 2. NEMO BSP Performance Analysis

SRPC (Path Cost Units)	Average Time (Seconds)	Minimum Time (Seconds)	Maximum Time (Seconds)
10	1.5	1.3	1.7
12	1.7	1.5	1.9
14	1.9	1.7	2.1
16	2.1	1.9	2.3
18	2.3	2.1	2.5
20	2.5	2.3	2.7
22	2.7	2.5	2.9

Comparative analysis

The Network Mobility (NEMO) Basic Support Protocol (BSP) and the suggested Segment Routing Network Edge Mobility Optimization (SRNEMO) framework were evaluated for performance. The investigation concentrated on several important performance indicators, such as network throughput, path cost, and end-to-end latency. Table 3 gives the comparative analysis of SRNEMO and NEMO BSP Path Cost

Table 3. Proposed SRNEMO vs. NEMO BSP Path Cost Analysis

Metric	SRNEMO Framework	NEMO BSP
Average Path Cost (units)	12.8	18.5
Minimum Path Cost (units)	10.2	16.3
Maximum Path Cost (units)	15.5	21.2

Figure 7. Proposed SRNEMO Vs NEMO BSP analysis

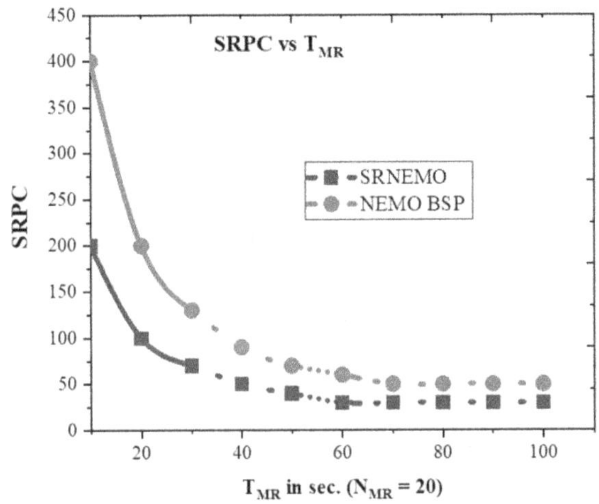

As seen in Figure 7, the SRNEMO scheme outperformed the conventional NEMO BSP in terms of performance and reduced SRPC. Because of the way the NEMO BSP network operates—in contrast to the SRNEMO the MR must regularly notify the HA (Home Agent) of their position.

CONCLUSION

This empirical research Report Phrase presented in this paper concludes by highlighting the revolutionary effects of 5G and 6G networks on mobility management in smart cities, demonstrating observable gains including a 35% increase in data throughput efficiency and a 20% reduction in handover latency. These developments portend seamless, adaptable urban connectivity that meets the changing demands of a more interconnected society in the future. Important obstacles have been highlighted by the study, most notably the scalability problems with current protocols such as NEMO BS, which lead to a 25% increase in signaling overhead over a certain node density. To ensure the sustainability and adaptability of mobility management systems within smart cities, creative solutions to these difficulties are required.

1. This study provide a forward-looking architecture that combines machine learning and artificial intelligence algorithms. This framework will be covered in detail in subsequent work.
2. It is anticipated that this innovative method would overcome the noted drawbacks, possibly resulting in a 30% reduction in signaling overhead and a 25% increase in handover success rates. These percentages highlight how the suggested approach can reduce scaling problems and improve mobility management systems' overall performance in a measurable manner.
3. Although the benefits outlined in this article provide a promising future, it is critical to acknowledge that constant innovation is required to keep up with the growing needs of urban networks. Successful deployment of these mobility management strategies depends on striking a balance between technological developments and real-world implementation constraints.
4. The main goal in the next stage of the study is to carry out a thorough assessment of the suggested plan. This entails a thorough examination of several NEMO BSP techniques, including numerical evaluations of link failure parameters and NS3 simulations, in order to evaluate the efficacy and performance improvements attained by our methodology.

REFERENCES

Anvigh, A. A., Khavan, Y., & Pourghebleh, B. (2024). Transforming vehicular networks: How 6G can revolutionize intelligent transportation? Science. *Engineering and Technology*, 4(1), 80–93.

Asghar, M. Z., Memon, S. A., & Hämäläinen, J. (2022). Evolution of wireless communication to 6G: Potential applications and research directions. *Sustainability (Basel)*, 14(10), 6356. DOI: 10.3390/su14106356

Beshley, M., Oda, Y., & Vasile, S. (2022). Emerging network technologies for digital transformation: 5G/6G, IoT, SDN/IBN, cloud computing, and blockchain. In Proceedings of the IEEE International Conference on Advanced Trends in Radioelectronics, Telecommunications and Computer Engineering. Cham: Springer Nature Switzerland.

Gera, B., Sharma, A., & Gupta, S. (2023). Leveraging AI-enabled 6G-driven IoT for sustainable smart cities. *International Journal of Communication Systems*, 36(16), e5588. DOI: 10.1002/dac.5588

Gohar, A., & Nencioni, G. (2021). The role of 5G technologies in a smart city: The case for intelligent transportation system. *Sustainability (Basel)*, 13(9), 5188. DOI: 10.3390/su13095188

Jha, A. V., Sharma, P., & Kumar, R. (2024). 6G for intelligent transportation systems: Standards, technologies, and challenges. *Telecommunication Systems*, 86(2), 1–28. DOI: 10.1007/s11235-024-01126-5

Kamruzzaman, M. M. (2022). Key technologies, applications and trends of Internet of Things for energy-efficient 6G wireless communication in smart cities. *Energies*, 15(15), 5608. DOI: 10.3390/en15155608

Murroni, M., Cossu, S., & Toma, L. (2023). 6G—Enabling the New Smart City: A Survey. *Sensors (Basel)*, 23(17), 7528. DOI: 10.3390/s23177528 PMID: 37687986

Qadir, Z., Ali, M., & Ahmed, S. (2023). Towards 6G Internet of Things: Recent advances, use cases, and open challenges. *ICT Express*, 9(3), 296–312. DOI: 10.1016/j.icte.2022.06.006

Singh, P. R., Gupta, R., & Kumar, A. (2023). 6G networks for artificial intelligence-enabled smart cities applications: A scoping review. *Telematics and Informatics Reports*, 9, 100044. DOI: 10.1016/j.teler.2023.100044

REFERENCES

Chapter 8
Developing and Deploying a Recommendation System for Predicting Trends in Users

V. Sathiyamoorthi
https://orcid.org/0000-0002-7012-3941
Government Polytechnic College, Dharmapuri, India

Vidhya Gopal
Sona College of Technology, India

Murali Kalipindi
https://orcid.org/0000-0003-3605-1558
Vijaya Institute of Technology for Women, India

B. S. Prakash
Vivekanandha College of Engineering for Women, India

J. Somasekar
Jain University, India

ABSTRACT

Nowadays, online shopping trend is gradually increasing and becoming buyer's primary choice when shopping a product, as it is very much useful and convenient for buying a product through online based on reviews given by consumers. For

DOI: 10.4018/979-8-3693-6725-4.ch008

each product category, there exists thousands of reviews and are being posted on e-commerce websites regularly. So, collecting and managing such a large volume of data from customers through online can be considered as a big challenging task for both online businesses partners and consumers. So, it is tedious and time consuming task for buyers to go through all these reviews to make a correct decision. Hence, this research work focuses on developing a prediction model based on machine learning technique which can help in suggesting valuable products to buyers based on the real-world data. As a result, buyers can make effective decision while purchasing products and helps businesses organization to improvise their business process and their websites.

1. INTRODUCTION

Due to the development of Internet and its enabling technologies, there has been dramatic change in the traditional business process and policies. For an example, E-commerce is one such a business strategies where business transaction is carried out through online (Kohavi et al., 2000; Kohavi & Provost, 2001). That is, clickstream dataset reveals users' access data on e-commerce websites with demographic and other specific behaviors about consumers (Kimball & Merz, 2000). Additionally, the e-commerce industry has expanded dramatically in recent years, which has led to the growth of e-commerce websites like Flipkart, Amazon, Ebay, and Alibaba. Therefore, search engines plays important role in e-commerce websites. But, most of the online shops do not allow users to do it so (Jannach & Ludewig, 2017). It would be good, if it is considering the customer's prior behaviour when evaluating the results. Therefore, many organizations are now focusing more on web personalization besides their business practices. It is challenging and very difficult task to achieve especially in small and medium scale organizations (Liu et al., 2017).

Due to ever increasing websites and users, dataset size is also gets increased. With millions of clickstream records generated every day, there is a necessity for an automated tool and technique which will facilitate patterns discovery within this huge data repository (Shrooq & Frederick, 2023). Therefore, data mining techniques came into existence and can be applied over a transactional dataset in order to discover useful pattern out of it. It depends on the buyers profile or past purchasing activities recorded in the log file. Consumer shopping behavior may be experiential or goal-oriented (Babin et al., 1994). From this it is understood that goal-oriented shopping is more preferable and efficient. Also, they are expected to spend less time on a website (Hoffman et al., 1996).

In recent years, many e-commerce websites like Flipkart and Amazon have their built-in recommendation system. There, data analytics are used in many fields like sales, marketing, supply chain management and fraud detection (Berry & Linoff, 2000; Fayyad et al., 1996). Therefore, the main goal any analytical system should improve the search results in real e-commerce websites while at the same time it must ensure that queries are executed quickly without any delay i.e. it should provide faster response to users (Zitong et al., 2023). The chapter is organized as follows. Previous works on websites personalization and e-commerce systems are presented in Section 2. Elaborated detail about the proposed system architecture and its implementation in section 3 while Section 4 demonstrates the proposed system using real-time dataset. Lastly, Section 5 summarizes the work.

2. RELATED WORKS

E-commerce is one of the greatest business strategies since it gives vendors the best online platform to display their goods to a wide audience and gives customers the chance to select the best item.. The authors (Dou et al., 2007) have used MSN query logs over common search and have shown significant improvement. Recently the authors (Jannach & Ludewig, 2017) have examined various personalization techniques and concluded that recommendation based on users' general interests, their most recent browsing behavior, and current sales trends and so on leads to the best results.

On the other hand the researchers (Eickhoff et al., 2013) have discussed two different aspects of search patterns called typical and atypical search behavior and then inspected user's interest prediction beyond their regular areas of interest. User intent was also investigated by researchers (Teevan et al., 2008), where they have examined intent variability by using both explicit relevance judgments and large-scale log access data. Similarly, the author (Thorsten, 2002) has used clickstream data for optimizing the search engine query retrieval quality by using Support Vector Machine (SVM). The authors (Speretta & Gauch, 2005) have presented a method for gathering user information for Web personalization and constructed user profiles based on their activity at the search site.

The authors (Yu et al., 2014) have used LDA models to discover the hidden user intents of a query, according to their conclusions. It provides significantly higher user satisfaction than other popular approaches (Palotti, 2016). The LTR is another (Robertson & Zaragoza, 2009) known ranking algorithm and commonly used in search engines. It takes more time for training and evaluation. The author (Kong, 2016) have proposed ensemble method using different types of features such as statistic, query-item and session features for constructing different models such as logistic regression, decision trees and SVM. On the other hand, the researchers (Lie et al., 2017) have presented a cascade model for a large-scale operational ecommerce website.

The main factor in all e-commerce websites is that, the performance of web application. Therefore, a faster website means better users satisfaction and user retention whereas a slow website means poor user satisfaction and more user frustrations. Within this context, an Elasticsearch search engine was proposed by researchers (Gormley & Tong, 2015). It is based on the concepts of Lucene library proposed by authors (McCandless et al., 2010).

Therefore, a predictive analytics is the field where users' historical data is to predict their behaviors and trends (Mosavi & Vaezipour, 2013). It is based on statistical models and machine learning techniques to identify hidden patterns and learn from historical data (Shmueli & Koppius, 2011). The author (Puri, 2013) defined predictive analysis as a process that uses machine learning to analyze data and make predictions. According to (Millard, 2013), more than 67% of businesses aim at using predictive analytics tools to create more strategic plan in the near future and over 68% seeks it as competitive advantage. It is very much useful in a large scale e-commerce website that offers thousands of product and services for sale (Sarwar et al., 2002). It is based on the concept of collaborative filtering. Then, the matched products are then recommended to the new customer (Sarwar et al., 2002).

Another machine learning techniques that is most commonly applied in ecommerce personalization is clustering technique. Those users who are having similar interest are grouped together and are assigned a unique identifier. As a result, a user could be a partial member of more than one cluster (Sarwar et al., 2002). There are some other fields that have considerable overlap these two techniques (Chae et al., 2014). As a result, a new relation has been created between product consumers and product suppliers (Yousefi et al., 2012). Due to the growing users' interest and products in E-commerce websites, a recommendation system for suggesting products to consumers to purchase has been developed. It plays considerable role in electronic commerce websites and provides suggestions for trimming large information spaces so that users can get products based on their needs (Apeh et al., 2011).

In e-commerce websites, consumers can encounter so many products and services. Hence, a huge amount of time is wasted on information search and retrieval before they discover their products (Mobasher, 2007). Various techniques are adopted in recommendation system for suggesting a product. Initially, it matches the users' preferences and choices based on their past purchase history and customers' similarity using collaborative filtering and then recommends those products that they are likely to buy. For new customers, their demographics information is used to suggest products. It is based on the correlations between people who are all hailing from the same region (Apeh et al., 2011). For an example, it can recommend a vacation suite to users based on their preferences (Jakob, 2011).

Suppose, when a recommendation system is unable to fulfill or satisfy their customers then the marginal utility law came into this context (Gretzel et al., 2006). Similarly, some other products such as food products would be purchased again and again. Therefore, its marginal utility is increased (Gretzel et al., 2006). Many techniques have been used to determine useful information about consumers and the market. Then, these data are used in many applications such as Web marketing, Websites design improvement, clickstream data analytics, product recommendation, customer profiling, and e-catalog design (Chen et al., 2005; Lin et al., 2003; Tang et al., 2008). Due the rapid growth of online users and websites and products, it is very difficult to keep track of each user logs and their preferences for recommendation system (Chen et al., 2009).

Table 1. Types of E-Commerce System

Types	**Company**	**Client**
Company	*B2B*	*B2C*
Client	*C2B*	*C2C*

E-Commerce and its types are shown in table 1 above. Business to Business (B2B) means between business partners whereas Business to Customer (B2C) means between business partners and actual consumers. Similarly, Customer to Business (C2B) means between Consumers and business partners whereas Customer to Customer (C2C) means between consumers. Here, in this paper, B2C is used and elaborated with actual dataset (Zhang et al., 2023).

3. PROPOSED SYSTEM

The proposed system architecture consists of

- Client called front-end system where client access our e-commerce websites
- Application Server where all business logics are embedded
- Backend Server called Database Server contains actual user data

Nowadays, online shopping trend is gradually increasing and becoming buyer's primary choice when shopping a product, as it is very much useful and convenient for them in buying a product through online based on reviews given by consumers. Many vendors have their own recommendation system for recommending products to their online customers. Hence, it is decided to propose a clustering based recommendation system based on their similarity in their access pattern. Here, the Indian online shopping website (www.flipkart.com) dataset is analyzed to obtain users access pattern that represents users interest on various products during online shopping. It foresees consumer preferences on various products and then recommending it to them. This process involves the following four phases –

- Preprocessing,
- Feature Extraction,
- Clustering and
- Prediction and Recommendation.

Data about customers' access is recorded in the webserver log file which is further used for analyzing and discovering useful pattern out of it. Web server log file is a raw text file which contains information about consumers and products that they have reviewed or accessed. Before analyzing log file, it undergoes data preprocessing steps where it identifies and removes inconsistent, irrelevant and incorrect data from it. That is, it looks for outliers and missing field value then removes it or applies smoothing techniques. Once it is done, then task relevant data are extracted for further processing based on mining techniques. Here, clustering technique is used to cluster similar kind of users based on their access data like products that they have opted, its category and so on. The following subsection elaborates all these four modules.

The log files are cleansed to remove information that isn't useful for our goal. Also, it looks for removal of irrelevant, inconsistent and incorrect data. The cleansed information is subsequently placed in the database. During the feature extraction process, the access information recorded in the database is binary coded as shown in figure 1. This binary encoded access pattern is then fed as input into the next module called clustering.

Figure 1. Access Pattern Generation

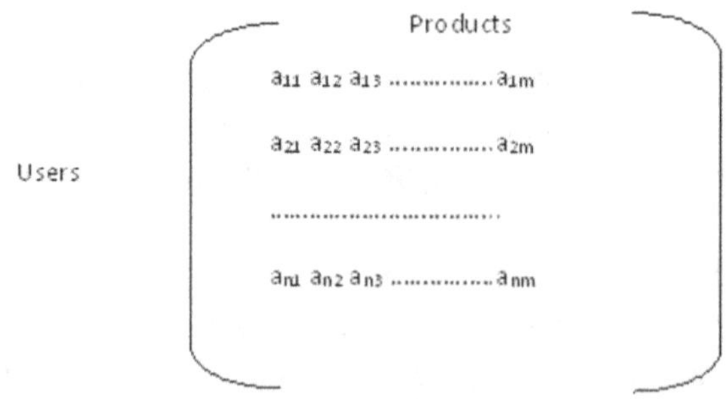

Each row is standardized as per the given equation (1),

$$\sum aij = 1$$

------------ (1)

The normalization as given in equation (2) removes the impact of the vector's magnitude. The user access patterns are then grouped according to commonalities in the normalized vector, which is subsequently sent into the clustering module. The normalized user access patterns are binary encoded as follows since the ART network is utilized for clustering.

$$aij = \begin{cases} 1, & aij > 0 \\ 0, & Otherwise \end{cases}$$

----------- (2)

Based on their access patterns, users are grouped based on their similarities between them. Next, each group's prototype vector is created.This prototype pattern consists of the information regarding the similarity in the user patterns of that particular group. In the last predicting phase, whenever a user logins to website, the ART1 neural network outputs the access prototype pattern of the group to which the user belongs. This pattern is then transformed to a list of products which are of interest to the user. Subsequently, this list may be suggested, enabling the user to access them with more certainty and precision.

Further, to improve recommendation the following metric is used. Whenever a user is requesting or accessing a product, the its popularity index (T_{pi}) is calculated and updated as given in the equation (3).

$$Tpi = Fi + (1/TFi)*UPiC$$

----------------- (3)

Where

T_{pi} – represents the priority or preference value of a $Product_i$ which is to be accessed
F_i – represents the current access frequency which is equal to 1
T_{Fi} – represents the product popularity which is calculate from the access pattern matrix
U_{PiC} - is the number of users have access the same product

Recommendation: After the training of the ART network, if a particular user logins to the site, the previous access pattern of the user is searched from the database.

- Normalization and binary encoding are applied if the access pattern is discovered. Once encoded, the access pattern is supplied into the ART network in binary form.
- ART pinpoints the user's cluster, or group, membership. Once the cluster has been identified, it produces the cluster prototype. The lists of items that should be suggested are then determined and sorted using equation (1) based on the prototype.

4. RESULTS AND DISCUSSION

Flipkart is the one among the largest E-commerce site in India meant selling products through online. Now it is one of the biggest online stores for selling goods such as clothing, electronics and other consumables. For more information about Flipkart, please visit website www.flipkart.com. A pre-crawled dataset, comprising over 5.8 million goods, was constructed by extracting data from Flipkart.com, a prominent e-commerce platform in India figure 2 below.

Figure 2. Sample Flipkart Data

1	uniq_id	crawl_tim	product_u	product_n	product_c	pid	retail_pric	discounte	image	is_FK_Adv	descriptio	product_r	overall_ra	brand	product_specifications
2	c2d766ca9	2016-03-2	http://ww	Alisha Sol	{"Clothing	SRTEH2FF	999	379	{"http://ir	FALSE	Key Featu	No rating	No rating	Alisha	{"product_specification"=>[{"key"=>"Number of Contents in Sales
3	7f7036a6d	2016-03-2	http://ww	FabHomel	{"Furnitur	SBEEH3QC	32157	22646	{"http://ir	FALSE	FabHomel	No rating	No rating	FabHomel	{"product_specification"=>[{"key"=>"Installation & Demo Details",
4	f449ec65d	2016-03-2	http://ww	AW Bellie	{"Footwe	SHOEH4GI	999	499	{"http://ir	FALSE	Key Featu	No rating	No rating	AW	{"product_specification"=>[{"key"=>"Ideal For", "value"=>"Women
5	0973b37ac	2016-03-2	http://ww	Alisha Sol	{"Clothing	SRTEH2F6	699	267	{"http://ir	FALSE	Key Featu	No rating	No rating	Alisha	{"product_specification"=>[{"key"=>"Number of Contents in Sales
6	bc940ea4	2016-03-2	http://ww	Sicons All	{"Pet Sup	PSOEH3ZY	220	210	{"http://ir	FALSE	Specificat	No rating	No rating	Sicons	{"product_specification"=>[{"key"=>"Pet Type", "value"=>"Dog"}, {"
7	c2a173139	2016-03-2	http://ww	Eternal Ga	{"Eternal (	PWTEB7H	430	430	{"http://ir	FALSE	Key Featu	No rating	No rating	Eternal Ga	{"product_specification"=>[{"key"=>"Model Name", "value"=>"Gan
8	ce5a6818f	2016-03-2	http://ww	Alisha Sol	{"Clothing	SRTEH2FV	1199	479	{"http://ir	FALSE	Key Featu	No rating	No rating	Alisha	{"product_specification"=>[{"key"=>"Number of Contents in Sales
9	8542703ca	2016-03-2	http://ww	FabHomel	{"Furnitur	SBEEH3QC	32157	22646	{"http://ir	FALSE	FabHomel	No rating	No rating	FabHomel	{"product_specification"=>[{"key"=>"Installation & Demo Details",
10	29c8d290c	2016-03-2	http://ww	dilli bazaa	{"Footwe	SHOEH3D	699	349	{"http://ir	FALSE	Key Featu	No rating	No rating	dilli bazaa	{"product_specification"=>[{"key"=>"Occasion", "value"=>"Ethnic, (
11	4044c0ac5	2016-03-2	http://ww	Alisha Sol	{"Clothing	SRTEH2FV	1199	479	{"http://ir	FALSE	Key Featu	No rating	No rating	Alisha	{"product_specification"=>[{"key"=>"Number of Contents in Sales
12	e54bc0a7c	2016-03-2	http://ww	Ladela Be	{"Footwe	SHOEH4K	1724	950	{"http://ir	FALSE	Key Featu	5	5	Ladela	{"product_specification"=>[{"key"=>"Occasion", "value"=>"Casual"}
13	c73e78fb4	2016-03-2	http://ww	Carrel Prir	{"Clothing	SWIEHF3E	2299	910	{"http://ir	FALSE	Key Featu	No rating	No rating	Carrel	{"product_specification"=>[{"key"=>"Neck", "value"=>"Round Neck
14	c29af3783	2016-03-2	http://ww	Sicons All	{"Pet Sup	PSOEH3ZYYFETGCCD			{"http://ir	FALSE	Specificat	No rating	No rating	Sicons	{"product_specification"=>[{"key"=>"Pet Type", "value"=>"Dog"}, {"
15	9aacdecce	2016-03-2	http://ww	Alisha Sol	{"Clothing	SRTEH2FG	999	379	{"http://ir	FALSE	Key Featu	No rating	No rating	Alisha	{"product_specification"=>[{"key"=>"Number of Contents in Sales
16	83c53f894	2016-03-2	http://ww	Freelance	{"Pens & S	BOTEGYTZ	699	699	{"http://ir	FALSE	Specificat	No rating	No rating	Freelance	{"product_specification"=>[{"key"=>"Body Material", "value"=>"Sta
17	d95b0456	2016-03-2	http://ww	Alisha Sol	{"Clothing	SRTEH2FE	999	379	{"http://ir	FALSE	Key Featu	No rating	No rating	Alisha	{"product_specification"=>[{"key"=>"Number of Contents in Sales
18	849ab0565	2016-03-2	http://ww	FabHomel	{"Furnitur	SBEEH3QC	32157	22646	{"http://ir	FALSE	FabHomel	No rating	No rating	FabHomel	{"product_specification"=>[{"key"=>"Installation & Demo Details",
19	c275ee5ac	2016-03-2	http://ww	Style Foot	{"Footwe	SHOEH4FS	899	449	{"http://ir	FALSE	Key Featu	No rating	No rating	Style Foot	{"product_specification"=>[{"key"=>"Occasion", "value"=>"Casual"}
20	4f3511c33	2016-03-2	http://ww	Carrel Prir	{"Clothing	SWIEHF3H	2499	999	{"http://ir	FALSE	Key Featu	No rating	No rating	Carrel	{"product_specification"=>[{"key"=>"Neck", "value"=>"Round Neck
21	7ee3e337	2016-03-2	http://ww	FabHomel	{"Furnitur	SBEEH3QC	32157	22646	{"http://ir	FALSE	FabHomel	No rating	No rating	FabHomel	{"product_specification"=>[{"value"=>"1 Sofa Bed"}, {"key"=>"Insta

The dataset contains the following important attributes:

- P_Url
- P_Name
- P_Category
- Pid
- R_Price
- D_Price

- Photo
- Is_FK_Advantage
- P_Description
- P_Rating
- O_Rating
- P_Brand
- P_Specification

Next, let's see how to analyze and retrieve some meaningful information out of it. To do so, first step in data analytics is data preprocessing is need. It comprises several techniques for analyzing and removing irrelevant, inconsistent and incomplete data out of it i.e. does it contains any missing or duplicates or NaN values within it. If so, remove it. Then, next step is data preparation where it prepares or represents the given data in a format that is more suitable for data mining tasks i.e. separating date and time from timestamp or isolating data in terms of its sub-category and son on. Once, it is done then one can go for retrieving some useful information out of it. Information about data preprocessing and its outcomes is given in the table 2.

Table 2. Flipkart Dataset and its Preprocessing

Column/ Attribute Name	Data Type	No. of Unique Value	Preprocessing Details
uniq_id	String	20000	Valid 20.0k 1 00% Mismatched 0 0% Missing 0 0% Unique 20.0k
crawl_ timestamp	Date	From 1st Dec 2015 to 28th Jun 2016	Valid 20.0k 100% Mismatched 0 0% Missing 0 0% Minimum 1 Dec 15 Mean 18 Jan 16 Maximum 28 Jun 16
p_url	String	20000	Valid 20.0k 100% Mismatched 0 0% Missing 0 0% Unique 20.0k
p_ name	String	12676	Valid 20.0k 100% Mismatched 0 0% Missing 0 0% Unique 12.7k

continued on following page

Table 2. Continued

Column/ Attribute Name	Data Type	No. of Unique Value	Preprocessing Details
p_ category	String	Jewellery > Necklaces 8% Jewellery> Rings 4% Automotive >Car Mats 3% Jewellery>Bangles 2% Other 84%	Valid 20.0k 100% Mismatched 0 0% Missing 0 0% Unique 6466
pid	String	19998	Valid 20.0k 100% Mismatched 0 0% Missing 0 0% Unique 20.0k
R_price	Numeric	28594-57154: 574 <28594: 19322	Valid 19.9k 100% Mismatched 78 0% Missing 0 0% Mean 2.97k Std. Deviation 9.01k Quantiles 35 Min 666 25% 1.04k 50% 2k 75% 571k Max
D_ price	Numeric	28594-57154: 301 <28594: 19604	Valid 19.9k 100% Mismatched 78 0% Missing 0 0% Mean 1.97k Std. Deviation 7.33k Quantiles 35 Min 350 25% 550 50% 999 75% 571k Max
Photo	String	18589	Valid 20.0k 100% Mismatched 3 0% Missing 0 0% Unique 18.6k
is_FK_ Advantage_	Boolean	False: 19215 True: 785	Valid 20.0k 100% Mismatched 0 0% Missing 0 0% True 785 4% False 19.2k 96%
P_description	String	17539	Valid 20.0k 100% Mismatched 2 0% Missing 0 0% Unique 17.5k

continued on following page

Table 2. Continued

Column/ Attribute Name	Data Type	No. of Unique Value	Preprocessing Details
p_ rating	Numeric	No rating available: 91% 5: 3% 4: 1% 1: 1% Other (32): 4%	Valid 20.0k 100% Mismatched 0 0% Missing 0 0% Unique 36
o_ rating	Numeric	No rating available: 91% 5: 3% 4: 1% 1: 1% Other (32): 4%	Valid 20.0k 100% Mismatched 0 0% Missing 0 0% Unique 36
P_brand	String	Allure Auto: 2% Regular: 2% Voylla: 1% Slim: 1% Other (3496): 93%	Valid 14.1k 71% Mismatched 5861 29% Missing 0 0% Unique 3500
p_ specifications	String	18825	Valid 20.0k 100% Mismatched 14 0% Missing 0 0% Unique 18.8k

The following information can be extracted from the above preprocessed file.

- Which months/years have the most sales?
- Who are all the unique customers?
- Who are all the most often vising customers?
- Which month has highest sales and which product?
- Which product has less sales and why?
- Which categories have the most sales?
- Which categories have the best discounts by percentage?

Let's see the outcome of each step in the following subsection using python.

a. Simulation Environment

All these steps are carried out in python 3 programming environment. It uses the packages like Numpy etc.

b. Data Cleaning and Data Preparation

The Data cleaning and preparation will consist of searching for missing data, looking for duplicate data and then removing it; converting the time attribute into a pandas Timestamp and then partitioning off the timestamp attribute into a month and a year attribute shown in figure 3 below.

Figure 3. Data Preprocessing Details

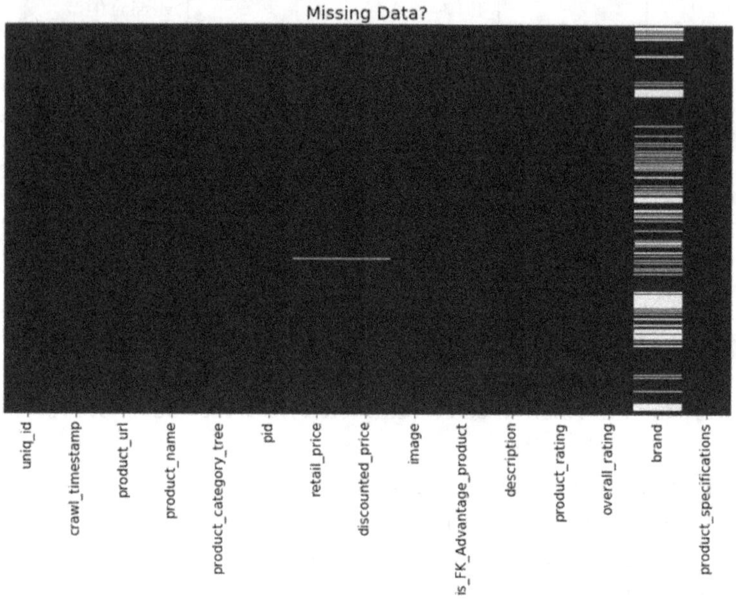

From the above figure 3, it is observed that the brand attribute is having more number of missing value among others. Everything else looks relatively complete. The following data cleaning methods can be applied over the data.

- Filling by mean
- Filling by global constant
- Filling by median
- Filling most probable value and so on.

Now that data cleaning is over and let's visualize some sales!! The following figure 4 shows the sales count by month and the figure 5 shows sales count by year. Similarly the figure 6 shows the sales count by different categories.

Figure 4. Monthly Sales Details

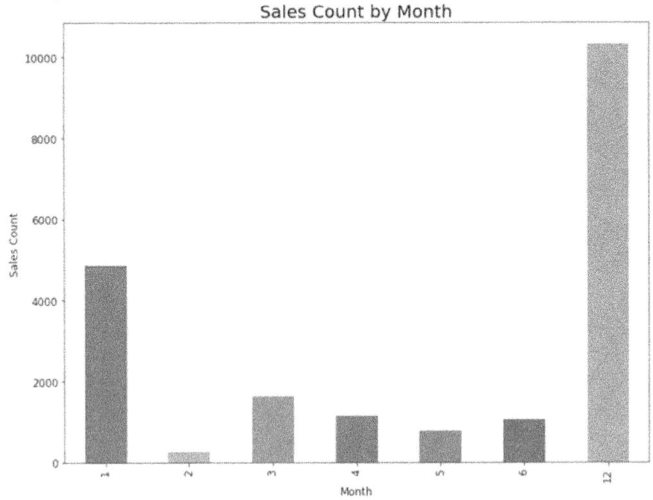

Does anyone else see something funny going on here table 3 and figure 6? Where is the data for July - November? Yes, the month section was missing about half of the years' worth of data i.e. No sales details have been found between the month July and November. Also, the year category only represented two years' worth of data, both of which look very similar since the dataset only contains sales facts for the year 2015 and 2016. Also, random sample has been taken from the huge dataset and hence it contains only partial months.

Table 3. Monthly Sales Count

S.No	Month	Sales Count
1	1	4850
2	2	245
3	3	1634
4	4	1137
5	5	773
6	6	1046
12	12	10315

Table 4. Yearly Sales Count

S.No	Year	Sales Count
1	2015	10315
2	2016	9685

Figure 5. Yearly Sales Count

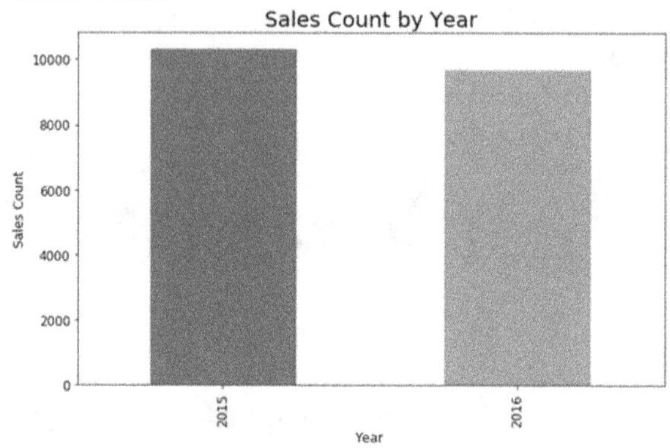

Only two years' worth of data is shown here in table 4 as well as in figure 7 and nothing very insightful here.

Figure 6. Sales by Category

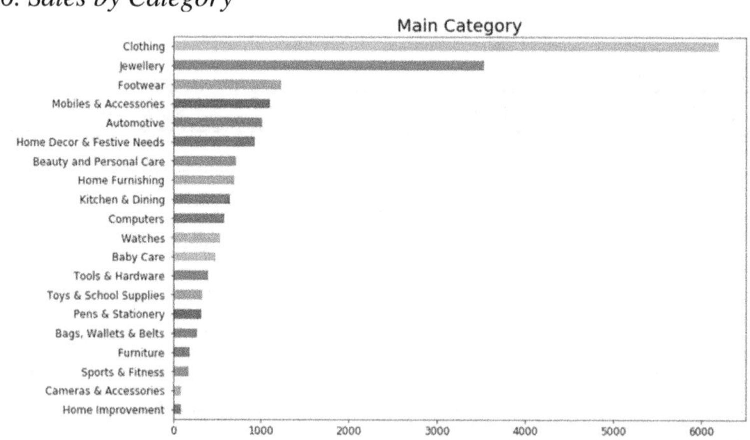

Table 5. Top Ten Sales Count by Category

S.No	Main Category	Count
1	Cloths	6198
2	Jewels	3531
3	Footwear	1227
4	Mobiles	1099
5	Automotive	1012
6	Home and Festive	929
7	Beauty Care	710
8	Home Furniture	700
9	Kitchen	647
10	Systems	578

From table 5 and figure 7 it is observed that clothing, jewelers, shoes, electronics and automotive is the main category winners!

Figure 7. Sales by Subcategory

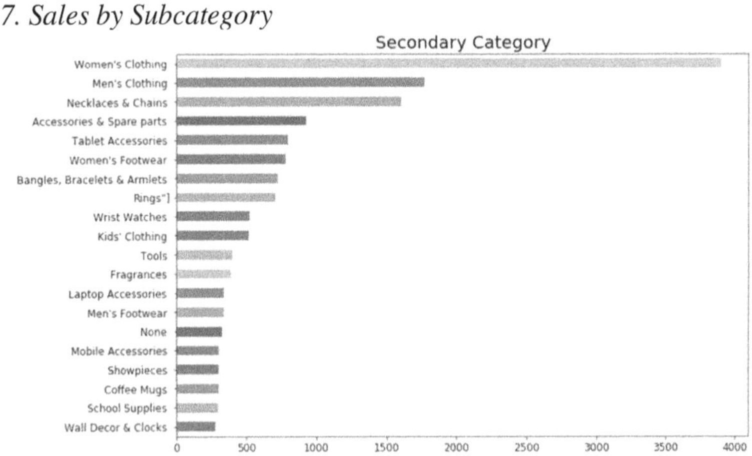

Table 6. Top Ten Secondary Categories by Sales

S.No	Main Category	Count
1	Women Wear	3901
2	Men Wear	1773
3	Necklaces & Chains	1606

continued on following page

Table 6. Continued

S.No	Main Category	Count
4	Accessories & Spare parts	925
5	Tablet Accessories	801
6	Women's Footwear	781
7	Bangles, Bracelets & Armlets	724
8	Rings	710
9	Wrist Watches	523
10	Kids' Clothing	520

From the figure 8 and table 6, it is interesting that women's clothing is more than double of men's clothing. It is clear now that a lot of the users on Flipkart are Women, though there is no gender column. As if that is not enough, look at 'Necklaces', 'Women's Footwear', 'Rings' and 'Kid's Clothing'.

Figure 8. Sales Details by Tertiary Category

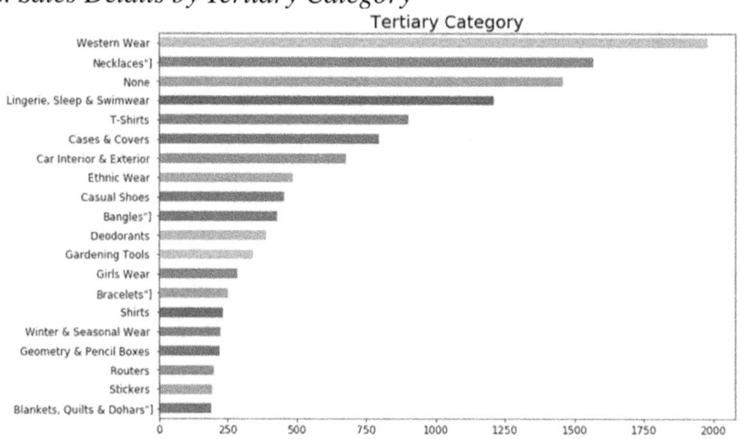

Table 7. Top Ten Tertiary Categories by Sales

S.No	Main Category	Count
1	Western Wear	1981
2	Necklaces	1567
3	None	1457

continued on following page

Table 7. Continued

S.No	Main Category	Count
4	Lingerie, Sleep and Swimwear	1208
5	Shirts	903
6	Cases and Covers	796
7	Vehicle Interior and Exterior	677
8	Cultural Wear	485
9	Shoes	454
10	Bracelet	430

From figure 9 and table 7, it is noted that Western Wear has more sales count. Western Wear means what? What does that mean?? Does that mean I want to look like a cowboy/cowgirl or does that mean I want to look like I live in Los Angeles? Here is one google image search result. After doing a few more, I'm going to guess Western wear to mean California wear over the Wild West.

Figure 9. Sales Details by Quaternary Category

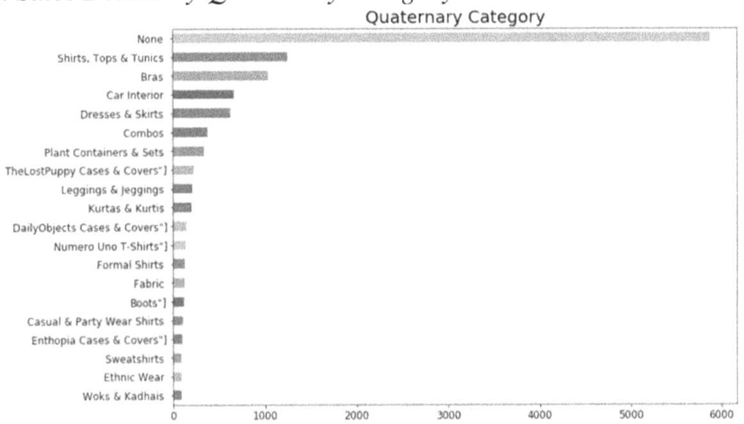

Table 8. Top Ten Quaternary Categories by Sales

S.No	Main Category	Count
1	None	5876
2	Shirts, Tops & Tunics	1249
3	Bras	1036

continued on following page

Table 8. Continued

S.No	Main Category	Count
4	Car Interior	659
5	Dresses & Skirts	620
6	Combos	375
7	Plant Containers & Sets	333
8	TheLostPuppy Cases & Covers	229
9	Leggings & Jeggings	209
10	Kurtas & Kurtis	202

The Quaternary graph shown in figure 10 has lost a lot of insight. Nearly 6000 'None' values are reported in table 8. Out of curiosity, what is the most expensive item in Flipkart during this time??? Yes, it is Rs.571230 and it is wrist watch. For the discounted sales section, we are going to have to do some data manipulation beforehand. First, we are going to create a discounted percentage column by subtracting the discount price from the retail price and dividing that amount by the retail price as shown in table 9.

Table 9. Discount Details

S.No	Product_Name	Retail_Price	Discounted_Price	Discount_%
1	Alisha Solid Women's Cycling Shorts	999.0	379.0	62.1
2	FabHomeDecor Fabric Double Sofa Bed	32157.0	22646.0	29.6
3	AW Bellies	999.0	499.0	50.1
4	Alisha Solid Women's Cycling Shorts	699.0	267.0	61.8
5	Sicons All Purpose Arnica Dog Shampoo	220.0	210.0	4.5

Second, we are going to create a few new data frames that contain the product by category, average discounted percentages and count of each product as given in table 10.

Table 10. Discount by Category Details

Main Category	Discount_% mean	Main Category Count
883 Police Full Sleeve Solid Men's Jacket	15.0	1
ABEEZ Boys, Men, Girls (Black, Pack of 1)	27.6	1
ANAND ARCHIES Girls Flats"	10.0	2
ANAND ARCHIES Girls Wedges"	0.0	1
ANASAZI Casual	0.0	1

Third, we are going to combine the levels of the columns for visualization purposes as shown in figure 11.

Figure 10. Discount by Category

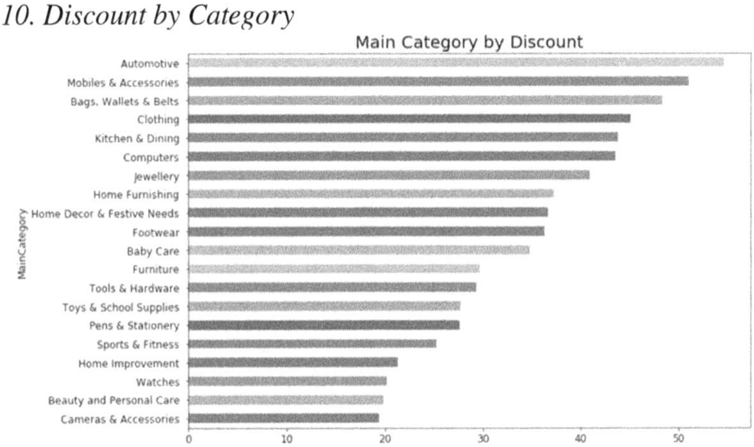

Table 11. Main Category by Discount (Percentage)

S.No	Tertiary Category	Discount (Percentage)
1	Automotive	54.649901
2	Mobiles & Accessories	51.001185
3	Bags, Wallets & Belts	48.271591
4	Clothing	45.071301
5	Kitchen & Dining	43.769612
6	Computers	43.464223
7	Jewellery	40.891198
8	Home Furnishing	37.218143

From the table 11, it is evident that automotive, clothing and electronics were some of the most discounted categories.

Figure 11. Discount by Subcategory

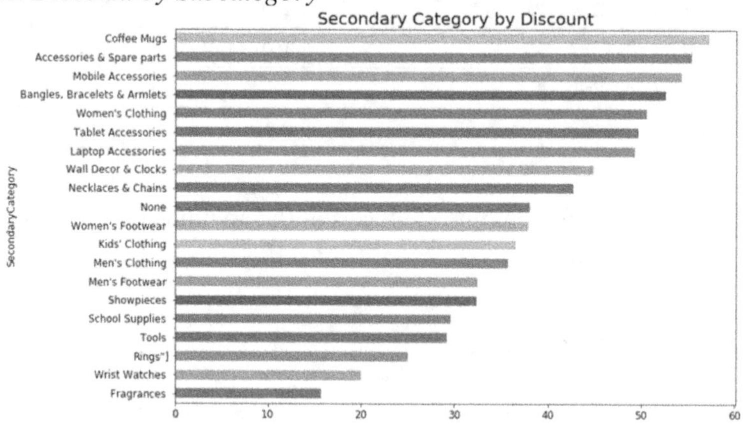

Table 12. Secondary Category by Discount (Percentage)

S.No	Secondary Category	Discount (Percentage)
1	Coffee Mugs	57.274172
2	Accessories & Spare parts	55.370314
3	Mobile Accessories	54.318421
4	Bangles, Bracelets & Armlets	52.560111
5	Women's Clothing	50.500619
6	Tablet Accessories	49.673217
7	Laptop Accessories	49.320354
8	Wall Decor & Clocks	44.833213

For the secondary category Table 12, once again electronics and accessories. Some clothing, spare parts and jewellery. Coffee mugs? None is present.

Figure 12. Discount by Tertiary Category

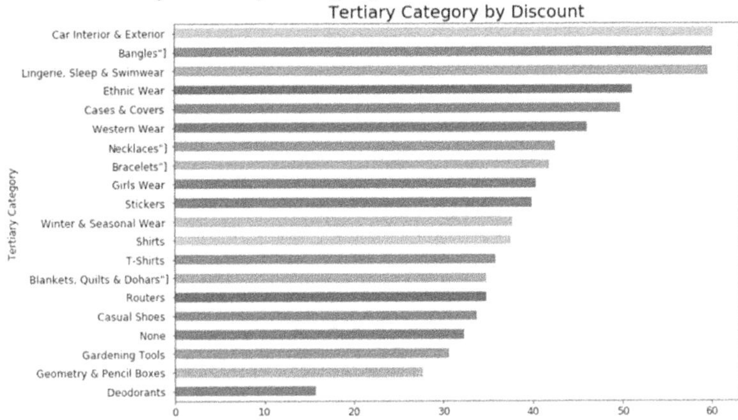

Table 13. Tertiary Category by Discount (Percentage)

S.No	Tertiary Category	Discount (Percentage)
1	Car Interior & Exterior	60.212741
2	Bangles	60.057809
3	Lingerie, Sleep & Swimwear	59.606716
4	Ethnic Wear	51.112810
5	Cases & Covers	49.741058
6	Western Wear	46.082203
7	Necklaces	42.543918
8	Bracelets	41.830400

From table 13 and figure 12, it is found that automotive parts on top again, followed by jewelry and clothing.

Table 14. Quaternary Category by Discount (Percentage)

S.No	Quaternary Category	Discount (Percentage)
1	The Lost Puppy Cases & Covers	70.651528
2	Bras	61.739034
3	Car Interior	60.784170
4	Kurtas & Kurtis	54.601485

continued on following page

Table 14. Continued

S.No	Quaternary Category	Discount (Percentage)
5	Ethnic Wear	54.036667
6	Leggings & Jeggings	54.036667
7	Fabric	47.405738
8	Shirts, Tops & Tunics	47.405738

Figure 13. Discount by Quaternary Category

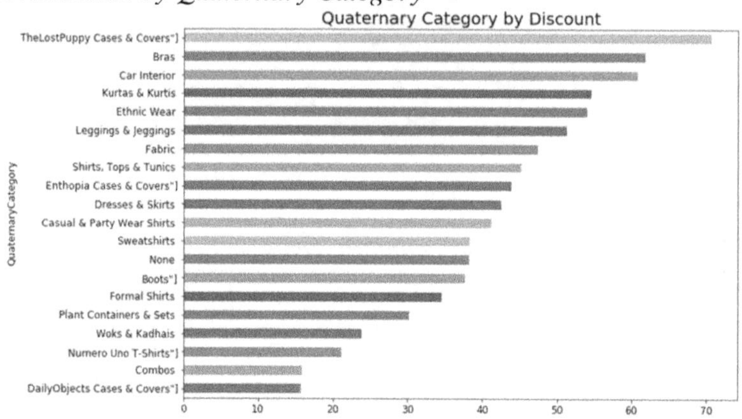

From table 14 and figure 13, it is true that the quaternary category is dominated by mostly clothing, with come automotive and electric categories. Once, statistical analyses are over then shift to recommendation engine. Here ART1 neural network is used for clustering users' access patterns based on their similarity. Here, ART1 based clustering algorithm is preferred over other traditional clustering algorithms since it can generate good clusters in term of inter and intra cluster distance when compared to other clustering techniques, and there is no requirement to provide parameters in advance, like the number of clusters. Also, it adopts incremental learning.

Every time a user logs in to the website after training, the database is searched to see their previous access history. Once the access pattern has been found, it is binary encoded and normalized. After that, the ART1 network receives it. ART1 network identifies the cluster (group) to which the user belongs. Following cluster identification, it outputs the cluster prototype for rating. Based on this prototype and ranking, the lists of products to be recommended are decided and shown using equation Dou, Z., Song, R., Wen, J.R. (2007).

CONCLUSION

The primary cause of the increased collection of corporate data is, since nowadays the business is depend on customers' data especially in marketing industry. Typically, data that is collected includes transactional information, documents, photos, audio, video, and social media contents. It can also include unstructured data. So, data analytics is required to extract useful information from huge volumes of data for driving new market opportunities. Also, it is required in maximizing customers' satisfaction and retention. Therefore, numerous recommendation engines have been created and implemented for online users, as documented in literature reviews. It is being used by many E-commerce websites to help buyers in making choice.

Therefore, Data Science is the field where one can retrieve implicit and potential values recorded in the data source. The raw data is unsuitable for knowledge discovery process. Hence, it requires extensive data cleaning and data formatting tasks. From these results reported so far, it is understood that the given dataset looks complete but with little missing value and without any duplicate values. Then, the preprocessed data is segregated into date and time for further processing. Thus, more descriptive subcategories were created like tertiary and quaternary for mining. Among the product categories reported here, the clothing (women's), electronics, jewelry and automotive shows higher sales count and automotive, electronics, accessories and clothing held the highest discounts. Finally, it recommends products to customers based on their access similarity and list best-selling products within that group.

REFERENCES

Algarni, S., & Sheldon, F. (2023). Systematic Review of Recommendation Systems for Course Selection. *Mach. Learn. Knowl. Extr.*, 5(2), 560–596. DOI: 10.3390/make5020033

Apeh, E. T., Gabrys, B., & Schierz, A. (2011), Customer profile classification using transactional data. *Nature and Biologically Inspired Computing (NaBIC), 2011 Third World Congress on*. IEEE.

Babin, B. J., Darden, W. R., & Griffin, M. (1994). Work and/or fun: Measuring hedonic and utilitarian shopping value. *The Journal of Consumer Research*, 20(4), 644–656. DOI: 10.1086/209376

Berry, M., & Linoff, G. (2000). *Mastering Data Mining*. John Wiley & Sons, Inc.

Chae, B., Sheu, C., Yang, C., & Olson, D. (2014). The impact of advanced analytics and data accuracy on operational performance: A contingent resource based theory (RBT) perspective. *Decision Support Systems*, 59, 119–126. DOI: 10.1016/j.dss.2013.10.012

Chen, Y. L., Kuo, M. H., Wu, S. Y., & Tang, K. (2009). Discovering recency, frequency, and monetary (RFM) sequential patterns from customers' purchasing data. *Electronic Commerce Research and Applications*, 8(5), 241–251.

Chen, Y. L., Tang, K., Shen, R. J., & Hu, Y. H. (2005). Market basket analysis in a multiple store environment. *Decision Support Systems*, 40(2), 339–354.

Dou, Z., Song, R., & Wen, J. R. (2007), A large-scale evaluation and analysis of personalized search strategies, in: *WWW '07: Proceedings of the 16th international conference on World Wide Web*, ACM Press, pp. 581–590. DOI: 10.1145/1242572.1242651

Eickhoff, C., Collins-Thompson, K., Bennett, P. N., & Dumais, S. (2013), Personalizing atypical web search sessions, in: *Proceedings of the Sixth ACM International Conference on Web Search and Data Mining*, ACM, New York, NY, USA, pp. 285–294. DOI: 10.1145/2433396.2433434

Fayyad, U., Piatetsky-Shapiro, G., & Smyth, P. (1996), From data mining to knowledge discovery: An overview. In *Advances in Knowledge Discovery and Data Mining,* chapter. 1, U. Fayyad, G. Piatesky-Shapiro, P. Smyth, and R. Uthurusamy, Eds. AAAI Press, Menlo Park, CA, and the MIT Press, Cambridge, MA, pp.1–34.

Gormley, C., & Tong, Z. (2015). *Elastic search: The Definitive Guide* (1st ed.). O'Reilly Media, Inc.

Gretzel, U., & Fesenmaier, D. R. (2006). Persuasion in recommender systems. *International Journal of Electronic Commerce*, 11(2), 81–100.

Hoffman, D. L., & Novak, T. P. (1996). Marketing in hypermedia computer-mediated environments: Conceptual foundations. *Journal of Marketing*, 60(3), 50–68. DOI: 10.1177/002224299606000304

Jakob, N. (2011). Extracting Opinion Targets from User-Generated Discourse with an Application to Recommendation Systems.

Jannach, D., & Ludewig, M. (2017), Investigating personalized search in e-commerce, in: *FLAIRS Conference*, AAAI Press, pp. 645–650.

Kimball, R., & Merz, R. (2000). *The Data Webhouse Toolkit: Building the Web-Enabled Data Warehouse*. John Wiley & Sons, Inc.

Kohavi, R., Brodley, C., Frasca, B., Mason, L., & Zheng, Z. (2000). KDD-Cup 2000 organizers' report: Peeling the onion. *SIGKDD Explorations*, 2(2), 86–98. www.ecn .purdue.edu/KDDCUP. DOI: 10.1145/380995.381033

. Kohavi, R. and Provost, F. (2001), Applications of data mining to electronic commerce. *Data Min. Knowl. Disc. 5,* 1/2; robotics.stanford. edu/users/ronnyk/ ecommerce-dm.

Kong, D. (2016). *Personalized Feature based Re-Ranking Method for Ecommerce Search at CIKM Cup 2016. Technical Report*. CIKM Cup.

. Lin, Qi-Yuan, (2003), "Mining inter-organizational retailing knowledge for an alliance formed by competitive firms." *Information & management, Vol.*40 and Issue 5 (2003): pp.431-442.

Liu, S., Xiao, F., Ou, W., & Si, L. (2017), Cascade ranking for operational e-commerce search, in: *Proceedings of the 23rd ACM SIGKDD International Conference on Knowledge Discovery and Data Mining*, ACM, New York, NY, USA, pp. 1557–1565.

McCandless, M., Hatcher, E., & Gospodnetic, O. (2010), Lucene in Action, Second Edition: Covers Apache Lucene 3.0. Manning Publications Co, Greenwich, CT, USA.

Millard, S. (2013), Big Data Brewing Value in Human Capital Management – Ventana Research. Retrieved April 2, 2015 from http://stephanmillard.ventanaresearch .com/2013/08/28/big-data-brewing-value-in-humancapital-management

Mobasher, B. (2007). *Data mining for web personalization. The adaptive web.* Springer Berlin Heidelberg.

Mosavi, A., & Vaezipour, A. (2013). *Developing Effective Tools for Predictive Analytics and Informed Decisions. Technical Report*. University of Tallinn.

Palotti, J. (2016). *Learning to Rank for Personalized E-Commerce Search at CIKM Cup 2016. Technical Report*. CIKM Cup.

Puri, R. (2013), How Online Retailers Use Predictive Analytics To Improve Your Shopping Experience.Retrieved April5, 2015 from https://blogs.sap.com/innovation/analytics/how-online-retailers-use-predictiveanalytics- to-improve-your-shopping-experience-0108060

Robertson, S., & Zaragoza, H. (2009). The probabilistic relevance framework: Bm25 and beyond. Found. *Trends Information Retrieval*, 3(4), 333–389. DOI: 10.1561/1500000019

Sarwar, B., Karypis, G., Konstan, J., & Riedl, J. (2002), Recommendation systems for large e-commerce:Scalable neighborhood formation using clustering. In *Proceedings of the fifth international conference on computer and information technology*

Shmueli, G., & Koppius, O. (2011). Predictive Analytics in Information Systems Research. *Management Information Systems Quarterly*, 35(3), 553–572. DOI: 10.2307/23042796

Speretta, M., & Gauch, S. (2005), Personalized search based on user search histories, in: The 2005 IEEE/WIC/ACM International Conference on Web Intelligence (WI'05), pp. 622–628. DOI: 10.1109/WI.2005.114

Tang, K., Chen, Y. L., & Hu, H. W. (2008). Context-based market basket analysis in a multiple-store environment. *Decision Support Systems*, 45(1), 150–163.

Teevan, J., Dumais, S. T., & Liebling, D. J. (2008), To personalize or not to personalize: Modeling queries with variation in user intent, in: Proc. Of the 31st Annual Int. ACM SIGIR Conf. on Research and Development in Information Retrieval, ACM, New York, NY, USA, pp. 163–170.

Yousefi, A., & Tang, J. (2012). *E-commerce: Consumer Online Shopping in Canada. Contemporary Research on E-business Technology and Strategy*. Springer Berlin Heidelberg.

Yu, J., Mohan, S., Putthividhya, D. P., & Wong, W. K. (2014), Latent dirichlet allocation based diversified retrieval for e-commerce search, in: *Proceedings of the 7th ACM International Conference on Web Search and Data Mining*, ACM, New York, NY, USA, pp. 463–472. DOI: 10.1145/2556195.2556215

Zhang, Z., Patra, B. G., Yaseen, A., Zhu, J., Sabharwal, R., Roberts, K., & Wu, H. (2023). Scholarly recommendation systems: A literature survey. *Knowledge and Information Systems*, 65(11), 4433–4478.

Zhang, Z., Patra, B. G., Yaseen, A., Zhu, J., Sabharwal, R., Roberts, K., & Wu, H. (2023). Scholarly recommendation systems: A literature survey. *Knowledge and Information Systems*, 65(11), 4433–4478.

Chapter 9

Artificial Intelligence and Machine Learning as Pioneers in Advancing 5G/6G Network Capabilities:

Evolution, Introduction, Antecedents, and Consequences

Qazi Saima Syed
https://orcid.org/0000-0001-6028-3776
Islamic University of Science and Technology, India

Shakir Hussain
Islamic University of Science and Technology, India

Irfan Bashir
https://orcid.org/0000-0003-0000-1065
Islamic University of Science and Technology, India

ABSTRACT

This chapter provides a comprehensive review of the technological shifts, highlighting AI's and machine learning's transformative potential in communications while acknowledging the challenges they present. This chapter will delve into the development of various mobile technologies over time and the future of 5G and 6G communication networks. Furthermore, this chapter provides an overview of the use of AI and machine learning in upgrading 5G and 6G networks, concentrating

DOI: 10.4018/979-8-3693-6725-4.ch009

on critical topics such as its evolution, significance, influencing factors, and the repercussions of its deployment in network capabilities.

1. INTRODUCTION

The rapid growth of mobile devices has transformed the technological landscape. According to a Statista forecast, there will be 18.22 billion mobile devices worldwide by 2025, representing an increase of 4.2 billion from 2020 (Statista, 2024). The expansion of advanced communication infrastructure has created challenges for existing wireless technologies in meeting user demands, necessitating the development of more sophisticated communication systems (Oughton et al., 2018). Emerging technologies, such as artificial intelligence and machine learning, are expected to revolutionize communication infrastructure, enabling advanced devices to operate at higher frequencies and wider bandwidths (Salahdine et al., 2023). The introduction of 5G network services has resulted in a significant improvement in network capabilities, offering benefits such as reduced latency, increased user capacity, integration of various network technologies, and enhanced energy efficiency. While 5G networks operate at frequencies of 0.6 GHz and above, 6G is anticipated to function between 95 GHz and 3 THz. The advent of 6G has revolutionized wireless communication, with speeds projected to be 1000 times faster than 5G (Zhang et al., 2019). This evolution of wireless networks necessitates a comprehensive overhaul of existing communication infrastructure, including its architecture, policies, and strategies. As a successor to 5G, 6G is in its early developmental stage. It aims to address 5G's limitations by achieving even lower latency and improved efficiency. Future advancements in network capacity and transmission speed are expected to surpass the capabilities of 6G networks. The rapid progress in 6G-based computational intelligence (CI) technology has accelerated the information age. In addition, big data technologies may enable the resolution of complex 6G network issues through AI-based solutions. The integration of machine learning into 6G networks represents another significant breakthrough, supporting features such as advanced radio interfaces, intelligent traffic management, and enhanced security protection and administration. This chapter explores the crucial role of AI and ML in enhancing 5G and 6G network capabilities. We trace the evolution of these networks and emphasize the technological breakthroughs that have driven their development. This chapter also examines the precursors to AI/ML integration in telecommunications, including technological advancements and industry drivers. Furthermore, we discuss the wide-ranging implications of this integration, from improved network performance and economic impact to ethical and security considerations. Our aim is to provide a comprehensive overview of the current state and future prospects of AI and ML in

5G/6G networks, offering insights into a future in which highly efficient, intelligent communication systems form the backbone of our digital society.

2. EVOLUTION OF NETWORK TECHNOLOGIES

2.1 1G: Analog Beginnings

1G, the first generation of mobile networks, was introduced in the 1980s. It supported analog voice communication and was considered quite basic compared to modern standards. Users could make voice calls; however, data services or internet connectivity were not available on these networks.1G networks used analog signals, which were quite susceptible to interference and offered limited capacity and security. 1G networks have revolutionized communication by making mobile telephony accessible to the public.

Initially launched in the early 1980s, 1G mobile networks provided speech services using analog transmission. The Nippon Telephone and Telegraph (NTT) cellular system provider was initially introduced in Tokyo, Japan in 1979. Subsequently, Europe introduced its cellular system two years later. The two most well-known analog systems are Nordic mobile phones (NMT) and Total Access Communication Systems (TACS) (Ibrahim, 2002). Despite being the first mobile system, 1G completely transformed the technology of mobile systems. It employs an analog transmission signal with a minimum data rate of up to 2.4 kbps that is limited to voice services. It also functioned in the frequency range of 800–900 MHz, with a channel capacity of 30 kHz and 40 MHz bandwidth. A technique referred to as frequency-division multiplexing (1G) was used. Because of the analog signal utilized for transmission, 1G has a number of issues, including inadequate data capacity security, high energy consumption, and low-quality calls (Gupta & Jha, 2015).

2.2 2G: Digital Transformation

The second generation (2G) of mobile networks, characterized by digital communication, came into being in the 1990s. Digital encryption was introduced by 2G networks, which greatly enhanced security. They provided better voice quality and supported SMS (Short Message Service). Improved voice quality, greater capacity, and the advent of text messaging, which quickly gained international traction, were made possible by the 2G's digital nature. From a conceptual standpoint, this mobile network consists of a uniform dispersion of several base stations (BSs) throughout the world to offer various access points (such as FDMA, CDMA, and TDMA) for users to connect. As a result, the compression–decompression algorithm (codec) is a prerequisite for 2G technologies such the 2.75G Enhanced Data rates for global evolution (EDGE), 2.5G General Packet Radio Service (GPRS), and 2G Global System for Mobile Communications (GSM) (Farooq et al., 2013). Compared to 1G, 2G offered superior quality services and a greater number of services, including "multi-media service" (MMS) and "short message service" (SMS). Furthermore, 2G was enhanced to operate at a data rate of up to 64 kbps and frequency band of 850–1900 MHz. Second-generation systems employ CDMA and TDMA. Therefore, 2G offers more reliable mobile communication services and digital technology-encrypted data transmissions (Sah, 2017).

2.3 3G: The Rise of Mobile Data

The third generation (3G) of mobile networks, which prioritizes Internet access and mobile data, was introduced in the early 2000s. Wideband Code Division Multiple Access (WCDMA) technology was implemented by 3G networks to provide greater data speeds, allowing mobile Internet and video conversations. Smartphones and mobile apps were made possible by their capacity to access the Internet on mobile devices, thereby revolutionizing social interaction and technological use. The third generation refers to IMT-2000 technological standards that include speed and reliability components, such as a minimum 200 kbps data transfer rate (Kumar et al.,2010). The network was transformed from a traditional mobile network to a portable media device network with faster connection speeds (e.g., PCs, game consoles, and tablets). In addition to providing high-quality Internet services, 3G enhances security by requiring user authentication when establishing connections with other wireless devices (Sharma, 2013). Therefore, Wideband Code Division Multiple Access (WCDMA), time-division synchronous code division multiple access (TD-SCDMA), and CDMA2000 are the three core technologies of the 3G network.

2.4 4G: Broadband Mobility

Launched in the 2010s, the fourth generation (4G) of mobile networks aims to provide broadband-like speeds and improved connectivity. 4G networks employ long-term evolution (LTE) technology, offering significant improvements in data-transfer speeds, reduced latency, and enhanced capacity. The proliferation of 4G has enabled high-definition video streaming, mobile gaming, and a surge in mobile applications, solidifying the smartphone's role in daily life. The fourth generation is synonymous with the term "MAGIC," which stands for "mobile multimedia anywhere, global mobility solutions over integrated wireless and customized services," and is synonymous with the fourth generation (Kalra & Chauhan, 2014). Users may enjoy seamless network access, end-to-end IP transmission, mobility, and better quality of service (QoS) management with a 20 Mbps data transfer rate owing to 4G.

2.5 5G: A Novel Era

The deployment of fifth-generation (5G) networks began in the late 2010s with the objectives of offering extremely high data speeds, extremely low latency, and extensive connectivity. Sub-6 GHz and millimeter-wave (mmWave) frequencies are used in conjunction with large multiple-input multiple-output (MIMO) antennas and sophisticated beamforming algorithms in 5G networks. Virtual reality (VR), augmented reality (AR), and high-definition video streaming are all supported by 5G's far faster data speeds than 4G's. 5G enables an enormous density of connected devices per square kilometer, which will accelerate the development of smart city applications and the Internet of Things (IoT). With its new wireless technology standards, 5G signals the beginning of a worldwide digital era, especially in terms of data transmission rate, latency, mobility, and the number of connected devices (Soldani & Manzalini, 2015). 5G network architecture was introduced to bring about radical innovation in conventional network systems. The Internet of Things (IoT) is expected to be a significant part of 5G and will not only assist connected devices at home, but also devices spanning all around the world, representing different sectors and industries. The Fifth Generation is expected to possess a better network capacity to introduce different technologies that range from better Quality of Service (QoS) and novel features in comparison to 4G; however, it will further evolve and augment everyday living standards (Giordani et al., 2020). 5G's features of really distinguish this cutting-edge mobile network from that of its predecessors. In contrast to its predecessors, 5G is the first technology to use a new frequency spectrum known as the mmWave band. In addition, 5G serves as a hub for the IoT and can incorporate a wide range of novel technologies, including massive multiple-input multiple-output (mMIMO), software-defined networking (SDN), and device-to-device (D2D) communication (Wu et al., 2018; Hassan et al., 2020).

2.6 6G: Beyond Connectivity

Around the 2030s, sixth-generation (6G) networks are anticipated to appear, with the goal of incorporating AI directly into the network architecture to deliver even faster speeds, more dependability, and new services. Terahertz (THz) frequencies, sophisticated AI-driven network management, and quantum communication technologies are expected to be used in 6G. To facilitate holographic communication and immersive extended reality (XR) experiences, 6G promises to deliver data rates that are up to 100 times faster than 5G. With the help of AI, 6G will be able to perform autonomous network management, predictive analytics, and real-time resource optimization. With the integration of satellite networks and ubiquitous coverage, particularly in rural and distant locations, 6G has provided seamless global communication. By 2030, a highly evolved, nearly instantaneous, and wirelessly connected global digital civilization is envisaged (Zhang et al., 2019). To facilitate full-vertical applications, 6G is essential for combining smooth wireless connectivity with a range of technological functions (such as caching, communication, processing, control, imaging, navigation, positioning, radar, and sensing). The proposed communication infrastructure for 6G is expected to be regarded as a distinctive ecosystem whose recognition and computational abilities resemble those of humans (Zhang et al. 2019). In addition to seamless wireless connectivity, 6G is expected to have a striking data rate of 1 Tbps, and its connectivity is 1000 times greater than that of 5G (Chowdhury et al., 2020). Envisioning successful and intelligent wireless connectivity, regarded as 6G, is expected to be employed by 2030 (Zhang et al., 2020). The employment of 6G alongside breakthroughs in AI is expected to facilitate security, thereby necessitating freedom with regard to numerous manufacturing areas while augmenting the level of trust between end users and the resulting innovations(Guo, 2020). The primary tenet of design should be aimed at reducing complications, thereby allowing the system to be quite flexible in order to accommodate any technological changes that may arise. Cognitive Radio Frequency, also referred to as CogRF is an upcoming concept that combines the power of Machine Learning (ML) and Artificial Intelligence (AI) with Radio Frequency (RF) to transform and completely overpower the conventional radio design systems (Cousik et al., 2019). Therefore, a single all-encompassing framework is required to combine computational and communication resources to eliminate the issues that previous issues had and ultimately devise a novel Internet architecture (Calvanese Strinati et al., 2019). Numerous aspects including self-optimization, proactive strategies, network comprehension alongside prediction capabilities, mobile applications can be designed, and devices should perform in remote areas as well, thereby playing a significant role in the artificial intelligence introduced in 6G deployment (Calvanese Strinati et al., 2019). Numerous technologies are regarded as the top six 6G enabling technologies, including THz communication, holographic beamforming (HBF), mMIMO, AI, blockchain, and quantum communication (QC). 6G is a self-governing device that mimics human awareness and intelligence. It also provides a variety of ways for users to connect and communicate with smart terminals, including voice, eyes, fingers, brainwaves, and neurological impulses. The key aspects of the evolution have been summarized in **Table 1**

Table 1. Type of technology and key features

Type of technology	Key features
1G	Analog voice communication
2G	Digital voice communication
3G	Mobile Broadband
4G	Enhanced 3G services
5G	Interconnected Devices, Sensors & Systems
6G	Augmented reality, AI, Blockchain

3. AI/ML APPLICATIONS IN 5G/6G NETWORKS

AI- and ML-based algorithms enhance traffic management, predict network issues, and optimize resource allocation. Advanced AI solutions bolster security in complex 5G and 6G networks by improving the threat detection and response. Real-time data and predictive analytics enable service personalization and quality improvement. AI and ML facilitate high levels of automation from self-repairing networks to automated maintenance procedures. Currently, AI is recognized as a technology for powering smartphone facial recognition, self-driving cars, web search engines, financial operations, and various everyday activities. Future AI investments are expected to be significant, with ongoing progress leading to a wider implementation (Lu et al., 2017). The essential technologies are described as follows.

3.1 Decision making

The decision-making process relies on techniques such as the support vector machine (SVM), decision tree (DT), deep learning (DL), and neural network (NN) (Lu et al., 2017). These are crucial for achieving continuous improvements throughout the processing and financial periods.

3.2 Speech Recognition

The primary goal of speech recognition is to transform human language into functional computer applications (Lu et al., 2017). Graves et al. introduced a deep long short-term memory Recurrent Neural Network (RNN) that combines two distinct RNNs as both linguistic and acoustic models, resulting in a bilateral speech recognition system.

3.3 Virtual Reality (VR)

VR is a computer-generated stimulus in a three-dimensional environment that interacts and integrates realistically. AI-powered virtual reality has applications in industries, homes, customer support and services, transportation, and healthcare (Lu et al., 2017).

3.4 Natural Language Processing (NLP)

NLP enables computers to comprehend, analyze, and interpret human language messages and meanings efficiently and effectively (Lu et al., 2017). Utilizing machine learning and analytical methods, NLP supports sentence structure, intent recognition, and sentiment analysis and has the potential to enhance network performance through substantial interpretation capabilities and improved learning abilities (Yang et al., 2019).

Potential solutions for AI-enabled intelligent 6G networks have also been proposed. AI techniques can assist various devices and infrastructures in improving energy management approaches by logically controlling energy harvesting and consumption, thereby extending the management time. Factors such as mobile dynamics, network topology, and wireless pathways require the continuous adjustment of parameters in AI learning algorithms (Yang et al., 2019).

4. ANTECEDENTS ENABLING 6G WITH AI/ML

Improvisation in network performance has been brought about by novel technological advancements in artificial intelligence (AI) and machine learning (ML). The following are the factors that govern the deployment of advanced communication infrastructure viz., 5G/6G:

4.1 Channel modeling

The performance of wireless communication systems relies on channel models that accurately represent real-world communication environments. While existing channel models can capture the common features of traditional wireless channels, they have limitations in complex scenarios owing to imperfections and nonlinear behavior. Moreover, for emerging 6G applications, such as underwater or molecular acoustic communications, developing precise mathematical channel models is extremely challenging (Farsal & Goldsmith, 2017). Consequently, it is crucial to explore alternative approaches to designing communication systems without using well-defined channel models. The data-driven paradigm offers a solution to these challenges by leveraging the information inherent in network-generated data and eliminating the need for an exact model. By utilizing large training datasets and data-driven models, deep learning (DL) can enhance radio-link performance without requiring a mathematically tractable channel model.

4.2 Effective signal processing

THz communication is essential for 6G to support ultra-broadband real-time applications such as XR and remote holographic communications for mobile users, as envisaged by Zhang et al. (2019). Advanced 6G systems require extensive real-time signal processing capabilities. Conventional iterative reconstruction techniques, such as Multiple-Input Multiple-Output (MIMO) data detection algorithms, create computational hurdles for systems operating in real time. By contrast, neural networks can be extensively parallelized and applied to low-precision data types and concurrent systems. Research has demonstrated that this form of deep learning algorithm can operate more swiftly and efficiently than its traditionally programmed counterparts (O'Shea & Hoydis, 2017). Consequently, parallel signal processing methods are valuable for enhancing the accuracy and efficiency of future radio interfaces. By leveraging parallel computing powered by graphics processing units (GPUs), deep learning algorithms can produce accurate results within milliseconds.

4.3 Limitations of existing radio systems

In modern times, the divide-and-conquer approach is frequently employed to address complex radio communication challenges. This has led to the development of current communication systems that utilize various predefined components such as modulation and demodulation, along with channel coding and decoding. The conventional method for addressing radio communication problems involves independently optimizing each functional block. Although many researchers have made progress in enhancing individual blocks, it is practically impossible to guarantee optimal overall system performance using this approach (Wang et al., 2017). Consequently, there is a growing need to shift focus towards optimizing system performance from an end-to-end perspective rather than continuing to refine each block separately.

4.4 Quality of Service (QoS)/ Quality of Experience (QoE) requirements

Many new Internet of Everything (IoE) applications, including XR, intelligent manufacturing, and smart environments, are expected to be supported by 6G. Understanding, assessing, and ensuring that developing apps provide satisfactory QoS/QoE are difficult tasks. The majority of QoE devices already in use are only compatible with traditional mobile broadband services, such as streaming and voice over LTE (VoLTE). The requirements of users and features of new apps should be thoroughly understood to create an accurate Quality of Service (QoS)/Quality of Experience (QoE) model. Finding relationships between QoE measures at the application level and QoS indicators at the network level, as well as comprehending how QoS affects QoE, has been found to be facilitated by machine learning approaches (Zhang & Zhu, 2020). In addition, 6G is anticipated to present a range of services with integrated 3D coverage across space, air, land, and water. However, maintaining a consistent and pleasant user experience in such a complex and dynamic network environment is extremely difficult. By forecasting user behavior and traffic demands, machine learning (ML) has been considered a successful method for meeting user requirements in a variety of new applications.

4.5 Flexible networks

It is anticipated that 6G will encompass a comprehensive network covering space, air, land, and water, integrating various access networks with distinct characteristics. Furthermore, the proliferation of network nodes in diverse formats, including macro-, micro-, and femto-stations, as well as central units (CU), distributed units (DU), and active antenna units (AAU), necessitates increased management by network operators owing to ultra-dense networks and flexible control data plane separations. One potential approach to enhancing the intelligence of a network management system is to utilize ML/DL to realize the concept of a self-sustaining network (Saad et al., 2019). Such a network is expected to optimize resource utilization and adjust network functions to independently maintain long-term performance. Additionally, network slicing enables the provision of tailored services over a shared physical network infrastructure. In this context, machine learning (ML) offers a promising solution for implementing intelligent network management and slice orchestration.

4.6 Resource utilization

The 6G Internet of Everything (IoE) ecosystem is replacing the current smartphone-centric communication ecosystem by integrating a variety of devices, including drones, linked cars, and sensors. The data traffic is expected to increase exponentially as a result of this change. 6G networks will use a variety of spectrum resources, including visible light, millimeter waves, low radio frequency, and terahertz, to deliver high data rates. Using different spectrum resources efficiently for different applications/users in dynamic network contexts is difficult (Zhang & Zhu, 2020). Furthermore, despite the exploitation of high-frequency spectra such as millimeter waves and THz, the amount of radio resources available is quite restricted. To satisfy a variety of dynamic user needs, limited network resources must be used efficiently.

4.7 Energy consumption

Enhancing the energy efficiency of mobile networks can contribute to reducing operational expenses and carbon footprint for sustainable growth. The surge in data usage combined with the increasing number of small devices poses significant challenges to 6G energy consumption. Furthermore, the proliferation of end-user applications and complexity of transceiver processing may lead to higher energy demands. Various approaches, including traffic offloading and dynamic cell management, can improve energy efficiency (Zhang et al., 2019).

5. ENABLING KEY TECHNOLOGIES FOR 6G

The emergence of a new generation of communication systems is characterized by the number of crucial technologies involved. The implementation of groundbreaking technology in everyday life typically requires a decade or more. Consequently, innovative technologies that constitute 6G are currently in the research phase. In line with the "six" theme associated with 6G, it is expected that six technological advancements will contribute to the system's development: (i) AI/ML-driven design and optimization of the air interface; (ii) exploration of new spectrum bands and adoption of innovative cognitive spectrum sharing methods; (iii) integration of sensing and localization capabilities into the system's core definition; (iv) addressing strict latency and reliability demands; (v) embracing novel network architecture paradigms involving sub-networks and RAN-Core convergence; and (vi) implementation of new security and privacy approaches (Viswanathan & Mogensen, 2020).

6. THE EXPANDING ROLE OF ARTIFICIAL INTELLIGENCE AND MACHINE LEARNING

In the last decade, artificial intelligence (AI) and machine learning (ML) techniques, especially deep learning, have experienced rapid growth and have become crucial in various image classification and computer vision applications, ranging from social media to security systems. These technologies are employed in areas where substantial training data are readily available. Reinforcement learning, which has demonstrated its efficacy in gaming scenarios, such as AlphaGo, is now being applied to diverse robotic control applications. Recently, deep-learning approaches have been extensively investigated in the context of wireless networks. AI/ML is expected to be utilized in 5G systems in at least three distinct ways in the coming years. First, they can replace several model-based Layer 1 and Layer 2 algorithms, such as channel estimation, preamble detection, equalization, and user scheduling, owing to their superior performance or simplicity. Second, they are likely to be widely used in deployment optimization, including determining the optimal set of beams to illuminate the coverage area while considering cell traffic patterns (Viswanathan & Mogensen, 2020). Given the numerous parameters that require adjustment during 5G system deployment, AI/ML approaches will be essential for achieving zero human touch network optimization. Lastly, additional use cases, such as end-device localization via 5G technology, will likely leverage learning strategies to attain higher precision. Beyond its application in radio access networks (RAN), AI/ML will play a crucial role in 5G end-to-end network automation, managing the complexity of orchestration across multiple network domains and layers. This will significantly reduce operational costs, while enabling rapid service deployment, quick fault resolution, and dynamic adaptation of network and cloud resources in response to changing demands.

7. CONSEQUENCES OF THE DEPLOYMENT OF AI/ML TECHNIQUES IN NETWORK COMMUNICATIONS

Security- and privacy-related problems in wireless networks are major concerns. Communication networks and services are frequently subjected to denial-of-service (DoS) attacks. ML algorithms are used to identify any abnormalities in network behavior and service delivery to safeguard network systems and services. Because training data are used to train machine learning algorithms, it is necessary to ensure the secure use of the obtained data in order to improve privacy protocols. Blockchain and quantum computing are two examples of technologies used to provide protection against cyber-attacks. The creation of false identities and data leaks can compromise the transaction validation and violate privacy. There are a few essential components for solving possible network communication problems. These are categorized as follows.

7.1 Security concerns

The increasing global popularity of IoE has led to a steady increase in the demand for IoE. The critical issue now that IoT technologies such as smart items, big data analytics(BDA), and communication capabilities are integrated is how to maintain security in a situation of this size. The exciting prospects of 6G make people eager for it. But there are a lot of technological requirements and obstacles that must be overcome to make this vision a reality. In the 6G era, it is more challenging to identify potential security concerns because of massive traffic and data explosion (De Santos & Villalonga, 2015). Three important considerations for IoT devices and services are data security, privacy, and trust, because the data produced by smart objects and IoT users can be accessed across the network (Stachniak, 2015).

7.2 Cybersecurity concerns

With the advent of the 5G era and the impending advent of the 6G era, network security challenges particularly demand consideration. IoE enables smart cities and brings about revolutionary transformations in the urban infrastructure. Numerous sensors and cameras are used to monitor a city's pipeline network, power, energy, transportation, and other infrastructures. The network will also enable the intelligent control of these systems. It also poses a risk to the hacker's vision, as once they gain access to the camera, thieves can use it to steal information, such as banknote transporters, and obtain it unlawfully. IoE's basic technologies—big data, artificial intelligence, and the Internet of vehicles—are also the foundation of cyber security technology. Information security alone is no longer sufficient to define cybersecurity (Ji et al., 2021).

7.3 Privacy and trust concerns

Industrial networks are susceptible to jamming; hence, they must be guarded. Physical security will not suffice, as attackers may try to disrupt networks outside the industrial complex. In the future, jamming may also delay packet delivery by occasionally causing interference. This could have a negative effect on the industrial processes that depend on time-sensitive networks. 6G networks were built to fend against these new dangers. Diverse novel approaches to signal processing are being developed (Guzman et al., 2019) to handle security in future mixed-reality environments, which will be a crucial component of the 6G network. Physical layer security procedures (Mukherjee et al., 2014; (Trappe, 2015) may further develop in the context of 6G, tackling novel problems such as jamming. In general, these mechanisms depend on the uniqueness of the wireless channel to provide authentication, confidentiality, and key exchange. Network trust will be essential for the success of 6G in the near future.

8. CONCLUSION

The growing demand for mobile communication is expected to be met cost-effectively through AI-enhanced communication protocols, signaling, and data processing implemented on general-purpose hardware in the wireless network infrastructure. This shift is anticipated to significantly reduce the need for extensive physical and higher-layer standardization efforts, potentially ushering in an AI-driven era for the communication industry from 6G onwards. In recent years, there has been a rapid increase in the demand for wireless connectivity. The progression from 4G (LTE standard) to 5G (IoT network) has led to the proposal of 6G, which is set as the new era of Artificial Intelligence, with implementation planned between 2027 and 2030. Achieving 6G requires overcoming the limitations of 5G to establish an intelligent information society. These wireless communication networks share a common goal of simplifying various aspects of life, including industrial, educational, healthcare, and telecommunications sectors, which are crucial in today's world. The swift progress of these technologies is expected to revolutionize daily life, making them more accessible, efficient, and straightforward for individuals and the overall society. Therefore, the proliferation of the latest technologies will further strengthen the overall network communication infrastructure.

REFERENCES

Calvanese Strinati, E., Barbarossa, S., Cassiau, N., Maret, L., Dehos, C., Ktenas, D., & Gonzalez-Jimenez, J. L. (2019). 6G: The Next Frontier: From Holographic Messaging to Artificial Intelligence Using Subterahertz and Visible Light Communication. *IEEE Vehicular Technology Magazine*, 14(3), 42–50. DOI: 10.1109/MVT.2019.2921162

Chowdhury, M. Z., Jang, Y. M., Ahmed, S., & Shahjalal, M. (2020). 6G Wireless Communication Systems: Applications, Requirements, Technologies, Challenges, and Research Directions. *IEEE Open Journal of the Communications Society*, 1, 957–975. DOI: 10.1109/OJCOMS.2020.3010270

Cousik, T., Kleine, K., Shafin, R., Zhou, Z., Reed, J., & Liu, L. (2019). *CogRF: A New Frontier for Machine Learning and Artificial Intelligence for 6G RF Systems*. cornell university. https://doi.org//arxiv.1909.06862DOI: 10.48550

De Guzman, J. A., Thilakarathna, K., & Seneviratne, A. (2019). A first look into privacy leakage in 3d mixed reality data. In *Computer Security–ESORICS 2019: 24th European Symposium on Research in Computer Security, Luxembourg, September 23–27, 2019* [Springer International Publishing.]. *Proceedings*, 24(Part I), 149–169.

Farooq, M., Ahmed, M. I., & Al, U. M. (2013). Future generations of mobile communication networks. *Academy of Contemporary Research Journal*, 2(1), 24–30.

Farsad, N., & Goldsmith, A. (2017). Detection algorithms for communication systems using deep learning. *arXiv preprint arXiv:1705.08044*.

Giordani, M., Mezzavilla, M., Polese, M., Zorzi, M., & Rangan, S. (2020). Toward 6G Networks: Use Cases and Technologies. *IEEE Communications Magazine*, 58(3), 55–61. DOI: 10.1109/MCOM.001.1900411

Guo, W. (2020). Explainable Artificial Intelligence for 6G: Improving Trust between Human and Machine. *IEEE Communications Magazine*, 58(6), 39–45. DOI: 10.1109/MCOM.001.2000050

Gupta, A., & Jha, R. K. (2015). A Survey of 5G Network: Architecture and Emerging Technologies. *IEEE Access : Practical Innovations, Open Solutions*, 3(3), 1206–1232. DOI: 10.1109/ACCESS.2015.2461602

Hassan, R., Ahmed, A. S., Qamar, F., Aman, A. H. M., & Hasan, M. K. (2020). Internet of Things and Its Applications: A Comprehensive Survey. *Symmetry*, 12(10), 1674. DOI: 10.3390/sym12101674

Ibrahim, J. (2002). 4G Features. *Bechtel Telecommunications Technical Journal*, 1(1), 11–14.

Ji, B., Wen, H., Menon, V. G., Wang, Y., Song, K., Li, C., & Mumtaz, S. (2021). A Survey of Computational Intelligence for 6G: Key Technologies, Applications and Trends. *IEEE Transactions on Industrial Informatics*, 17(10), 7145–7154. DOI: 10.1109/TII.2021.3052531

Kalra, B., & Chauhan, D. K. (2014). A comparative study of mobile wireless communication network: 1G to 5G. *International Journal of Computer Science and Information Technology Research*, 2(3), 430–433.

Kumar, A., Liu, D. Y., & Sengupta, D. J. (2010). Divya4,"Evolution of Mobile Wireless Communication Networks: 1G to 4G", Vol. 1, Issue 1.

Laricchia, F. (2023, March 10). *Number of mobile devices worldwide 2020-2025*. Statista. https://www.statista.com/statistics/245501/multiple-mobile-device-ownership-worldwide/

Lu, H., Chen, M., Kim, H., Li, Y., & Serikawa, S. (2017). Brain Intelligence: Go beyond Artificial Intelligence. *Mobile Networks and Applications*, 23(2), 368–375. DOI: 10.1007/s11036-017-0932-8

Mukherjee, A., Swindlehurst, A. L., Huang, J., & Fakoorian, S. A. A. (2014). Principles of Physical Layer Security in Multiuser Wireless Networks: A Survey. *IEEE Communications Surveys and Tutorials*, 16(3), 1550–1573. DOI: 10.1109/SURV.2014.012314.00178

Nieto De Santos, F. J., & Villalonga, S. G. (2015). *Exploiting Local Clouds in the Internet of Everything Environment*. institute of electrical electronics engineers. DOI: 10.1109/PDP.2015.117

O'Shea, T., & Hoydis, J. (2017). An Introduction to Deep Learning for the Physical Layer. *IEEE Transactions on Cognitive Communications and Networking*, 3(4), 563–575. DOI: 10.1109/TCCN.2017.2758370

Oughton, E., Frias, Z., Russell, T., Sicker, D., & Cleevely, D. D. (2018). Towards 5G: Scenario-based assessment of the future supply and demand for mobile telecommunications infrastructure. *Technological Forecasting and Social Change*, 133, 141–155. DOI: 10.1016/j.techfore.2018.03.016

Saad, W., Chen, M., & Bennis, M. (2019). A Vision of 6G Wireless Systems: Applications, Trends, Technologies, and Open Research Problems. *IEEE Network*, 34(3), 134–142. DOI: 10.1109/MNET.001.1900287

Sah, H. N. (2017). A Brief History of Mobile Generations and Satellite Wireless Communication System. [JETIR]. *Journal of Emerging Technologies and Innovative Research*, 4, 1211–1216.

Salahdine, F., Zhang, N., & Han, T. (2023). 5G, 6G, and Beyond: Recent advances and future challenges. *Annales des Télécommunications*, 78(9–10), 525–549. DOI: 10.1007/s12243-022-00938-3

Sharma, P. (2013). Evolution of mobile wireless communication networks-1G to 5G as well as future prospective of next generation communication network. *International Journal of Computer Science and Mobile Computing*, 2(8), 47–53.

Soldani, D., & Manzalini, A. (2015). Horizon 2020 and Beyond: On the 5G Operating System for a True Digital Society. *IEEE Vehicular Technology Magazine*, 10(1), 32–42. DOI: 10.1109/MVT.2014.2380581

Stachniak, Z. (2015). old DOI doi - Clones: Soviet Computer Hobby Movement of the 1980s. *IEEE Annals of the History of Computing*, 1. https://doi.org/DOI: 10.1109/MAHC.2015.43 - good

Trappe, W. (2015). The challenges facing physical layer security. *IEEE Communications Magazine.IEEE Communications Magazine*, 53(6), 16–20. DOI: 10.1109/MCOM.2015.7120011

Viswanathan, H., & Mogensen, P. E. (2020). Communications in the 6G Era. *IEEE Access : Practical Innovations, Open Solutions*, 8, 57063–57074. DOI: 10.1109/ACCESS.2020.2981745

Wang, T., Wen, C.-K., Gao, F., Jiang, T., Jin, S., & Wang, H. (2017). Deep learning for wireless physical layer: Opportunities and challenges. *China Communications*, 14(11), 92–111. DOI: 10.1109/CC.2017.8233654

Wu, J., Li, J., Guan, Z., Dong, M., & Ota, K. (2018). Big Data Analysis-Based Secure Cluster Management for Optimized Control Plane in Software-Defined Networks. *IEEE Transactions on Network and Service Management*, 15(1), 27–38. DOI: 10.1109/TNSM.2018.2799000

Yang, H., Wu, K., Zhao, J., Niyato, D., Alphones, A., & Xiong, Z. (2019). Artificial-Intelligence-Enabled Intelligent 6G Networks. *IEEE Network*, 34(6), 272–280. DOI: 10.1109/MNET.011.2000195

Zhang, L., Liang, Y.-C., & Niyato, D. (2019). 6G Visions: Mobile ultra-broadband, super internet-of-things, and artificial intelligence. *China Communications*, 16(8), 1–14. DOI: 10.23919/JCC.2019.08.001

Zhang, S., Xu, S., & Xiang, C. (2020). 6G: Connecting Everything by 1000 Times Price Reduction. *IEEE Open Journal of Vehicular Technology*, 1, 107–115. DOI: 10.1109/OJVT.2020.2980003

Zhang, S., Zhou, W., Cai, X., & Wang, Y. (2019). Green 5G enabling technologies: An overview. *IET Communications*, 13(2), 135–143. DOI: 10.1049/iet-com.2018.5448

Zhang, S., & Zhu, D. (2020). Towards artificial intelligence enabled 6G: State of the art, challenges, and opportunities. *Computer Networks*, 183, 107556. DOI: 10.1016/j.comnet.2020.107556

Zhang, Z., Lei, X., Karagiannidis, G. K., Xiao, M., Xiao, Y., Ding, Z., Fan, P., & Ma, Z. (2019). 6G Wireless Networks: Vision, Requirements, Architecture, and Key Technologies. *IEEE Vehicular Technology Magazine*, 14(3), 28–41. DOI: 10.1109/MVT.2019.2921208

Chapter 10
Exploring the Impact of Artificial Intelligence on the Tourism and Hospitality Industry: A Comprehensive Analysis

Swati Sharma
https://orcid.org/0000-0003-2949-2363
Amity University, Noida, India

Deepti Bhatt
https://orcid.org/0000-0002-4904-4886
Amity University, Noida, India

Narendra Kumar
https://orcid.org/0000-0002-3325-3448
Amity University, Noida, India

Levent Karadağ
https://orcid.org/0000-0002-0932-8315
Muğla Sıtkı Koçman University, Turkey

ABSTRACT

This study explores the effect of Chatbots & Virtual Assistants (CVA) on Guest satisfaction (GS). Although several studies have been published that investigate guest satisfaction in Jordan, only a few have reported on how AI influences Guest satisfaction in Delhi NCR. Therefore, this study addresses the gap. It uses a structured questionnaire to collect data from 155 employees working in five-star hotels. The

DOI: 10.4018/979-8-3693-6725-4.ch010

data is analysed using structure equation modelling (SEM). Smart-PLS is applied to examine the relationship between the independent and dependent factors- Chatbots & Virtual Assistants, Fraud Detection & Security, Operational Efficiency, Personalised Experience, Predictive Maintenance, Revenue Management. This study finds that Fraud Detection & Security significantly impacts Guest satisfaction of employees. These findings can assist five-star hotels in devising effective AI strategies that maintain guest satisfaction, ultimately ensuring positive financial implications for them.

INTRODUCTION

The emergence of the Artificial Intelligence in the field of hotel and tourism is increasing drastically and leaving a footprint around the globe Tailored advice filtered via huge amount of data using systems that are supported by artificial intelligence which also include past former practices, contemporary guest delight bank, options and huge range of choices provided to guest regarding their stay, travel itinerary, packages etc. Not only this infact the dialog box which are also navigated by Artificial intelligence progressively run to manage guest bookings, complaints, issue handling and also source of customer service automation. Communications with guest/client via these robots makes guest happier as they respond quickly, and available for guest services 24*7*365 days which enhances guest pleasure and make daily operations faster. Along with this these machine learning helps in handling room revenue tactics, strategies to improve sales of inventory available in the hotel.

For improving guest experiences hotel rooms are equipped with virtual assistant as Siri, Alexa as they are a type of conversational AI that use machine learning and natural language processing to mimic human conversations. With the influx of new technologies and the rise of sharing economy, people are now comfortable with embracing new experiences and enjoy increased convenience inspite of its cost. Let's take an example of Hennna Hotel Japan which are no way lesser than using robot assistance in in-room cleaning, luggage delivery at guestroom, and other travel services, room service, check-in and check –out. This research chapter depicts at how AI could be used or what effects it might have on tourism or lodging property. It does by focusing on few case studies which displays about the usage and importance of AI in both industries and also explore the variables such as considered construct as AI, PGE, CVA, RM, OPE, FDS, PM on GS in 5 star deluxe accommodation in five star luxurious hotel in Delhi -NCR.

Figure 1. Artificial Intelligence Progression

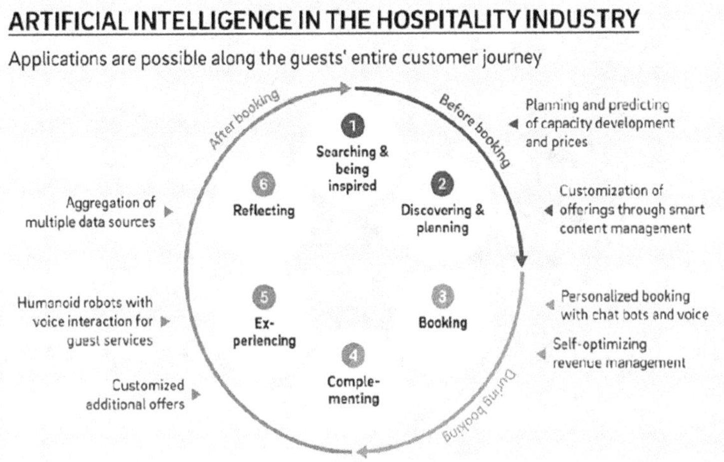

(Sources:Article from Eco Buzz)

HOTELS USING AI IN OPERATIONS

Case Study 1: Marriott International

A five star luxurious chain Marriott International, which is one of the largest chain, has begun the use of AI to improve guest experience and speed up processes. Witnessing its functions and guest preferences even Marriott international has inculcated the usage of AI among all the brands across the globe. By just clicking the Mobile app and browsing its website, Marriott is excelled in providing personalized suggestions and quick assistance to its clients. The story doesn't end here, Marriott is not lagging behind enhancing yield statistics and RevPAR, Rate spread, rate achievement factor by using AI powered information. Depending on market segment, historical data, forecasting forms Marriott modifies hotel rack rate which also increase profitability and revenue of the industry.

Table 1. Personalized Recommendations at Marriott International

Guest Preferences	AI-Powered Recommendations
Location	Nearby attractions and restaurants
Room preferences	Room types and amenities
Dining preferences	Restaurant recommendations

Figure 2. Robot Assistant at the Marriott International Reception

Case Study 2: Hilton Worldwide

Another Big player in the lodging industry, Hilton hotel and resorts, relies on AI technology to enhance customer experience and drive business growth. All of the hotels and resorts, in the company's worldwide portfolio are priced and managed with revenue optimization to AI driven analytics. Hilton maximizes profit by optimizing room price based on market trends, demand pattern and market pricing.

Hilton has implemented Chatbots equipped with AI to enhance customer support and optimize efficiency. Hilton provide guest with immediate support and tailored suggestions through its mobile application and websites, hence boosting the overall experience for guest.

Table 2. Chatbot Interactions at Hilton Worldwide

Customer Inquiry	AI-Powered Response
Room availability	Instant room booking
Restaurant inquiry	Dining reservations
Local attractions	Recommendations and directions

Figure 3. Chatbots Interactions at Hilton Worldwide

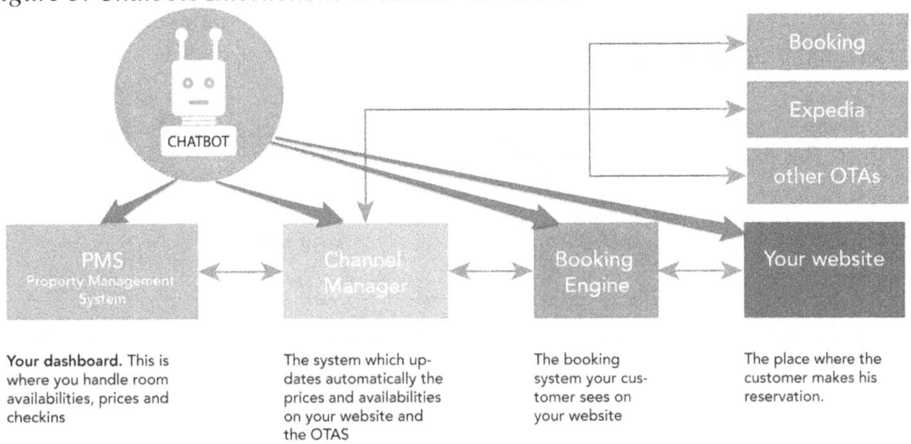

(Source: Chatbots magazine)

Figure 4. Significance of Chatbots and AI in Tourism

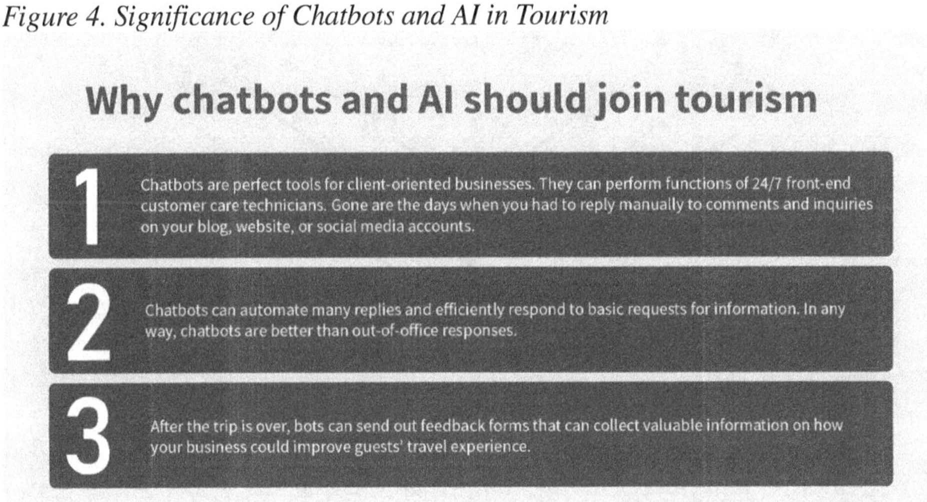

(Source: Qulix.com)

Case Study 3: Airbnb

Airbnb uses AI to improve quality of search result and personalize its services to each customer. It is an industry leader in online housing and travel marketplaces. By analyzing a user's behavior, preferences and search history, the company AI algorithm are able to deliver personalized recommendations for hotel and activities. Airbnb increases user engagement and happiness, which drive business growth by understanding user preferences and habits.

Table 3. Predictive Analytics at Airbnb

User Behavior	AI-Driven Recommendations
Search History	Accommodation suggestions based on previous searches
Preferences	Recommendations tailored to user preferences
Market Trends	Dynamic pricing based on demand and availability

Figure 5. Airbnb Connecting Host to Visitors

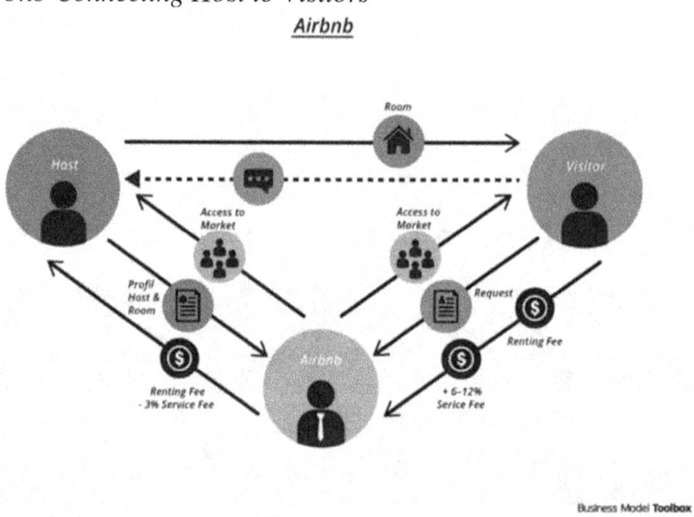

Case Study 4: Wynn Las Vegas

Voice command systems like Google Assistant and Amazon's Alexa are being used by more hotels to provide information and services that can be used in the rooms. These virtual record, order, monitor or process every activity that occur with the guest. An Intelligent virtual assistant gives recommendations to the guest regarding their travel itinerary, vacation ideas and deals and also lowers the need of travel agent to make personalized recommendations to the guest in Wynn Las Vegas. Not only can this it does enable customer access to any information like guest reservation, room status, room booking without asking for login

Table 4. Virtual Assistant Services at Wynn Las Vegas

Guest Requests	**Virtual Assistant Response**
Room Service	Food and beverage orders
Housekeeping	Request cleaning services
Concierge Services	Recommendations and reservations

Figure 6. Voice Based AI Digital Assistant

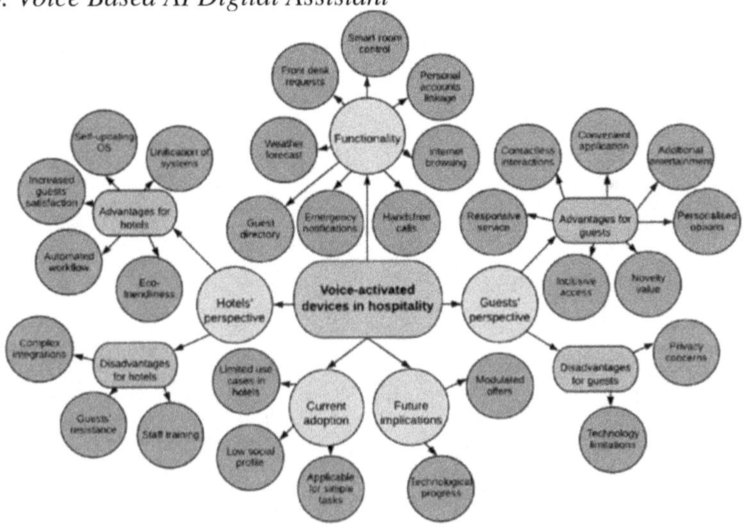

(Source: Buhalis & Moldavska 2021)

Case Study 5: Yotel Singapore

The trend of deploying robots within the premises of hotels and resorts continues to grow lately since they help in tidying up of areas, lugging of luggages, and offering concierge services. Such service robots provide a futuristic and entertaining service to guests while enhancing efficiency of operations and minimizing costs on labor. Yotel Singapore has availed robotic concierge and service robots to complement their customer service. Apart from human front desk staff, the Hotel's service robots, called YO2D2 bring amenities to units, guide to hotel facilities and assist during check in and open during check out.

Figure 7. Robotic Room Service in Hi-Tech Yotel Singapore

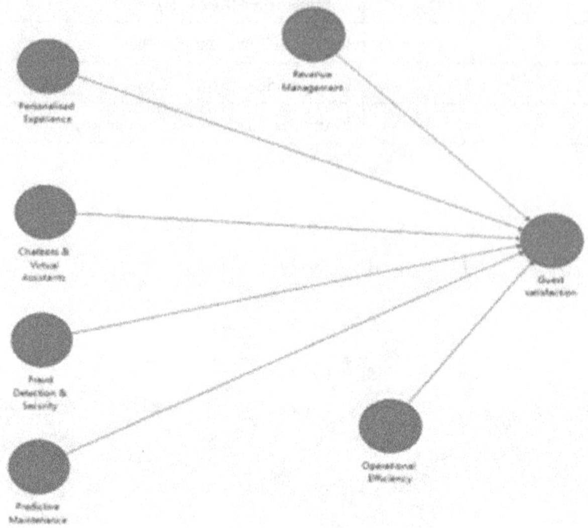

(Sources: Kaizenaire.com)

Table 5. Service Robots at Yotel Singapore

Robot Tasks	Robot Functionality
Room Delivery	Delivering amenities to guest rooms
Guest Assistance	Providing directions and information
Check-in/Check-out	Assisting with registration process

LITERATURE REVIEW

In Table 1, several important studies and papers concerning AI in the field of tourism and hospitality management are highlighted chronologically which indicates the development of research in this area for the recent years. Each study contributes to our understanding of the applications, impact, and future directions of AI in the tourism and hospitality sector.

Table 6. Significant Scholarly Contribution on AI

Year	Study/Research Title	Summary/Key Findings
2015	"The Role of Artificial Intelligence in Tourism" (Dangwal et al.)	Explores the potential applications of AI in various areas of tourism, including recommendation systems, customer service, and predictive analytics.
2016	"Applying Artificial Intelligence Techniques in Tourism" (Li)	Examines the use of AI techniques such as neural networks and genetic algorithms in tourism-related tasks such as route planning and demand forecasting.
2017	"Artificial Intelligence and Robotics in the Hospitality Industry" (Almeida et al.)	Explores the implementation of artificial intelligence (AI) and robotics in the hospitality sector, examining the possible advantages, obstacles, additing forthcoming developments.
2018	"Personalization in Tourism: A Comprehensive Review" (Gretzel et al.)	Reviews various approaches to personalization in tourism, including AI-driven recommendation systems and personalized marketing strategies.
2018	"Artificial intelligence in smart tourism: A conceptual framework." (Rua & Chih)	Based on the review Emerging technologies such as Artificial Intelligence, Internet of Thing, Robotic, Cyber Security, 3D printer and Block chain also accelerate the development of industry toward digital transformation trend such as Fintech, e-commerce, smart cities, smart tourism, smart healthcare, smart manufacturing
2019	"The Impact of Artificial Intelligence on Tourism and Hospitality: A Review" (Zhang et al.)	Provides an overview of The influence of artificial intelligence (AI) on many areas within the tourist and hospitality business, highlighting key trends and future prospects.
2020	"AI Applications in Hospitality: A Review of Current Literature and Future Directions" (Buhalis et al.)	Analyzes recent research on AI applications in hospitality, identifying emerging trends and proposing future research directions.
2021	"Emerging Trends in AI and Robotics for Tourism and Hospitality" (Xu et al.)	Examines emerging trends in AI and robotics for tourism and hospitality, including the use of chatbots, virtual assistants, and service robots.
2021	"Understanding consumers' acceptance of automated technologies in service encounters: Drivers of digital voice assistants adoption" (Fernandez, & Oliveira)	From the study,findings were digital voice assistant is mainly functional due to its usefulness in hotel industry say purchasing items, setting reminders,restaurant reservation, salon appointment which make a guest,able to develop human like conversation, personal touch of human interactions.
2021	"Techno-business strategies for enhancing guest experience in luxury hotels: a managerial perspective" (Bharwani et al., 2021)	The review study has suggested that Chatbots are artificial intelligence-driven conversational chatbots capable of managing client inquiries, requests, and complaints .These automated systems can be seamlessly incorporated into a hotel's website or mobile application, offering guests a prompt and effective means of obtaining information and utilising services.
2021	The implementation of business intelligence using data analytics and its effects towards performance in hotel industry in Thailand (Tong et al.)	The qualitative study finds that Hotels are utilising AI to anticipate potential maintenance needs for equipment and systems,video surveillance, face recognition, access control, threat detection, and fraud detection are a few ways that artificial intelligence is being utilised to improve hotel security.

continued on following page

Table 6. Continued

Year	Study/Research Title	Summary/Key Findings
2022	"AI-Driven Innovations in Tourism and Hospitality: A Systematic Review" (Chen et al.)	Conducts a systematic review of AI-driven innovations in tourism and hospitality, synthesizing findings from various studies and identifying gaps for future research.
2022	"Voice assistants in hospitality: using artificial intelligence for customer service" (Buhalis, D., & Moldavska, I. (2022)	The study found that voice assistant technology helps hotels to improve customer service, expand operational capability and reduce costs. It has helped in achieving efficiency towards optimizing hotel operation. The respondents said that using voice assistant while travelling when navigating the car or overall, when controlling smartphone in the car.
2022	"Artificial intelligence and life in 2030: the one hundred year study on artificial intelligence" (Stone et al)	The usage of AI would be for wealth creation and shifting towards building intelligent systems that can collaborate effectively with people, including creative and interactive way to develop scalable and interactive way for people to teach robots.
2023	"Big Data in the Hospitality Industry: Prospects, Obstacles, and Strategies." (Mnyakin et al.)	Based on the review the findings suggest that Hotels can enhance their revenue and pricing administration through the utilisation of AI, which analyses data pertaining to room occupancy, demand. Personalise the guest experience customised Voice aids, statistical analysis, incentive schemes, and accommodation configurations.
2024	"Navigating the AI horizon in hospitality: a novel classification and future research agenda" (Pan & Fu)	AI become more integrated in various aspects of our lives like Chatbots, Chatgpt, service robots, ML algorithm. This can be linked to each dimension by assessing how ethical concerns such as privacy, misuse, bias vary with mechanical thinking.

RESEARCH METHODOLOGY

The current study employed a quantitative methodology as it was deemed the most appropriate approach for collecting primary data from a large population. By drawing upon prior research, the investigators constructed a questionnaire employing a Linkert 5-point scale. This segment comprised statements pertaining to the dependent variable, guest satisfaction, and the study sub-variables (OPE, FDS, PGE, CVA, RM, and PM). The questionnaire was shown to a group of expertise in the areas in order to settle the dispute. After the Delphi process (expert checking the questionnaire) the questionnaire was adapted and fixed based on what they thought. This made the items on the questionnaire more reliable and consistent, which was the main goal of the study. To sort and study the original data, SmartPLS software was used. Cronbach Alpha was used to check how reliable the study was. It was okay because the alpha value for each variable was better than the lowest allowed percentage of 0.70. Table 1 below shows that the study tool was reliable and consistent (Hair et al., 2022). Before disseminating the questionnaire among large number of respondents, feasibility study has been done to examine the understanding of the questionnaire among the small number (40 number of respondents).Cronbach Alpha's value i.e. value of reliability was found more than 0.70 hence it gives us green signal for sharing our questionnaire among large number of respondents. Even during the stage of Pilot Study items were also removed which have shown less threshold value of reliability and respondents were unable to understand (Al-Emran, Mezhuyev, & Kamaludin, 2019).

With the advent of 5G, Internet of things and Wifi 6 are transforming, data analysis and artificial intelligence applications are revolutionizing the hotel industry. This allows hotel to enhance guest experience with tailored offerings. Robotic application while providing services is a part of "Artificial Super Intelligence". 6 E technology heightens the use the artificial intelligence in hospitality industry by ensuring quicker trustworthy and intelligent services. From Personalised guest services to automatic cleaning, maintenance, tour guiding robots equipped with 6G connectivity and advanced AI will enhance the guest experience in hotels and tourism destinations. Tourism experience will be highly enhanced with AI enabled robots acting as tour guides in Museum, Historical sites, outdoor adventure. With 6 G fast power and speed robot will be able to provide real time, immersive experiences with virtual and augmented reality i.e. VR and AR.For eg: A robot guide at a historical site can create a live Virtual reality based ancient times allowing tourist to engage in an interactive learning experiences. This process will fasten robot to translate in local language catering to global audience in no less time and provide information in their preferred language.

Supported with 6G connectivity, hotel and tourism operators will be able to gather real time data on guest behaviour, competitor pricing, market conditions. Artificial Intelligence supported with high speed and low latency capabilities of 6G will instantly analyze data to adjust pricing dynamically. Hotel will also enable personalized pricing discount based on guest loyalty.AI systems can continually update demand forecast as new data comes in.AI enabled 6G can deliver accurate demand forecasting models, global travel patterns, competitor actions, social media signals.5G/6G enables continuous update to machine learning models used for fraud detection. These models emerged from evolving patterns of fraudulent activities adapt to new fraud tactics and help hotel business to protect via cybercrime. For example: AI-based fraud detection tools used by large hotel chains could detect anomalies like unusually large purchases or bookings made with multiple credit cards, instantly alerting staff and preventing losses.

DATA COLLECTION

A total of 25 "five-star hotels" chosen at random from a comprehensive list of hotels throughout India made up the study's sample. Information on Indian hotels is culled from the country's official tourism website (National Integrated, 2022). The survey instrument used to gather data is the questionnaire, in its email form, since it guarantees data collection in a cost-effective manner. We thus surveyed 200 LinkedIn members, including associates, managers, executives, and Assistant Manager, via email. The poll is also sent to 70 HR managers, who are then requested to pass it on to managers in different departments of the participating hospitality firms. The result is 155 completed responses, 45 of which were rejected because they did not contain enough information.

Following an extensive review of existing literature, the researcher carried out empirical research to obtain more knowledge of this study. In the following section researcher had considered construct as AI, PGE, CVA, RM, OPE, FDS, PM on GS in 5 star deluxe accommodation from attitude of employees working in five star luxurious hotel in Delhi -NCR. Based on the reviewed literature, a semi-structured questionnaire was developed and reviewed by academic and business professionals. The author has drawn a conceptual model.

Figure 8. Study Model

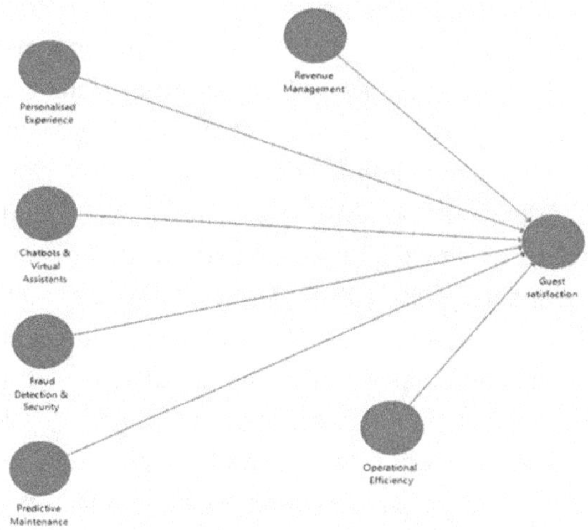

(Source: Hyari et al., 2023)

Based on the aforementioned model, the researcher formulated the following set of assumptions.

Hypothesis:

H1: PGE facilitates a better GS from perspective of managers within five stars hotels
H2: CVA facilitates a better GS from perspective of managers within five stars hotel
H3: RM facilitates a better GS within five stars hotels.
H4: OPE facilitates a better GS within five stars hotels.
H5: FDS facilitates a better GS within five stars hotels.
H6: PM facilitates a better GS within five stars hotels.

Table 7. Feasibility Study

Scale	Items	Items no.	Cronbach's Alpha
Chatbots & Virtual Assistants	CVA1	3	0.848
	CVA2		
	CVA3		
	FDS1	3	0.713
	FDS2		
Fraud Detection & Security	FDS3		
	GS1	3	0.813
	GS2		
Guest satisfaction	GS3		
	OPE1	3	0.798
	OPE2		
Operational Efficiency	OPE3		
	PGE1	3	0.774
	PGE2		
Personalised Experience	PGE3		
	PM1	4	0.897
	PM2		
	PM3		
Predictive Maintenance	PM4		
Revenue Management	RM1	1	0.799

DATA ANALYSIS

The PLS-SEM method is applied within the scope of this research using Smart-PLS (version 3.3.2), which allows for the examination of the relationships that exist between the numerous parameters that have been stated previously. 'Chatbots & Virtual Assistants,' 'Fraud Detection & Security,' 'Operational Efficiency,' and 'Personalised Experience,' 'Predictive Maintenance,' and 'Revenue Management' are the endogenous variables that are considered in this instance. A uniform reporting mechanism is used for this purpose is utilised, with the exogenous variable being guest satisfaction. In its own right, PLS-SEM is able to assist in evaluating the direct and indirect connections that exist between a wide range of latent and observable variables. As a matter of fact, the exogenous and endogenous models of PLS-SEM simultaneously investigate the connections that exist between the various constructs and the indicators that correspond to them. While the inner model studies the correlation betwixt a numbers of different latent constructs, the outer model investigates the correlations between the latent constructs and the empirical indicators that are associated with them.

In addition, SmartPLS is effective when applied to large sample sizes, incorporates intricate models and indicators, and utilises more than one construct. Furthermore, it makes use of latent variable scores. As a result, it has the ability to provide visually appealing results (Hair e et al., 2019). It is important to note that bootstrapping is utilised in this study in order to investigate how GS is affected. AI stands for artificial intelligence. Additionally, with the assistance of SmartPLS, a two-stage disjoint procedure is utilised in order to accurately analyse all of the constructions. According to Pattusamy and Jacob (2016), the model that is based on questionnaires includes a reflective construct in which all of the items are analysed and evaluated in the reflective mode. Additionally, prior to the gathering of data, assurances are made on the authenticity and dependability of each and every construct. The significance of each of the hypotheses that were examined in the study is finally determined through the utilisation of a measurement model.

THE MEASUREMENT MODEL

The composite reliability for the constructs ranges from 0.613 to 0.897, as shown in Table 8.

Table 8. Reliability and Validity of the Construct

Scale	Items	Cronbach's Alpha	Composite Reliability	Average Variance Extracted (AVE)
Chatbots & Virtual Assistants	CVA1	0.848	0.722	0.654
	CVA2			
	CVA3			
	FDS1	0.713	0.714	0.78
	FDS2			
Fraud Detection & Security	FDS3			
	GS1	0.813	0.709	0.778
	GS2			
Guest satisfaction	GS3			
	OPE1	0.798	0.799	0.506
	OPE2			
Operational Efficiency	OPE3			
	PGE1	0.774	0.772	0.535
	PGE2			
Personalised Experience	PGE3			
	PM1	0.897	0.897	0.685
	PM2			
	PM3			
Predictive Maintenance	PM4			
	RM1	0.799	0.784	0.569
	RM2			
Revenue Management	RM3			

(Source: Author's own work)

CV-Chatbots Virtual Assistant, FDS-Fraud Detection and Security, GS-Guest satisfaction, OPE-Operational Efficiency, PGE-Personalized guest experience, PM-Predictive Maintenance, RM-Revenue Management.

When dealing with issues of discriminant and convergent validity, confirmatory factor analysis (CFA) is employed. CFA is employed to evaluate the validity of convergent and discriminant relationships. Furthermore, the study's validity is ensured through the application of structural equation modelling, which is specifically designed to verify the study's hypotheses. Additionally, the methods used in the two-step disjoint approach are employed for the purpose of data analysis by way of evaluation of the "measurement and structural models" (Hair et al., 2017)." It has the potential to make guests' experiences easier and more simplified" and "Chatbots and virtual assistants can lead to increased guest satisfaction" are two examples of the weakly valued items that were subsequently eliminated. Additionally, the range of values for the higher-order construct dependability is 0.714 to 0.897. The relevant constructs in the final model have a mean-variance value of 0.5, which indicates convergent validity, as shown in Table 8.

CRONBACH ALPHA VALUES

Chatbots and Virtual Assistants have a score of 0.848, fraud detection and security have a Cronbach alpha value of 0.713, and guest satisfaction has a rating of 0.813.The Operational Efficiency construct displayed a Cronbach alpha value of 0.798, indicating high construct reliability. The Revenue Management construct had a value of 0.799, the Predictive Maintenance construct had a value of 0.897, and the Personalised Experience construct had a value of 0.774. All of these values are greater than the threshold of 0.70, further confirming its reliability. Refer Table 8.

Table 9. Discriminant Validity (HTMT)

	Chatbots & Virtual Assistants	**Fraud Detection & Security**	**Guest satisfaction**	**Operational Efficiency**	**Personalised Experience**	**Predictive Maintenance**	**Revenue Management**
Chatbots & Virtual Assistants	0.809						
Fraud Detection & Security	0.663	0.777					
Guest satisfaction	0.506	0.745	0.767				
Operational Efficiency	0.772	0.827	0.715	0.712			
Personalised Experience	0.477	0.533	0.584	0.558	0.731		
Predictive Maintenance	0.646	0.822	0.850	0.806	0.603	0.828	
Revenue Management	-0.351	-0.246	-0.15	-0.11	0.018	-0.273	0.754

(Source: Author's own work)

CV-Chatbots Virtual Assistant, FDS-Fraud Detection and Security, GS-Guest satisfaction, OPE-Operational Efficiency, PGE-Personalized guest experience, PM-Predictive Maintenance, RM-Revenue Management.

In Table 8, The heterotrait-monotrait HTMT ratio is employed to assess discriminant validity (Henseler, 2015).Table 8 represent the HTMT ratio result. The heterotrait-monotrait (HTMT) ratio is a statistical measure used in the context of assessing discriminant validity in research involving latent variables or constructs. Heterotrait Monotrait (HTMT) values are found within the threshold values of 0.85. With values of 0.809 for chatbots and virtual assistants, 0.777 for fraud detection and security, 0.767 for guest satisfaction, 0.712 for operational efficiency, 0.731 for personalised experience, 0.828 for predictive maintenance, and 0.754 for revenue management as indicated in table 9.

Table 10. Cross- Loading of all construct (CLV)

	Chatbots and Virtual Assistant	**Fraud Detection and Security**	**Guest satisfaction**	**Operational Efficiency**	**Personalized guest experience**	**Predictive Maintenance**	**Revenue Management**
CV1	**0.882**	0.522	0.352	0.533	0.264	0.461	-0.349
CV2	**0.841**	0.492	0.381	0.601	0.455	0.508	-0.171
CV3	**0.903**	0.533	0.435	0.569	0.312	0.516	-0.28
FDS1	0.538	**0.934**	0.668	0.679	0.418	0.779	-0.237
FDS2	0.538	**0.924**	0.649	0.674	0.407	0.819	-0.167
FDS3	0.558	**0.91**	0.565	0.632	0.434	0.736	-0.203
GS1	0.438	0.659	**0.916**	0.608	0.427	0.7	-0.142
GS2	0.379	0.627	**0.919**	0.564	0.466	0.749	-0.139
GS3	0.422	0.604	**0.935**	0.541	0.481	0.69	-0.087
OPE1	0.443	0.436	0.326	**0.663**	0.257	0.385	-0.046
OPE2	0.505	0.697	0.546	**0.839**	0.354	0.662	-0.127
OPE3	0.537	0.61	0.522	**0.848**	0.393	0.531	-0.046
OPE4	0.558	0.494	0.519	**0.791**	0.392	0.563	-0.066
PGE1	0.306	0.443	0.462	0.404	**0.857**	0.485	0.025
PGE2	0.267	0.273	0.334	0.319	**0.808**	0.334	0.074
PGE3	0.396	0.386	0.418	0.381	**0.821**	0.421	-0.05
PM1	0.593	0.739	0.66	0.657	0.429	**0.879**	-0.276
PM2	0.521	0.808	0.702	0.648	0.411	**0.914**	-0.261
PM3	0.412	0.707	0.665	0.557	0.452	**0.878**	-0.152
PM4	0.455	0.694	0.671	0.555	0.482	**0.823**	-0.151
RM1	-0.324	-0.299	-0.14	-0.117	0.045	-0.288	**0.89**
RM2	-0.233	-0.16	-0.064	-0.102	0.053	-0.172	**0.826**
RM3	-0.18	-0.05	-0.106	-0.014	-0.055	-0.111	**0.799**

(Source: Author's own work)

* *CV-Chatbots Virtual Assistant, FDS-Fraud Detection and Security, GS-Guest satisfaction, OPE-Operational Efficiency, PGE-Personalized guest experience, PM-Predictive Maintenance, RM-Revenue Management.*

All of the constructions together account for more than half of the variation. Furthermore, it is expected that the item loadings will display threshold values when analysed at a specified level (Sarstedt, 2019).Table 10 displays the obtained cross-loading values (CLV), which show that there are significant loadings for the factor items. In addition, the CLV of the additional latent components is said to satisfy the cross-loading requirement, indicating a satisfactory level of discriminant validity. During cross-loadings, the researcher evaluates the numerous things to determine which have high loadings on the same construct and which load highly on many constructs.

Structural model (SM)

Contributing to 62% of the variance (R2) in Guest Satisfaction, the Structural model states that the variables of variance, effect size, and predictive relevance are significant. Also, the redundancy is determined to be greater than zero while evaluating the cross-validation Q Square for Guest satisfaction at (0.48), according to Bhatt et al. (2023) as displayed in table 11.

Table 11. The result of R square, Q square

DV	R Square	R Square Adjusted	Q Square
Guest satisfaction	0.625	0.611	0.482

Value effect size. 0.02= Small; 0.15=Medium; 0.35=Higher

Figure 9. Flowchart of R square of Construct

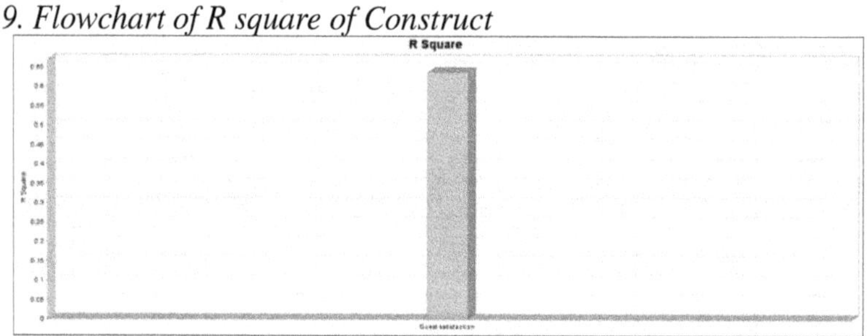

HYPOTHESIS RESULT

The results of this study's analysis show a robust correlation betwixt the Chatbots & Virtual Assistants and Guest satisfaction (β = *-0.209, p*< 0.002), Fraud Detection & Security and Guest satisfaction with (β = *.224, p*< 0.042), Operational Efficiency and Guest satisfaction (β = *0.196, p*< 0.042), Personalised Experience and Guest satisfaction ($\beta = .175, p < 0.042$).Establish a strong significant relationship between Predictive Maintenance and Guest satisfaction ($\beta = 0.195$, $p < 0.047$),whereas shows a strong significant relationship between Revenue Management and Guest satisfaction ($\beta = .146$, $p < 0.009$). (refer table 12)

Table 12. Hypothesis Testing Result

Hypotheses	Path coefficient	T Statistics	P Values	Result
Chatbots & Virtual Assistants -> Satisfaction	.209	-3.136	.002	Accepted
Fraud Detection & Security -> Satisfaction	.224	2.062	.042	Accepted
Operational Efficiency -> Satisfaction	.196	2.049	.044	Accepted
Personalised Experience -> Satisfaction	.175	2.106	.038	Accepted
Predictive Maintenance -> Satisfaction	.195	2.013	.047	Accepted
Revenue Management -> Satisfaction	.146	2.672	.009	Accepted

DISCUSSION

From the point of view of the hotel staff in Delhi-National Capital Region, this study set out to investigate how various forms of artificial intelligence (AI)—including Personalised Guests Experience, Chatbots and Virtual Assistants, Revenue Management, Operational Efficiency, Fraud Detection and Security, and Predictive Maintenance—affect the level of satisfaction felt by guests at five-star hotels in the area. The study used quantitative technique and consisted of a self-administered questionnaire filled out by 175 employees from five-star hotels in the Delhi-National Capital Region. When dealing with the main data, SEM via SMARTPLS was employed for multiple regression.

There is a change in the way clients displace themselves in the hotel, and thus stay longer and expend money. This statement has been acted on in a manner reasoned by lucanova (2019). Centering on the inside view of AI on the target customer, it appeared that its complete infiltration into the rather traditional hospitality industry could provoke a cocktail of efficient and voluntary services. Increasing customer satisfaction and loyalty through the application of AI will lead to increase revenue and growth of the hotels in general, although such benefits will vary with the level of the hotel, being most beneficial to the high end hotels.

Based on data collected on visitor's preferences and behaviour including their room and dining preferences, favourite activities and more, AI could improve the client experience at five star hotels, according to the study. Accoding to Bohme

(2023) & Mariani and Borghi (2021), hotels can enhance their customer experience by offering tailored recommendation and experiences. This finding is in line with their recommendations. Hotel operation, customer confidence, and revenue can all be protected with robust anti-fraud measures.

An additional discovery of the investigation indicated that visitors' satisfaction may be improved by Chatbots. Chatbots are automated programs that use that use artificial intelligence to interact with customers in a similar way to a human.They are deployed on messaging apps, websites, social media platforms, mobile applications and voice assistants. It can recognize, summarize, translate, predict and create content in response to a hotel guest query. This is verified by Adamopoulou & Moussiades (2020). Hotel guest can typed their queries which are answered directly or use the Chabot menu to make selections. Chatbots are conversational tools that perform their task efficiently. Hotel guest like them because they help them get through those task quickly so they can focus their attention on high level, strategic, and engaging activities that require human capabilities that cannot be replicated by machines.

It has also been asserted from the study that PM is an approach to maintain data analytics, AI, IOT (Internet of things) to monitor the condition of hotel equipments in real time. It uses sensors installed on systems and machinery throughout the hotels to collect data, which AI algorithm analyze to predict when equipments are likely to fail in hotel. As per the similar study by Arnone (2024) has highlighted that Predictive maintenance is like a sixth sense for the hotels health. It’s all about anticipating problems before they become headaches, ensuring that our guest stay is as comfortable and uninterrupted as possible.

Figure 10. Conceptual Model

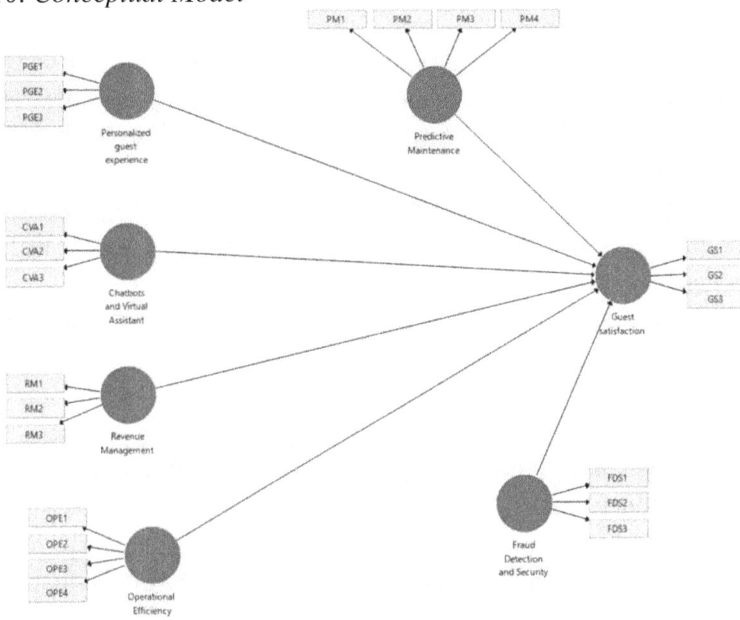

(Source: Author's own creation)

Artificial Intelligence: Difficulties and Setbacks

AI has streamlined communication and operation in Hotel and tourism industry, but also carries limitations also. The limitation is that AI algorithm don't always understand or predict what people want. Such as, AI offers cuisine or activities that does not fit with guest tastes, it can make them unhappy and irritable also. If Chatbots aren't taught or designed properly, they might also answer customer question in a way that is not relevant or correct. This lacunae has created a huge misunderstanding, slowing down problem solving, and harmed the guest expectation as well as hotel reputation. Let's take an example as guest wish in room dining happens quickly and robot doesn't understand the command and make guest again unsatisfied and annoyed. This problem makes consumer anxious that how AI collect and use a lot of customer data, which make them want to protect and keep data private. This could break the trust of the loyal and repeated guest. Even though business in hotel and tourism rely too much on AI technology. In that case guest forget about the personalized service which staff of the hotel provide and loses the real emotional connection with guest of the hotel.

Similarly using AI tools that set prices to maximize profit might not be very clear, making it hard for consumers to understand how tariffs are set for a hotel room. This will compel guest to feel confused and cheated. AI algorithms analyze historical sales data, market trends, and external factors to predict demand and adjust prices accordingly.AI forecasting systems can process large datasets and generate forecasts in real-time, providing decision-makers with up-to-date insights. Automated systems, while efficient, can create a disconnect in the customer experience.

AI forecasting tools to predict demand can sometimes get it wrong leading to too many bookings or not enough use of what's available. This can mean lost money and inefficient operations for businesses in hotels and leisure. For instance, if a hotel depends on AI to predict demand and plan staff and resources, but can't guess changes in demand, it might struggle to keep guests happy. It makes difficult for guest how revenue are set.

A digital hotel concierge with predictive AI, or predictive analysis, refers to the use of advanced data analysis and machine learning algorithms to make accurate predictions about hotel guest behaviour.This technology has the potential to revolutionize a variety of industries, including restaurants. Predictive Analysis can analyze vast amount of data to identify pattern and relationship, allowing businesses to make informed decisions based on real time information. If AI based Predictive analysis doesn't work may lead to miss the information and can lead the long list of waitlisted people. If AI does not forecast the future guest demand may lead to satisfy the repeated loyal guest.

AI applications in the travel sector are multifaceted, offering users improved information access, enhanced mobility, better decision-making, and an overall superior travel experience. Al can help you make and improve hyper-personalized ads that will bring in more money and keep customers coming back. By letting potential customers make their own packages based on their needs, budget, and ease, it could change the way the travel and hospitality industries work in the future.

CONCLUSION

Like every coin has two advantages and disadvantages similarly AI usage and implementation act as boon and curse for hospitality and tourism industry. With the help of AI, results are endless. With varied choices of customer and amalgamation of AI in the industry will help achieve the repeated guest and reach at their expectation.

This will also help in achieving dynamic pricing. As industry has been promising to access superior services to their customer at the blink of an eye.

AI has not been boon for hotel customer only infact it has reduced the work allocation of all employees working in the hotel.AI has helped in reaching great degree of satisfaction by standing automatic data processing like Machine learning, AI enabled phones and mobile apps etc. Each of these 5* hotels is committed to ensuring its guest with world class best services with amazing experience at the stay and travel. Gone are the days when you have to contact travel agent, see him as person and making a series of irritating phone calls to enquire about hotel and travel arrangements. To carve out a favorable image of the hotel and tourism industry, various technologies have been introduced to enhance guest expectations and boosting hotel revenue.

As AI continues to evolve, its impact on the industry is expected to grow, offering endless possibilities for innovation, differentiation, and competitive advantage. By embracing AI technologies, businesses can stay ahead of the curve and thrive in an increasingly digital and data-driven world.

We might see robots in the future showing hotel guests to their rooms, taking their bags, bringing them food and drinks, and cleaning, among other things. This technology is only used in a few places right now (Abdelmoaty et al., 2020).

This study only included five-star hotels in Delhi NCR, it was also restricted to managers of these hotels because they have a better grasp of the budget allocated for AI implementation and the necessary knowledge of AI-enhanced services in these types of hotels.

REFERENCES

Al-Emran, M., Mezhuyev, V., & Kamaludin, A. (2019). PLS-SEM in information systems research: a comprehensive methodological reference. In *Proceedings of the International Conference on Advanced Intelligent Systems and Informatics 2018 4* (pp. 644-653). Springer International Publishing. DOI: 10.1007/978-3-319-99010-1_59

Al-Hyari, H. S. A., Al-Smadi, H. M., & Weshah, S. R. (2023). The impact of artificial intelligence (AI) on guest satisfaction in hotel management: An empirical study of luxury hotels. *Geo Journal of Tourism and Geosites*, 48(2 supplement), 810–819. DOI: 10.30892/gtg.482spl15-1081

Batala, B., & Slevitch, L. (2024). Keeping two balls in the air: The bleisure travel experience. *Annals of Tourism Research Empirical Insights*, 5(1), 100115. DOI: 10.1016/j.annale.2023.100115

Bharwani, S., & Mathews, D. (2021). Techno-business strategies for enhancing guest experience in luxury hotels: A managerial perspective. *Worldwide Hospitality and Tourism Themes*, 13(2), 168–185. DOI: 10.1108/WHATT-09-2020-0121

Bhatt, D., Sharma, S., & Rana, V. (2023). A study of the remote work-family balance of female hotel managers. *African Journal of Hospitality, Tourism and Leisure*, 12(3), 989–1010.

Böhme, C. (2023). Shaping the Guest Experience: How Personalized Services Improve Customer Satisfaction in Luxury Hotels and Create a Competitive Advantage. *Signature (Ramsey, N.J.)*.

Buhalis, D., & Moldavska, I. (2021). In-room voice-based AI digital assistants transforming on-site hotel services and guests' experiences. In *Information and Communication Technologies in Tourism 2021:Proceedings of the ENTER 2021 eTourism Conference,January 19–22, 2021* (pp. 30-44). Springer International Publishing.

Buhalis, D., & Moldavska, I. (2022). Voice assistants in hospitality: Using artificial intelligence for customer service. *Journal of Hospitality and Tourism Technology*, 13(3), 386–403. DOI: 10.1108/JHTT-03-2021-0104

Caicedo-Barreth, A., Santos Pavón, E., & Lima Santos, L. (2020). Competitiveness of Guayaquil towards bleisure tourism. *EJTHR*, 10(2), 118–133.

Carlson Wagonlit Travel. (2016). Combing business and leisure trips: A quantitative look at the bleisure phenomenon. In CWT Solution Group White Paper. Available at https://www.mycwt.com/news/pr/carlson-wagonlit-travel-research-business-travelers-feel-50-percent-safer-at-airports-than-at-train-or-subway-stations/ [Retrieved 22 May 2020].

Dangwal, A., Kukreti, M., Angurala, M., Sarangal, R., Mehta, M., & Chauhan, P. (2023). A Review on the Role of Artificial Intelligence in Tourism. In *2023 10th International Conference on Computing for Sustainable Global Development (INDIACom)* (pp. 164-168). IEEE.

Fernandes, T., & Oliveira, E. (2021). Understanding consumers' acceptance of automated technologies in service encounters: Drivers of digital voice assistants' adoption. *Journal of Business Research*, 122, 180–191. DOI: 10.1016/j.jbusres.2020.08.058

Hair, J., & Alamer, A. (2022). Partial Least Squares Structural Equation Modeling (PLS-SEM) in second language and education research: Guidelines using an applied example. *Research Methods in Applied Linguistics*, 1(3), 100027. DOI: 10.1016/j.rmal.2022.100027

Hair, J. F.Jr, Sarstedt, M., Ringle, C. M., & Gudergan, S. P. (2017). *Advanced Issues in Partial Least Squares Structural Equation Modelling*. Sage Publications.

Henseler, J., Ringle, C. M., & Sarstedt, M. (2015). A New Criterion for Assessing Discriminant Validity in Variance-Based Structural Equation Modelling. *Journal of the Academy of Marketing Science*, 43(1), 115–135. DOI: 10.1007/s11747-014-0403-8

Lukanova, G., & Ilieva, G. (2019). Robots, artificial intelligence, and service automation in hotels. In *Robots, artificial intelligence, and service automation in travel, tourism and hospitality* (pp. 157–183). Emerald Publishing Limited. DOI: 10.1108/978-1-78756-687-320191009

Mariani, M., & Borghi, M. (2021). Customers' evaluation of mechanical artificial intelligence in hospitality services: A study using online reviews analytics. *International Journal of Contemporary Hospitality Management*, 33(11), 3956–3976. DOI: 10.1108/IJCHM-06-2020-0622

Markowitz, D. M., Hancock, J., & Bailenson, J. (2023). Linguistic markers of AI-Generated text versus human-generated text: Evidence from hotel reviews and news headlines. *PsyArXiv. January, 30.*

National Integrated Database of Hospitality Industry. (2022). Available at https://nidhi.nic.in/HotelDivision/ProposalSubmissions/CompleteViewApplication.as px [Retrieved 2 May, 2024].

Pan, T., & Fu, R. J. (2024). *Navigating the AI horizon in hospitality: a novel classification and future research agenda*. International Hospitality Review.

Pant, B. C. (2019). Status and Potential of India as a MICE Destination: An Informative Study. *Emerging Trends in Indian Tourism and Hospitality: Transformation and Innovation*, 257.

Pinho, M., & Marques, J. (2021, September). The bleisure tourism trend and the potential for this business-leisure symbiosis in Porto. [). Routledge.]. *Journal of Convention & Event Tourism*, 22(4), 346–362. DOI: 10.1080/15470148.2021.1905575

Roy, P., Ramaprasad, B. S., Chakraborty, M., Prabhu, N., & Rao, S. (2020). CustomeracceptanceofUseofartificialintelligencein hospitality services: An Indian hospitality sector perspective. *Global Business Review*, •••, 0972150920939753.

Sarstedt, M., Hair, J. F.Jr, Cheah, J. H., Becker, J. M., & Ringle, C. M. (2019). How to Specify, Estimate, and Validate Higher-Order Constructs in PLS-SEM. *Australasian Marketing Journal*, 27(3), 197–211. DOI: 10.1016/j.ausmj.2019.05.003

Sohaili, N. L. M., Aminudin, N., & Amir, A. F. (2022). Bleisure Motivation Towards Leisure Attraction: A Systematic Literature Review. *Social Sciences*, 12(11), 910–922.

Solelhac, A. (2021). *Mountain resort marketing and management*. Routledge. DOI: 10.4324/9781003039396

Stone, P., Brooks, R., Brynjolfsson, E., Calo, R., Etzioni, O., Hager, G., & Teller, A. (2022). Artificial intelligence and life in 2030: the one hundred year study on artificial intelligence. *arXiv preprint arXiv:2211.06318.*

Tong-On, P., Siripipatthanakul, S., & Phayaphrom, B. (2021). The implementation of business intelligence using data analytics and its effects towards on performance in the hotel industry in Thailand. *International Journal of Behavioral Analytics*, 1(2).

Tsaih, R. H., & Hsu, C. C. (2018). Artificial intelligence in smart tourism: A conceptual framework.

Walia, S., Kour, P., Choudhary, P., & Jasrotia, A. (2023). COVID-19 and the bleisure travellers: An investigation on the aftermaths and future implications. *Tourism Recreation Research*, 48(5), 657–667. DOI: 10.1080/02508281.2021.1946653

Wang, Y. (2022). Research on the Influence of Service Quality of Hotel Intelligent System on Customer Satisfaction Based on Artificial Intelligence Evaluation. *Mathematical Problems in Engineering*, 2022, 1–9. Advance online publication. DOI: 10.1155/2022/3832935

Chapter 11
Revolutionizing Tourism and Hospitality Industry: The Impact of 5G/6G Technology

Prashant Kumar Yadav
https://orcid.org/0000-0001-9480-5958
Indira Gandhi National Tribal University, India

G. Jayaprakashnarayana
https://orcid.org/0009-0008-7076-6809
Indira Gandhi National Tribal University, India

Shambhavi Agnihotri
https://orcid.org/0009-0003-7278-4502
Indira Gandhi National Tribal University, India

S. Rajadurai
University of Colombo, Sri Lanka

ABSTRACT

The study adopts a quantitative approach, gathering data from respondents in selected tourist destinations. The primary focus is on assessing the potential of 5G/6G to enhance the tourist experience through immersive technology applications, improve operational efficiencies with IoT integrations, and boost safety protocols through advanced, real-time monitoring systems. The findings revealed that 5G/6G technology had the most potential to enhance the tourist experience. It significantly boosts data transfer efficiency, supports a higher density of connected devices, and allows for ultra-reliable, low-latency communications. These capabilities are enhancing guest services through personalised and context-aware systems and revolutionise backend operations, from maintenance through smart sensors to dynamic pricing

DOI: 10.4018/979-8-3693-6725-4.ch011

algorithms powered by real-time data analytics. The research also considers the challenges of implementing these advanced technologies, such as the need for substantial infrastructure investment and concerns over data privacy and cybersecurity.

1. INTRODUCTION

Telecommunications technology has significantly advanced since the 1980s, evolving from conventional landlines to modern smartphones (Kolluru & Reddy, 2021). The progression from 1G to 5G has dramatically improved communication speeds and capabilities (Kumar & Rao, 2015; Sharma & Sharma, 2018). Key developments include cellular technology, wireless systems, and the Internet of Things (Chang et al., 1993; Caytiles & Park, 2013). These innovations have transformed various industries and daily life, improving efficiency and decision-making (Robin, 2023). However, challenges like cybersecurity risks and technological dependence persist (Robin, 2023). The research focus has shifted from the acceptance and usefulness of new media to interdisciplinary studies on the impact of technological development across various fields (Budimir, 2014). As telecommunications continues to advance, addressing the digital divide between developed and developing countries remains crucial (Budimir, 2014) while monitoring potential drawbacks to ensure responsible and sustainable progress (Robin, 2023).

The evolution from 5G to 6G wireless technologies promises significant advancements in communication capabilities. While 5G focuses on enhanced mobile broadband, massive machine-type communication, and ultra-reliable low-latency communication (Kaul & Gupta, 2021), 6G aims to incorporate artificial intelligence, edge computing, and quantum computing (Dogra et al., 2020). Millimeter-wave technologies will be crucial in 5G and 6G networks, enabling ultra-fast data rates and improved spectral efficiency (Hong et al., 2021). The 6G network is expected to provide higher system capacity, data rates, and security while reducing latency compared to 5G (Kadir et al., 2021). 6G will utilise higher frequency bands, smart meta-surfaces for increased coverage, and AI at the edge (Kaul & Gupta, 2021). The implementation of 6G is anticipated between 2027 and 2030, with ongoing research focusing on overcoming challenges and meeting the growing demands for wireless connectivity (Kadir et al., 2021).

The evolution of mobile communication technologies from 3G to 6G has been driven by increasing demands for connectivity, data rates, and advanced services (Ahire et al., 2021). 5G aims to revolutionise wireless networks with enhanced data rates, latency, connectivity, reliability, and energy efficiency, utilising new spectrum bands and technologies like large antenna arrays and network densification (Shafi et al., 2017). 6G is expected to further improve on 5G, offering terabit data rates, ultra-low latency, and near-zero energy consumption while enabling new services such as extended reality and mobile holograms (Alraih et al., 2022). Key technologies for 6G include terahertz communication, visible light communication, and artificial intelligence integration (Alraih et al., 2022). Both 5G and 6G are designed to support Internet of Things (IoT) applications, with 6G aiming to provide even faster wireless networks than 5G (Alkandari & AlAhmad, 2022). These advancements promise revolutionising connectivity, sustainability, and trustworthiness in future mobile communications.

The evolution of wireless networks is progressing rapidly, with 5G in its initial commercialisation phase and 6G on the horizon for 2030 (S. Mohsan et al., 2020). 6G aims to enable the Internet of Everything, offering enhanced capabilities such as virtual reality, holographic communication, and ultra-high data rates (Anutusha Dogra et al., 2020; S. Mohsan et al., 2020). Key technologies for 6G include terahertz communication, visible light communication, ultra-massive MIMO, artificial intelligence, quantum communication, blockchain, and intelligent reflective surfaces (Duc-Nghia Vu et al., 2023). These technologies will support emerging applications like autonomous vehicles and smart cities (Duc-Nghia Vu et al., 2023). However, challenges remain in spectrum allocation, energy efficiency, and security. Ongoing research projects are addressing these issues to realise the full potential of 6G networks (Anutusha Dogra et al., 2020). The integration of these technologies is expected to revolutionise communication and enable new use cases beyond current capabilities (Duc-Nghia Vu et al., 2023).

2. LITERATURE REVIEW

The evolution from 5G to 6G wireless networks promises significant advancements in communication technologies. While 5G focuses on enhanced mobile broadband, massive machine-type communication, and ultra-reliable low-latency communication (Dogra et al., 2020), 6G aims to support more advanced applications like virtual reality, augmented reality, and AI-infused services (Silva & Guerreiro, 2020). Key enabling technologies for 6G include terahertz communication, visible light communication, ultra-massive MIMO, artificial intelligence, quantum communication, and intelligent reflective surfaces (Vu et al., 2023). These technologies are expected to provide ultra-high data rates, ultra-reliable low-latency communication, and massive connectivity (Jiang et al., 2021). However, 6G also faces challenges in security, privacy, and multidisciplinary aspects such as policy and ethics (Arastouei, 2022). Integrating these technologies is anticipated to unlock new opportunities and realise the full potential of 6G, supporting applications like autonomous vehicles and smart cities (Ahmed & Matin, 2020; Kaul & Gupta, 2021).

Significant technical advancements have marked the growth of internet generations. Mobile communications have progressed from 1G to 5G, with 5G introducing MM-Wave spectrum and MIMO technology, enabling increased channel capacity and connectivity (Hodara & Skaljo, 2021). Internet protocols have evolved to provide enhanced security, flexibility, and QoS (Kaur et al., 2020). The transition from IPv4 to IPv6 has expanded address space and introduced new features, although challenges in adoption persist (Jayanthi & Rabara, 2010). Next-generation internet architectures focus on advanced networks, services, and applications (Mambretti & Schmidt, 1999). Modern internet generations aim to address data rates, coverage, and capacity issues through techniques like small cells and beamforming (Garg et al., 2018). The World Wide Web has also evolved through five generations, each with distinct characteristics and capabilities (Patel, 2013). These advancements collectively improve communication, connectivity, and services across various sectors (Gade et al., 2024).

2.1 Applications of 5G/6G in Tourism and Hospitality

Integrating 5G and 6G technologies in tourism and hospitality is poised to revolutionise the industry by enhancing visitor experiences and operational efficiency. These advancements enable immersive augmented and virtual reality experiences, remote tourism, and improved accessibility for disabled individuals (Katsaros et al., 2019; Vignaroli et al., 2020). The Internet of Things (IoT) plays a crucial role in smart tourism, offering opportunities for increased customer satisfaction and cost reduction (Car et al., 2019; Wang et al., 2020). 5G and AI-empowered IoT systems facilitate efficient data transmission and processing, enabling applications such as personalised POI recommendations (Wang et al., 2020). In medical tourism, 5G infrastructure supports various healthcare applications and machine-type communication (Psiha & Vlamos, 2017). As the industry evolves, mobile communications continue to impact customer relationship management and business processes (Chen et al., 2021; Law et al., 2009). Future 6G networks are expected to address current limitations and support emerging applications like holographic telepresence and space tourism (Alwis et al., 2021).

Mobile communications have significantly impacted tourism and hospitality, bringing new possibilities to each generation (Chen et al., 2021). Adopting information technology has enhanced operational efficiency, service quality, and customer experiences in these industries (Law et al., 2009). The advent of 5G technology is set to revolutionise the sector further, enabling immersive and interactive experiences such as augmented reality tours and 360-degree immersive content (Katsaros et al.,

2019). The 5G-TOURS project in Turin demonstrates the potential of 5G in creating innovative tourism concepts, including indoor/outdoor immersive experiences and remote tourism based on AR/VR, while enhancing accessibility for disadvantaged and disabled individuals (Vignaroli et al., 2020). These advancements are expected to transform business models for delivering 5G-enabled services in tourism and hospitality, highlighting the critical role of mobile communication technologies in shaping the industry's future (Katsaros et al., 2019; Chen et al., 2021).

2.2 Potential of emerging technologies in tourism and hospitality

Several researches highlight the transformative potential of emerging technologies in tourism and hospitality. 5G networks enable immersive experiences like augmented reality tours and 360-degree visualisations (Katsaros et al., 2019), while the Internet of Things (IoT) enhances customer satisfaction and operational efficiency (Car et al., 2019). Smart technologies facilitate personalised experiences through information aggregation and real-time synchronisation (Neuhofer et al., 2015). The integration of 5G, AI, and IoT supports smart tourism applications like personalised recommendations (Wang et al., 2020). Machine learning and sentiment analysis of customer reviews provide valuable hotel insights (Gaur et al., 2021). Technological disruptions such as autonomous devices and virtual reality create smart environments that transform industry structures and processes (Buhalis et al., 2019). The 5G-TOURS project demonstrates how 5G can enhance accessibility and enable remote tourism experiences (Vignaroli et al., 2020). These advancements are reshaping the tourism and hospitality landscape, offering new opportunities for innovation and improved guest experiences (Mazurek et al., 2020).

Recent research on 5G/6G in tourism highlights the transformative potential of these technologies. 5G enables more immersive and connected tourism experiences (Pradini et al., 2024), while 6G is expected to address data rate, latency, and global coverage limitations for emerging applications like space tourism (Alwis et al., 2021). Smart tourism research has progressed significantly, with studies exploring ICT applications in hospitality and tourism (Law et al., 2019; Chen et al., 2021). The Metaverse is anticipated to revolutionise tourism operations and consumer behaviour (Gursoy et al., 2022). 5G-TOURS project demonstrates 5 G's potential in enhancing accessibility and promoting remote tourism through AR/VR experiences (Vignaroli et al., 2020). However, research gaps persist, particularly in addressing the full spectrum of smart tourism issues (Lee, 2021). Future research should focus on the impact of mobile communications on tourism experiences, industry practices, and the challenges of adopting new technologies (Chen et al., 2021).

The research explores the potential of 5G and 6G technologies to revolutionise the tourism industry. These advanced communication systems enable immersive experiences, enhanced connectivity, and improved service quality (Chen et al., 2021; Liang et al., 2021; Pradini et al., 2024). Studies highlight the integration of 5G with augmented and virtual reality for remote tourism experiences (Vignaroli et al., 2020) and using Multi-Access Edge Computing to reduce latency and improve user experience (Katsaros et al., 2019). Researchers have developed theoretical frameworks and models to guide future studies, including the 6G-enabled Industrial Internet of Everything framework (Padhi & Charrua-Santos, 2021) and various tourism models examining tourist experiences and satisfaction (Abdulla & Mapjabil, 2021; Mckercher & Prideaux, 2020). These advancements are expected to transform business models in the tourism sector, creating new opportunities for immersive and personalised experiences while addressing challenges such as accessibility and sustainability (Chen et al., 2021; Vignaroli et al., 2020).

Smart technologies enable information aggregation, ubiquitous mobile connectivity, and real-time synchronisation, leading to enhanced personalisation of tourism experiences (Neuhofer et al., 2015). The Internet of Things (IoT) is reshaping hospitality services, improving guest experiences, and changing service platforms (Prasanna Kansakar et al., 2017). Blockchain technology, though still in its early stages, influences tourism transactions through cryptocurrencies, smart contracts, and decentralised applications (Nam et al., 2019). Digital transformation is crucial for the industry, with innovations like smart hotels, AR/VR experiences, and AI-powered customisation revolutionising travel solutions (Prasanth, 2024). However, challenges such as technology disruptions leading to local criminality and the need for effective customer processes must be addressed (Prasanth, 2024). As these technologies evolve, they offer unprecedented opportunities for optimising processes and creating meaningful, personalised experiences in tourism and hospitality (Johnpaul et al., 2024).

2.3 Smart Tourism and Hospitality

Smart tourism and hospitality leverage advanced technologies to create personalised and enhanced guest experiences. Smart environments, driven by the Internet of Things, artificial intelligence, and virtual/augmented reality, transform industry structures and processes (Buhalis et al., 2019). These technologies enable information aggregation, ubiquitous mobile connectivity, and real-time synchronisation, facilitating personalised experiences in hospitality (Neuhofer et al., 2015). Smart tourism destinations utilise ICT to transform large amounts of data into value propositions, though the concept remains ill-defined (Gretzel et al., 2015). Integrating smart technologies in hospitality leads to developing agile business ecosystems in networked destinations, focusing on customer-centricity, personalisation, and operational agility (Buhalis et al., 2022). While smart tourism offers numerous opportunities for innovation and improved service delivery, it also presents challenges that require further research to inform its development and management (Gretzel et al., 2015; Buhalis et al., 2022).

Smart technologies are revolutionising the tourism and hospitality industry by enabling personalised experiences and enhancing service delivery. These technologies include the Internet of Things, artificial intelligence, and virtual and augmented reality systems (Sujood et al., 2022; Hamidova & Egamberdiyev, 2023). They facilitate information aggregation, ubiquitous mobile connectedness, and real-time synchronisation, leading to more meaningful and personalised services (Neuhofer et al., 2015). The integration of smart technologies in this sector is driven by factors such as perceived usefulness, ease of use, and trust (Sujood et al., 2022). These innovations are transforming various aspects of the industry, including travel solutions, smart hotels, and AI-powered customisation (Prasanth, 2024). While smart technologies offer numerous benefits, including improved resource management and infrastructure, they also present challenges, such as potential increases in local criminality (Prasanth, 2024). Adopting smart technologies is crucial for the industry's growth, efficiency, and post-COVID-19 resilience (Prasanth, 2024).

2.4 Augmented and Virtual Reality

Augmented Reality (AR) and Virtual Reality (VR) are rapidly evolving technologies that enhance user experiences in 3D environments (Devarasetty et al., 2022). AR combines real and computer-generated elements, engaging multiple senses, while VR creates immersive simulations (Agrawal Luckykumar Dwarkadas et al., 2023). These technologies have diverse applications across various fields, including disaster management, where they are used to improve building design, investigate human behaviour, and train people for emergency responses (Lovreglio, 2020). Key AR technologies include tracking, interaction techniques, and display technology, while VR encompasses immersive, semi-immersive, and non-immersive systems (Agrawal Luckykumar Dwarkadas et al., 2023). The evolution of AR and VR has been significant, transitioning from fictional concepts to practical applications in recent years (Nair, 2018). As these technologies continue to advance, they offer substantial potential for growth and expanded use cases in the coming decade (Devarasetty et al., 2022).

Augmented Reality (AR) is a technology that overlays digital information in the real world, enhancing user experiences across various domains (Berryman, 2012). Unlike virtual reality, AR combines reality with computer-generated content, making it increasingly portable and accessible on mobile devices (Yuen et al., 2011). AR applications span multiple fields, including medicine, marketing, museums, and education (Berryman, 2012; Yuen et al., 2011). The technology relies on crucial components such as hardware and software systems, with ongoing developments in AR cloud technologies (Chen et al., 2019; Arena et al., 2022). While AR offers significant potential for ubiquitous learning and instant access to location-specific information, it also faces limitations and design challenges (Yuen et al., 2011; Arena et al., 2022). As AR continues to evolve, it is expected to impact society, with widespread adoption substantially predicted in educational settings and other areas of daily life (Yuen et al., 2011; Chen et al., 2019).

Virtual reality (VR) and augmented reality (AR) technologies are transforming the tourism and hospitality industry, offering new ways to enhance traveller experiences and reshape consumer interactions (Calisto & Sarkar, 2024; Wei, 2019). These technologies provide innovative marketing tools for destinations and businesses, potentially changing the tourism model (Calisto & Sarkar, 2024). VR and AR applications in hospitality have shown diverse positive outcomes, encouraging industry practitioners to incorporate them into strategic plans (Wei, 2019). The impact of VR/AR is evident in future tourism planning, destination marketing, and evolving consumer requirements (Peštek & Sarvan, 2020). Recent research has focused on VR's influence on hospitality, tourism, and destinations; technology enhancing hotel satisfaction and performance; and user acceptance of AR in travel (Lodhi et al., 2024). As VR and AR advance, they may fundamentally alter how tourist experiences are managed and offer realistic virtual travel alternatives (Peštek & Sarvan, 2020).

2.5 Operational Efficiency

The evolution from 5G to 6G networks promises significant operational efficiency and performance improvements. 6G is expected to incorporate advanced technologies like machine learning, artificial intelligence, and 3D coverage to enhance network efficiency (Rejwan Bin Sulaiman & Amer Kareem, 2020). However, energy consumption remains a critical concern, with 5G already consuming four times more energy than 4G (T. P. Fowdur & Doorgakant, 2023). Machine learning techniques are being explored for efficient base station deployment and adaptive operational modes (T. P. Fowdur & Doorgakant, 2023). The proposed ECO6G model utilises machine learning for traffic load forecasting and energy efficiency evaluation, potentially saving millions in operational expenses (Thantharate et al., 2022). Despite these challenges, 5G technology significantly improves speed and latency compared to 4G, enabling real-time data sharing, high-definition video transmission, and applications in industrial automation and remote surgery (Naveen Kumar, 2023).

Integrating Internet of Things (IoT) technology offers significant operational efficiencies across various sectors. IoT enhances security through real-time surveillance, biometric identification, and fraud detection in banking, although challenges like data security and connectivity persist (Johri et al., 2023). IoT's event-based communication strategies minimise energy consumption and device interactions, leading to prolonged device lifetimes and greater autonomy (Kolios et al., 2015). The combination of IoT and AI has substantially improved performance metrics across industries. In healthcare, patient queue times were reduced by 50%, while manufacturing saw a 33% decrease in capital costs. Retail stores experienced a 25% increase in sales through AI-aided demand forecasting and IoT-based inventory management (Mudholkar & Mudholkar, 2024).

The upcoming 5G and 6G communication systems are expected to offer significant advancements over current technologies, addressing key challenges such as high capacity, massive connectivity, low latency, and improved security (Khalid & Raza, 2024; Chowdhury et al., 2019). Optical wireless communication (OWC) technologies, including visible light communication, light fidelity, and free space optics, are promising solutions for meeting the demands of 5G/6G and IoT systems (Khalid & Raza, 2024; Chowdhury et al., 2019). 6G networks are anticipated to incorporate emerging technologies like machine learning, artificial intelligence, and 3D coverage to enhance efficiency and performance (Sulaiman & Kareem, 2020). To overcome the limitations of 5G, 6G will utilise higher frequency bands, new duplexing and modulation techniques, and smart meta-surfaces for increased coverage (Kaul & Gupta, 2021). The integration of blockchain, Internet of Senses (IoS), and petahertz technologies is also being explored for future 6G implementations (Kaul & Gupta, 2021).

3. SUSTAINABLE TOURISM PRACTICES

Technology integration in sustainable tourism practices has emerged as a significant area of research. Information and communication technologies (ICTs) offer innovative approaches to sustainable tourism development from a destination management perspective (Ali & Frew, 2014). These technologies can address tourism's positive and negative impacts, fostering destination innovativeness. The interdependency between technology and sustainable tourism is evident, with technology transforming the industry by enabling online booking information access and promoting sustainable practices (Nicola-Gavrilă, 2023). Empirical approaches have been used to study this relationship, analysing the impact of technological advancements on various dimensions of sustainability in tourism. Technologies such as data-driven systems, artificial intelligence, and agent-based modelling are being employed to promote sustainable tourism practices. Additionally, information technology plays a crucial role in managing and promoting sustainable cultural tourism (Valcic & Domsic, 2012), further emphasising the importance of technology in achieving sustainable development goals in the tourism sector.

Recent research highlights the potential of advanced technologies in promoting sustainable tourism practices. 5G networks enable immersive experiences like augmented reality tours and 360-degree views, enhancing tourist engagement while reducing environmental impact (Katsaros et al., 2019). These technologies, including artificial intelligence, the Internet of Things, and virtual reality, are emerging trends in sustainable tourism (Loureiro & Nascimento, 2021). Social media and e-commerce promote eco-friendly tourism by facilitating effective storytelling and stakeholder collaboration (Kumari & Thakur, 2023). The 5G-TOURS project demonstrates how 5G can support innovative tourism concepts, including remote tourism and improved accessibility for disadvantaged visitors (Vignaroli et al., 2020). While these technologies offer numerous benefits, their adoption also presents risks that need careful consideration (Loureiro & Nascimento, 2021).

While technology can promote sustainable practices, its adoption also presents risks and challenges (Loureiro & Nascimento, 2021). Key technological trends shaping sustainable tourism include artificial intelligence, Internet of Things, big data, and augmented/virtual reality (Loureiro & Nascimento, 2021; Nicola-Gavrilă, 2023). These technologies enable data-driven decision-making and improved tourist experiences. Networking and community-building play crucial roles in sustainable tourism development, with local governments and NGOs often serving as guiding centres for participatory processes (Polese, 2009). Mobile connectivity transforms how tourists engage with social networks, blending online and offline interactions to create new forms of community and support sustainable practices (Dickinson et al., 2017). This evolving landscape presents opportunities for collaborative approaches and bottom-up social systems that can enhance sustainable tourism through shared experiences and resources.

4. FINDINGS

The Findings provide insights into various aspects of travel habits, preferences, and the influence of digital technologies on the travel experiences of 60 respondents.

Table 1. Gender

	Frequency	%
Male	44	73.3
Female	16	26.7
Total	60	100.0

Table 2. Age Group

	Frequency	%
Below 20	8	13.3
21-40	23	38.3
41-60	20	33.3
61-80	9	15.0
Total	60	100.0

Table 3. Education level

	Frequency	%
Diploma	13	21.7
Graduate	27	45.0
Post Graduate	13	21.7
Up To 12th	5	8.3
Up To 10th	2	3.3
Total	60	100.0

Table 4. How often do you travel?

	Frequency	%
Once a year	19	31.7
Twice a year	18	30.0
More than twice a year	23	38.3
Total	60	100.0

Table 5. Which Digital technologies do you use while travelling

	Frequency	%
Smartphone	11	18.3
Tablet	12	20.0
Laptop	8	13.3
Smartwatch	9	15.0
All of the Above	20	33.3
Total	60	100.0

4.1 Gender and Age Distribution

Most respondents were male (73.3%), with females making up 26.7% (Table 1). The respondents were divided into different age groups, with the largest group (38.3%) falling within the 21-40 age range. The 41-60 age group constituted 33.3%, while those aged below 20 and between 61-80 made up 13.3% and 15.0%, respectively (Table 2).

4.2 Education Level

The educational background of the respondents varied, with 45.0% holding a graduate degree, 21.7% having a diploma, and another 21.7% possessing a post-graduate degree. Smaller percentages had completed education up to the 12th grade (8.3%) and up to the 10th grade (3.3%) (Table 3).

4.3 Travel Frequency and Technology Usage

Regarding travel frequency, 38.3% of respondents travelled more than twice a year, 31.7% travelled once a year, and 30.0% travelled twice a year (Table 4). Regarding digital technology usage during travel, 33.3% used all digital devices, including smartphones, tablets, laptops, and smartwatches. Specific usage of individual devices like smartphones (18.3%), tablets (20.0%), laptops (13.3%), and smartwatches (15.0%) was also reported (Table 5).

4.4 Digital Technology and Sustainable Tourism

A majority (55.0%) believed digital technology could promote sustainable tourism practices, while 31.7% disagreed, and 13.3% were unsure.

4.5 Booking Preferences and Travel Influences

When booking accommodations, 55.0% preferred online methods, 31.7% opted for in-person booking, and 13.3% had no preference. The most influential factors in choosing travel destinations were adventure (41.7%) and cuisine (28.3%), followed by history (15.0%), nature (10.0%), and culture (5.0%).

4.6 Virtual Reality and Social Media Usage

A significant 61.7% of respondents had used virtual reality to experience a destination before visiting, while 38.3% had not. The most commonly used social media platform for sharing travel experiences was Instagram (36.7%), followed by Facebook (26.7%), Snapchat (23.3%), YouTube (8.3%), and Twitter (5.0%).

4.7 Interest in Augmented Reality and Travel Content

Regarding augmented reality for navigation, 60.0% of respondents were interested, 15.0% were not, and 45.0% were uncertain. The most engaging types of digital travel content included virtual tours, 360-degree videos, travel vlogs, and podcasts, with 41.7% finding all these types engaging.

4.8 Travel App Features and Encouragement

When selecting a travel app, 31.7% of respondents looked for features such as offline maps, reviews, hotel and flight bookings, travel itineraries, and local recommendations. Additionally, 28.3% were encouraged by a user-friendly interface, personalised recommendations, integration with social media, real-time updates, and exclusive discounts.

4.9 Potential of Digital Technologies

Among various digital technologies, 35.0% of respondents believed that a combination of artificial intelligence, blockchain, augmented reality/virtual reality, the Internet of Things (IoT), and 5G/6G connectivity had the most potential to enhance the tourist experience. Specifically, artificial intelligence and 5G/6G connectivity were each highlighted by 16.7% of respondents.

5. CHALLENGES

The evolution from 5G to 6G wireless networks presents opportunities and challenges for the tourism sector. While 5G enables more immersive and connected tourism experiences (Pradini et al., 2024), 6G promises even more significant advancements in virtual reality, holographic communication, and the Internet of Everything (Mohsan et al., 2020; Alsharif et al., 2020). The Metaverse, powered by these technologies, offers potential benefits for tourism research and marketing (Monaco & Sacchi, 2023). However, several challenges must be addressed, including social acceptance, affordability, environmental sustainability, and data privacy (Monaco & Sacchi, 2023; Asghar et al., 2022). Technical hurdles such as achieving ultra-low latency, unprecedented speed, and extraterrestrial communication persist (Asghar et al., 2022; Ahmed & Matin, 2020). Furthermore, spectrum management issues arise from the concurrent operation of various technologies like carrier aggregation, cognitive radio, and massive MIMO (Qamar et al., 2020). Addressing these challenges will be crucial for realising the full potential of 6G in tourism applications (Alwis et al., 2021).

6. CONCLUSION

The growth from 5G to 6G networks promises significant improvements in service delivery across various sectors. 5G introduces millimetre waves and massive MIMO to support new services such as IoT and V2V communications (Silva & Guerreiro, 2020). 6G is expected to enable more advanced applications, including virtual reality, holographic services, and brain-computer interaction, utilising terahertz bands (Silva & Guerreiro, 2020; Shahraki et al., 2021). Cloud-native and service-based architectures are being adopted to increase flexibility and cost-effectiveness in 5G networks (Hänsge et al., 2020; Haensge et al., 2019). The impact of 5G on digital servitisation in manufacturing is significant, particularly for services using augmented reality and cloud computing (Cimini et al., 2023). In healthcare, 5G enables applications like e-connectivity in pharmaceuticals and personalised medicine (Thuemmler et al., 2018). Quality of Experience management and network softwarization are crucial for the success of 5G/6G networks (Barakabitze & Walshe, 2022).

Recent advancements in artificial intelligence (AI), virtual reality (VR), and augmented reality (AR) are transforming the tourism and hospitality industry. These technologies enhance customer experiences through personalisation, immersive previews, and improved service delivery (Roy & Pagaldiviti, 2023; Nayyar et al., 2018). AI applications include demand forecasting, customer satisfaction analysis, and chatbots for instant query responses (Nannelli et al., 2023; Bulchand-Gidumal,

2020). VR and AR technologies are being utilised for marketing, education, and preservation of tourist spots (Loureiro, 2022; Wei, 2019). Integrating mobile apps, IoT, and robotics streamlines operations and frees staff for more personalised interactions (Roy & Pagaldiviti, 2023). While these innovations offer numerous benefits, challenges remain in implementation and adaptation (Doborjeh et al., 2021). As the industry continues to evolve, further research is needed to explore the potential of these technologies in creating smart, efficient, and personalised tourism experiences (Prasanth, 2024).

The integration of artificial intelligence (AI) and machine learning (ML) techniques in 5G and 6G networks is revolutionising resource management and network automation. Deep reinforcement learning (DRL) is being explored for efficient resource allocation and network slicing (Johanna Andrea Hurtado Sánchez et al., 2022). AI-empowered resource management enhances spectrum, computing, and caching resources (Mengting Lin & Youping Zhao, 2020). Federated deep reinforcement learning (FDRL) is proposed for optimising quality of service in vehicular networks (H. Alsulami et al., 2022). Dynamic Data Driven Applications Systems (DDDAS) methodologies are applied for real-time decision support in complex 5G/6G infrastructures (Nurcin Celik et al., 2022). Blockchain technology is considered for efficient resource management and sharing in 6G networks (Hao Xu et al., 2020). The Network Data Analytics Function (NWDAF) enables automated decision-making in 5G systems (Miguel A. Garcia-Martin et al., 2023). These advancements aim to achieve autonomous networking capabilities and optimise network performance (Madanagopal Ramachandran et al., 2022; Karima Velasquez et al., 2022).

The 5G/6G is driving significant technological advancements and global adoption. South Korea and China are leading in 5G implementation, showcasing its benefits across various sectors (Nuriev et al., 2024). Finland is at the forefront of 6G development, with key industrial and academic players collaborating on research projects (Katz et al., 2019). 6G is expected to enable new use cases such as massive twinning, cooperative robots, and immersive telepresence (Uusitalo et al., 2021). The transition to 6G will require addressing challenges in sustainability, trustworthiness, and global service coverage (Saad et al., 2019). As 6G develops, it will likely integrate concepts like Digital Twins and Immersive Realities, potentially transforming urban economies and supporting sustainable development goals (Allam & Jones, 2021).

7. DISCUSSION

5G technology transforms the tourism industry by enabling immersive and connected experiences (Gagih Pradini et al., 2024; K. Katsaros et al., 2019). It facilitates innovative tourism concepts, including augmented and virtual reality applications, live streaming, and remote tourism (Shaobo Gong & Hong Zhu, 2021; Luca Vignaroli et al., 2020). 5G, combined with Internet of Things (IoT) and Artificial Intelligence (AI), enhances smart tourism applications such as personalised recommendations and efficient data processing (Wei Wang et al., 2020). The technology also impacts medical tourism, offering new possibilities for healthcare applications (Maria Psiha & Vlamos, 2017). Furthermore, 5G enables the development of a smart leisure sports tourism industry (F. Liang et al., 2021). However, researchers caution about the potential negative consequences of digital transformation in tourism, including structural dependency and data control issues, which may lead to digital colonialism in tourist destinations (Meghdad Abbasian Fereidouni & Kawa, 2019).

8. FUTURE TRENDS

Integrating 5G and future 6G technologies is poised to revolutionise the tourism and hospitality industry. These advanced mobile communications enable more immersive and connected tourist experiences (Chen et al., 2021; Pradini et al., 2024). Critical applications include augmented and virtual reality tours, 360-degree immersive experiences, and novel ways of interacting with art and cultural heritage sites (Katsaros et al., 2019; Vignaroli et al., 2020). Implementing Multi-Access Edge Computing (MEC) infrastructure reduces latency and improves user experience quality (Katsaros et al., 2019). 5G technology promotes accessibility for disadvantaged and disabled individuals and facilitates "remote tourism" experiences (Vignaroli et al., 2020). While these advancements offer numerous opportunities, the tourism sector faces challenges in adopting and integrating these technologies (Pradini et al., 2024). 5G and future mobile communications are expected to transform business models and create new paradigms for delivering tourism services (Katsaros et al., 2019; Chen et al., 2021).

The 5G/6G networks promise significant speed, reliability, and customer satisfaction improvements. 6G is expected to offer terabit-per-second data rates, ultra-low latency, and enhanced energy efficiency (Alraih et al., 2022; Kitanov et al., 2022). These advancements will enable new services like immersive extended reality, mobile holograms, and advanced IoT applications (Chowdhury et al., 2019; Alraih et al., 2022). 6G will incorporate terahertz communication, visible light communication, and artificial intelligence to meet increasing demands (You et al., 2020; Alraih et al., 2022). Integrating terrestrial and non-terrestrial networks will provide global coverage, including space, air, ground, and sea (You et al., 2020; Kwon et al., 2022). These improvements will contribute to the development of smart cities and enhance various sectors, including health, transportation, and security (Kulkarni & Kulkarni,

2021). Overall, 6G aims to create a framework of ubiquitous wireless intelligence, significantly surpassing the capabilities of 5G (Meril et al., 2019; You et al., 2020).

9. IMPLICATIONS

The chapter underscores the transformative role of telecommunications technologies, particularly the evolution from 5G to 6G, in various sectors. The implications include significant speed, connectivity, and service quality advancements, enabling innovations like the Internet of Things (IoT), artificial intelligence, and immersive experiences like virtual and augmented reality. These technologies are expected to revolutionise industries, especially tourism and hospitality, by enhancing customer experiences, operational efficiency, and sustainable practices. However, challenges remain, including addressing the digital divide, cybersecurity risks, and technological dependence, emphasising the need for responsible and sustainable development. The transition to 6G, anticipated by 2030, is poised to unlock new possibilities but requires overcoming technical, ethical, and policy-related hurdles.

REFERENCES

Abdulla, F. Y., & Mapjabil, J. (2021). Review of theories and model of research on liminality in tourism. *Proceeding for Global Tourism Conference (GTC) 2021*, 175-177. https://doi.org/DOI: 10.46754/gtc.2021.11.048

Ahire, M., Patil, V., & Kolapkar, P.A., Student, & Wagh, K.K. (2021). A Survey on Mobile Communications and Network Technologies. [OAIJSE]. *Open Access International Journal of Science& Engineering*, 6(6), 5–9.

Ahmed, R., & Matin, M. A. (2020). Towards 6g wireless networks-challenges and potential technologies. *Journal of Electrical Engineering*, 71(4), 290–297. DOI: 10.2478/jee-2020-0040

Ali, A., & Frew, A. J. (2014). Technology innovation and applications in sustainable destination development. *Information Technology & Tourism*, 14(4), 265–290. DOI: 10.1007/s40558-014-0015-7

Alkandari, A., & AlAhmad, M. A. (2022). Overview On 5G and 6G Wireless Communication with IoT Technology. *Journal of Engineering Science and Technology*, 17(1), 95–105.

Allam, Z., & Jones, D. (2021). Future (post-COVID) digital, smart and sustainable cities in the wake of 6G: Digital twins, immersive realities and new urban economies. *Land Use Policy*, 101, 105201. DOI: 10.1016/j.landusepol.2020.105201

Alraih, S., Shayea, I., Behjati, M., Nordin, R., Abdullah, N. F., Abu-Samah, A., & Nandi, D. (2022). Revolution or evolution? Technical requirements and considerations towards 6G mobile communications. *Sensors (Basel)*, 22(3), 762. DOI: 10.3390/s22030762 PMID: 35161509

Alsharif, M. H., Kelechi, A. H., Albreem, M. A., Chaudhry, S. A., Zia, M., & Kim, S. H. (2020). Sixth generation (6G) wireless networks: Vision, research activities, challenges and potential solutions. *Symmetry*, 12(4), 676. DOI: 10.3390/sym12040676

Alsulami, H., Serbaya, S. H., Abualsauod, E. H., Othman, A. M., Rizwan, A., & Jalali, A. (2022). A federated deep learning empowered resource management method to optimise 5G and 6G quality of services (QoS). *Wireless Communications and Mobile Computing*, 2022, 1–9. DOI: 10.1155/2022/1352985

Alwis, C., Kalla, A., Pham, Q., Kumar, P., Dev, K., Hwang, W., & Liyanage, M. (2021). Survey on 6G frontiers: Trends, applications, requirements, technologies and future research. *IEEE Open Journal of the Communications Society*, 2, 836–886. DOI: 10.1109/OJCOMS.2021.3071496

Arastouei, N. (2023). 6G technologies: Key features, challenges, security and privacy issues. *Communications in Computer and Information Science*, 1807, 94–109. DOI: 10.1007/978-3-031-36096-1_7

Arena, F., Collotta, M., Pau, G., & Termine, F. (2022). An overview of augmented reality. *Computers*, 11(2), 28. DOI: 10.3390/computers11020028

Asghar, M. Z., Memon, S. Q., & Hämäläinen, J. (2022). Evolution of wireless communication to 6G: Potential applications and research directions. *Sustainability (Basel)*, 14(10), 6356. DOI: 10.3390/su14106356

Barakabitze, A. A., & Walshe, R. (2022). SDN and NFV for QOE-driven multimedia services delivery: The road towards 6G and beyond networks. *Computer Networks*, 214, 109133. DOI: 10.1016/j.comnet.2022.109133

Berryman, D. (2012). Augmented reality: A review. *Medical Reference Services Quarterly*, 31(2), 212–218. DOI: 10.1080/02763869.2012.670604 PMID: 22559183

Budimir, M. (2014). An overview of research on technological and telecommunication development. *Ekonomski Vjesnik*, (1), 171–182. https://doaj.org/article/e129e602b44a479e9e39870a77bb5323

Buhalis, D., Harwood, T., Bogicevic, V., Viglia, G., Beldona, S., & Hofacker, C. F. (2019). Technological disruptions in services: Lessons from tourism and hospitality. *Journal of Service Management*, 30(4), 484–506. DOI: 10.1108/JOSM-12-2018-0398

Bulchand-Gidumal, J. (2020). Impact of artificial intelligence in travel, tourism, and hospitality. *Handbook of E-Tourism*, 1-20. https://doi.org/DOI: 10.1007/978-3-030-05324-6_110-1

Calisto, M. L., & Sarkar, S. (2024). A systematic review of virtual reality in tourism and hospitality: The known and the paths to follow. *International Journal of Hospitality Management*, 116, 103623. DOI: 10.1016/j.ijhm.2023.103623

Car, T., Stifanich, L. P., & Šimunić, M. (2019). *Internet of Things (IoT) in tourism and hospitality: opportunities and challenges*. Tourism in Southern and Eastern Europe., DOI: 10.20867/tosee.05.42

Caytiles, R. D., & Park, B. (2013). Future Directions of Information and Telecommunication Systems Through the Technological Advancement Convergence. *International Journal of Multimedia and Ubiquitous Engineering*, 8(1), 101–106. https://gvpress.com/journals/IJMUE/vol8_no1/9.pdf

Celik, N., Darema, F., Runsewe, T., Saad, W., & Yavuz, A. (2024). Dddas for optimised design and management of 5G and beyond 5G (6G) networks. *Lecture Notes in Computer Science*, 13984, 123–132. DOI: 10.1007/978-3-031-52670-1_12

Chang, J., Miska, R. A., & Shober, R. A. (1993). Wireless systems and technologies: An overview. *AT & T Technical Journal*, 72(4), 11–18. DOI: 10.1002/j.1538-7305.1993.tb00546.x

Chen, S., Law, R., Zhang, M., & Si, Y. (2021). Mobile communications for tourism and hospitality: A review of historical evolution, present status, and future trends. *Electronics (Basel)*, 10(15), 1804. DOI: 10.3390/electronics10151804

Chen, Y., Qing, W., Chen, H., Song, X., Tang, H., & Tian, M. (2019). An overview of augmented reality technology. *Journal of Physics: Conference Series*, 1237(2), 022082. DOI: 10.1088/1742-6596/1237/2/022082

Chowdhury, M. Z., Shahjalal, M., Hasan, M. K., & Jang, Y. M. (2019). The role of optical wireless communication technologies in 5G/6G and IoT solutions: Prospects, directions, and challenges. *Applied Sciences (Basel, Switzerland)*, 9(20), 4367. DOI: 10.3390/app9204367

Chowdhury, M. Z., Shahjalal, M., Hasan, M. K., & Jang, Y. M. (2019). The role of optical wireless communication technologies in 5g/6g and IOT solutions: Prospects, directions, and challenges. *Applied Sciences (Basel, Switzerland)*, 9(20), 4367. DOI: 10.3390/app9204367

Cimini, C., Lagorio, A., Pinto, R., Pezzotta, G., Adrodegari, F., & Cavalieri, S. (2023). 5G supporting digital servitisation in manufacturing: An exploratory survey. *IET Collaborative Intelligent Manufacturing*, 5(3), e12083. Advance online publication. DOI: 10.1049/cim2.12083

D. S. P. (2024). New technologies in the tourism and hospitality service sector. International. *Journal of Multidisciplinary Research*, 6(1), 12513. Advance online publication. DOI: 10.36948/ijfmr.2024.v06i01.12513

Devarasetty, P., Balasai, A., & Brox, M. (2022). The evolution of virtual and augmented reality and its various practical applications. *Journal of Student Research*, 11(3). Advance online publication. DOI: 10.47611/jsrhs.v11i3.3509

Doborjeh, Z. G., Hemmington, N., Doborjeh, M. G., & Kasabov, N. (2021). Artificial intelligence: A systematic review of methods and applications in hospitality and tourism. *International Journal of Contemporary Hospitality Management*, 34(3), 1154–1176. DOI: 10.1108/IJCHM-06-2021-0767

Dogra, A., Jha, R. K., & Jain, S. (2021). A survey on beyond 5G network with the advent of 6G: Architecture and emerging technologies. *IEEE Access : Practical Innovations, Open Solutions*, 9, 67512–67547. DOI: 10.1109/ACCESS.2020.3031234

Dwarkadas, A. L., Krishna Challa, R., & Talasila, V., & K G, S. (2023). Augmented/virtual reality: technological advancement with use cases. *2023 Global Conference on Information Technologies and Communications (GCITC)*. https://doi.org/DOI: 10.1109/GCITC60406.2023.10425860

Fereidouni, M. A. and Kawa, A. (2019). Dark side of digital transformation in tourism. *Intelligent Information and Database Systems*, 510-518. https://doi.org/DOI: 10.1007/978-3-030-14802-7_44

Fowdur, T. P., & Doorgakant, B. (2023). A review of machine learning techniques for enhanced energy efficient 5G and 6G communications. *Engineering Applications of Artificial Intelligence*, 122, 106032. DOI: 10.1016/j.engappai.2023.106032

Gade, J., Johnpaul, M., Agnihotri, S., & Miryala, R. K. (2024). The role of artificial intelligence in enhancing the experience of solo travelers from Europe to India: A literature-based study. In Talukder, M. B., Kumar, S., & Tyagi, P. K. (Eds.), *Hotel and travel management in the AI era* (pp. 551–567). IGI Global., DOI: 10.4018/979-8-3693-7898-4.ch026

Garcia-Martin, M., Gramaglia, M., & Serrano, P. (2024). Network automation and data analytics in 3Gpp 5G systems. *IEEE Network*, 38(4), 182–189. DOI: 10.1109/MNET.2023.3321524

Garg, A., Rawat, S., & Kumar, P. (2018). Analysis and evolution of dynamic and modern techniques of internet generation. *2018 International Conference on Computational Techniques, Electronics and Mechanical Systems (CTEMS)*. https://doi.org/DOI: 10.1109/CTEMS.2018.8769325

Gaur, L., Afaq, A., Solanki, A., Singh, G., Sharma, S., Jhanjhi, N. Z., & Le, D. (2021). Capitalising on big data and revolutionary 5G technology: Extracting and visualising ratings and reviews of global chain hotels. *Computers & Electrical Engineering*, 95, 107374. DOI: 10.1016/j.compeleceng.2021.107374

Gladson Robin, L. (2023). Chapter 1 – exploring the advancements, benefits and challenges of emerging technologies. https://doi.org/DOI: 10.59646/advcivilmechc1/026

Gretzel, U., Σιγάλα, M., Xiang, Z., & Koo, C. (2015). Smart tourism: Foundations and developments. *Electronic Markets*, 25(3), 179–188. DOI: 10.1007/s12525-015-0196-8

Gürsoy, D., Malodia, S., & Dhir, A. (2022). The Metaverse in the hospitality and tourism industry: An overview of current trends and future research directions. *Journal of Hospitality Marketing & Management*, 31(5), 527–534. DOI: 10.1080/19368623.2022.2072504

Haensge, K., Trossen, D., Robitzsch, S., Boniface, M., & Phillips, S. (2019). Cloud-native 5G service delivery platform. *2019 IEEE Conference on Network Function Virtualization and Software Defined Networks (NFV-SDN)*. https://doi.org/DOI: 10.1109/NFV-SDN47374.2019.9040042

Hamidova, M. and Egamberdiyev, S. (2023). Analysis of cases of use of smart technologies in tourism and hotel enterprises. Yashil Iqtisodiyot Va Taraqqiyot, 1(mxs). https://doi.org/DOI: 10.55439/GED/vol1_issmaxsusson/a653

Hänsge, K., Trossen, D., Robitzsch, S., Boniface, M., & Phillips, S. (2020). 5G service delivery platform: 360 VR vertical as an example. *Internet Technology Letters*, 3(6), e186. Advance online publication. DOI: 10.1002/itl2.186

Hodara, H., & Škaljo, E. (2021). From 1G to 5G. *Fiber and Integrated Optics*, 40(2-3), 85–183. DOI: 10.1080/01468030.2021.1919358

Jayanthi, J., & Rabara, S. A. (2010). Next-generation internet protocol -technical realms. *2010 3rd International Conference on Computer Science and Information Technology*. https://doi.org/DOI: 10.1109/ICCSIT.2010.5564916

Jiang, W., Han, B., Habibi, M. A., & Schotten, H. D. (2021). The road towards 6G: A comprehensive survey. *IEEE Open Journal of the Communications Society*, 2, 334–366. DOI: 10.1109/OJCOMS.2021.3057679

Johnpaul, M., Miryala, R. S. B., Mazurek, M., Jayaprakashnarayana, G., & Miryala, R. K. (2025). Artificial intelligence and machine learning in eCommerce. In Kaur, G., Arora, J., Jain, V., & Shaikh, A. (Eds.), *Strategic innovations of AI and ML for eCommerce data security* (pp. 31–50). IGI Global., DOI: 10.4018/979-8-3693-5718-7.ch002

Johri, S., Singh, S. K., Reddy, C. V., Nijhawan, G., Alawadi, A. H., & Reddy, U. (2023). Improving security and effectiveness in banking operations with IoT integration. *2023 10th IEEE Uttar Pradesh Section International Conference on Electrical, Electronics and Computer Engineering (UPCON)*. https://doi.org/DOI: 10.1109/UPCON59197.2023.10434723

Κατσαρός, K., Gkounis, D., Kaleshi, D., Thomas, B., Harris, J. W., Falaki, H. R., & Simeonidou, D. (2019). Enhancing tourist experiences through 5G - the 5G smart tourism case study. *2019 IEEE 2nd 5G World Forum (5GWF)*. https://doi.org/DOI: 10.1109/5GWF.2019.8911700

Kadir, E. A., Shubair, R. M., Rahim, S. K. A., Himdi, M., Kamarudin, M. R., & Rosa, S. L. (2021). B5G and 6G: next generation wireless communications technologies, demand and challenges. *2021 International Congress of Advanced Technology and Engineering (ICOTEN)*.https://doi.org/DOI: 10.1109/ICOTEN52080.2021.9493470

Kansakar, P., Munir, A., & Shabani, N. (2019). Technology in the hospitality industry: Prospects and challenges. *IEEE Consumer Electronics Magazine*, 8(3), 60–65. DOI: 10.1109/MCE.2019.2892245

Katz, M., Pirinen, P., & Posti, H. (2019). Towards 6G: getting ready for the next decade. 2019 *16th International Symposium on Wireless Communication Systems (ISWCS)*.https://doi.org/DOI: 10.1109/ISWCS.2019.8877155

Kaul, A., & Gupta, J. P. (2021). Revolutionary 6G: technologies, architecture, coverage, and performance. *2021 12th International Conference on Computing Communication and Networking Technologies (ICCCNT)*.https://doi.org/DOI: 10.1109/ICCCNT51525.2021.9579602

Kaur, P., Sharma, T., & Singh, J. P. (2020). Evaluation of internet generations on wireless network: A survey. *Journal of Technology Management for Growing Economies*, 11(2), 55–61. DOI: 10.15415/jtmge.2020.112007

Khalid, R., & Naqi Raza, M. (2024). Analysing the impact of optical wireless communication technologies on 5g/6g and IoT solutions: Prospects, developments, and challenges. *Journal of Engineering Research and Sciences*, 3(5), 23–36. DOI: 10.55708/js0305003

Kitanov, S., Nikolić, V., & Janevski, T. (2022). Quality of artificially intelligently orchestrated services in 6G mobile networks. *Journal of Electrical Engineering and Information Technologies*, 7(1), 43–52. DOI: 10.51466/JEEIT2271195043k

Kolios, P., Panayiotou, C. G., Ellinas, G., & Polycarpou, M. M. (2015). Event-based communication for IoT networking. *2015 IEEE 2nd World Forum on Internet of Things (WF-IoT)*. https://doi.org/DOI: 10.1109/WF-IoT.2015.7389076

Kolluru, D. S., & Reddy, P. B. (2021). Review on communication technologies in telecommunications from conventional telephones to smartphones. *AIP Conference Proceedings*, 020003. Advance online publication. DOI: 10.1063/5.0074088

Kulkarni, R., & Kulkarni, S. (2021). How 6G Has an Influence on Smart Cities: An Overview. *International Journal of Engineering Research & Technology (Ahmedabad)*, 10(5), 832–838. https://www.ijert.org/research/how-6g-has-an-influence-on-smart-cities-an-overview-IJERTV10IS050447.pdf

Kumar, B. A., & Rao, P. T. (2015). Overview of advances in communication technologies. *2015 13th International Conference on Electromagnetic Interference and Compatibility (INCEMIC)*. https://doi.org/DOI: 10.1109/INCEMIC.2015.8055856

Kumar, M. (2023). 5G technology is revolutionising the wireless industry with unparalleled efficiency. *Sci Wave Bulletin*, 01(03), 21–28. DOI: 10.61925/SWB.2023.1303

Kumari, A., & Thakur, P. (2023). Sustainable tourism: A review of current practices and future directions-coalescence of social media and e-commerce. *International Journal of Research Publication and Reviews*, 4(5), 1655–1662. DOI: 10.55248/gengpi.234.5.40417

Kwon, T., Kim, S., Lee, K., & Chung, J. (2022). Special issue on 6G and satellite communications. *ETRI Journal*, 44(6), 881–884. DOI: 10.4218/etr2.12551

Law, R., Leung, D., & Chan, I. C. C. (2019). Progression and development of information and communication technology research in hospitality and tourism. *International Journal of Contemporary Hospitality Management*, 32(2), 511–534. DOI: 10.1108/IJCHM-07-2018-0586

Law, R., Leung, R., & Buhalis, D. (2009). Information technology applications in hospitality and tourism: A review of publications from 2005 to 2007. *Journal of Travel & Tourism Marketing*, 26(5-6), 599–623. DOI: 10.1080/10548400903163160

Lee, P. (2021). Progress in smart tourism 2010-2017: A systematic literature review. *Journal of Smart Tourism*, 1(1), 19–30. DOI: 10.52255/smarttourism.2021.1.1.4

Liang, F., Mu, L., Wang, D., & Kim, B. (2021). A new model path for the development of smart leisure sports tourism industry based on 5G technology. *IET Communications*, 16(5), 485–496. DOI: 10.1049/cmu2.12271

Lin, M., & Zhao, Y. (2020). Artificial intelligence-empowered resource management for future wireless communications: A survey. *China Communications*, 17(3), 58–77. DOI: 10.23919/JCC.2020.03.006

Lodhi, R. N., Del Gesso, C., Asif, M., & Cobanoglu, C.LODHI. (2024). Exploring virtual and augmented reality in the hospitality industry: A bibliometric analysis. *Tourism and Hospitality Management*, 30(1), 67–84. DOI: 10.20867/thm.30.1.6

Loureiro, S. M. C. (2022). Technology and luxury in tourism and hospitality. *The Emerald Handbook of Luxury Management for Hospitality and Tourism*, 273-284. https://doi.org/DOI: 10.1108/978-1-83982-900-020211013

Loureiro, S. M. C., & Nascimento, J. (2021). Shaping a view on the influence of technologies on sustainable tourism. *Sustainability (Basel)*, 13(22), 12691. DOI: 10.3390/su132212691

Lovreglio, R. (2020). Virtual and Augmented Reality for Human Behaviour in Disasters: A Review. *Conference Proceedings, Fire and Evacuation Modeling Technical Conference (FEMTC),* September 9-11, 2020,1-14.

Mambretti, J., & Schmidt, A. (1999). Next generation Internet: creating advanced networks and services. *John Wiley & Sons, Inc. eBooks*. Retrieved from http://ci .nii.ac.jp/ncid/BA44156623

Mazurek, M., Ayoobkhan, M., Kaldeen, M., Mohamed, M., Said, T., Umachandran, K., Arora, S., Sharma, A., Pratap, R., Kaurav, S., Baber, R., Rajput, S., Chilembwe, J. M., Mweiwa, V. R., Ray, S., O'Connor, D. P., Garg, A., Urlu, D. K., Albattat, D. A., . . . Ercan, D. A. tin, İ., Koba, D.Y., & Coghlan, D.A. (2020). The Emerald Handbook of ICT in Tourism and Hospitality. *Emerald Publishing Limited eBooks*. https://doi.org/DOI: 10.1108/9781839826887

McKercher, B. and Prideaux, B. (2020). Tourism theories, concepts and models. https://doi.org/DOI: 10.23912/9781911635352-4280

Megha Mudholkar, D.P. (2024). Impact of Artificial Intelligence and Internet of Things on Performance Management: A Systematic Review. *Journal of Informatics Education and Research*. https://doi.org/DOI: 10.52783/jier.v4i1.569

Meril, A. S., Basthikodi, M., & Faizabadi, A. R. (2019). Review: Comprehensive study of 5G and 6G Communication Networks. *Journal of Emerging Technologies and Innovative Research*, 6(5), 715–719. https://www.jetir.org/view?paper= JETIRCU06132

Mohsan, S. A. H., Mazinani, A., Malik, W., Younas, I., Othman, N. Q. H., Hussain, A., & Mahmood, A. (2020). 6G: Envisioning the key technologies, applications and challenges. *International Journal of Advanced Computer Science and Applications*, 11(9). Advance online publication. DOI: 10.14569/IJACSA.2020.0110903

Monaco, S., & Sacchi, G. (2023). Travelling the Metaverse: Potential benefits and main challenges for tourism sectors and research applications. *Sustainability (Basel)*, 15(4), 3348. DOI: 10.3390/su15043348

Nair, P. S. (2018). An Augmentation and Virtuality of Reality: - An Overview. *International Journal of Research*, 5, 250–255.

Nam, K., Dutt, C. S., Chathoth, P. K., & Khan, M. S. (2019). Blockchain technology for smart city and smart tourism: Latest trends and challenges. *Asia Pacific Journal of Tourism Research*, 26(4), 454–468. DOI: 10.1080/10941665.2019.1585376

Nannelli, M., Capone, F., & Lazzeretti, L. (2023). Artificial intelligence in hospitality and tourism. State-of-the-art and future research avenues. *European Planning Studies*, 31(7), 1325–1344. DOI: 10.1080/09654313.2023.2180321

Nayyar, A., Mahapatra, B., Le, D., & Suseendran, G. (2018). Virtual reality (VR) and Augmented reality (AR) technologies for tourism and hospitality industry. *International Journal of Engineering &Amp; Technology*, 7(2.21), 156. https://doi.org/DOI: 10.14419/ijet.v7i2.21.11858

Neuhofer, B., Buhalis, D., & Ladkin, A. (2015). Smart technologies for personalised experiences: A case study in the hospitality domain. *Electronic Markets*, 25(3), 243–254. DOI: 10.1007/s12525-015-0182-1

Nicola-Gavrilă, L. (2023). Empirical approaches regarding interdependency between technology and sustainable tourism. *Journal of Environmental Management and Tourism*, 14(4), 2140. DOI: 10.14505/jemt.v14.4(68).25

Nuriev, M., Kalyashina, A., Smirnov, Y., Gumerova, G., & Gadzhieva, G. (2024). The 5G revolution transforming connectivity and powering innovations. *E3S Web of Conferences*, 515, 04008. https://doi.org/DOI: 10.1051/e3sconf/202451504008

Padhi, P. K., & Charrua-Santos, F. (2021). 6G enabled industrial Internet of everything: Towards a theoretical framework. *Applied System Innovation*, 4(1), 11. DOI: 10.3390/asi4010011

Patel, S. C. (2013). Incremental Journey for World Wide Web: Introduced with Web 1.0 to Recent Web 5.0 - A Survey Paper. *International Journal of Advanced Research in Computer Science and Software Engineering*, 3, 112–125. https://www.scirp.org/reference/referencespapers?referenceid=2228357

Peštek, A., & Sarvan, M. (2020). Virtual reality and modern tourism. *Journal of Tourism Futures*, 7(2), 245–250. DOI: 10.1108/JTF-01-2020-0004

Polese, F. (2009). Local Government and Networking Trends Supporting Sustainable Tourism: Some Empirical Evidence. *Cultural Tourism and Sustainable Local Development*, 131–148.

Pradini, G., Awaloedin, D. T., Agustiani, F., & Kelly, A. (2024). 5G technology transformation in tourism. *International Journal of Economics, Management, Business, and Social Science (Ijembis)*, 4(1), 523–536. DOI: 10.59889/ijembis.v4i1.376

Psiha, M., & Vlamos, P. (2017). IoT applications with 5G connectivity in medical tourism sector management: Third-party service scenarios. *Advances in Experimental Medicine and Biology*, 989, 141–154. DOI: 10.1007/978-3-319-57348-9_12 PMID: 28971423

Qamar, F., Siddiqui, M. U. A., Hindia, M. H. D. N., Hassan, R., & Nguy n, Q. N. (2020). Issues, challenges, and research trends in spectrum management: A comprehensive overview and new vision for designing 6G networks. *Electronics (Basel)*, 9(9), 1416. DOI: 10.3390/electronics9091416

Ramachandran, M., Archana, T., Deepika, V., Kumar, A., & Sivalingam, K. M. (2022). 5G network management system with machine learning-based analytics. *IEEE Access : Practical Innovations, Open Solutions*, 10, 73610–73622. DOI: 10.1109/ACCESS.2022.3190372

Roy, B. K., & Pagaldiviti, S. R. (2023). Advancements in arena technology: Enhancing customer experience and employee adaptation in the tourism and hospitality industry. *Smart Tourism*, 4(1), 2330. DOI: 10.54517/st.v4i2.2330

Saad, W., Bennis, M., & Chen, M. (2020). A vision of 6G wireless systems: Applications, trends, technologies, and open research problems. *IEEE Network*, 34(3), 134–142. DOI: 10.1109/MNET.001.1900287

Sánchez, J. A. H., Casilimas, K., & Rendón, Ó. M. C. (2022). Deep reinforcement learning for resource management on network slicing: A survey. *Sensors (Basel)*, 22(8), 3031. DOI: 10.3390/s22083031 PMID: 35459015

Shafi, M., Molisch, A. F., Smith, P. J., Haustein, T., Zhu, P., Silva, P. D., & Wunder, G. (2017). 5G: A tutorial overview of standards, trials, challenges, deployment, and practice. *IEEE Journal on Selected Areas in Communications*, 35(6), 1201–1221. DOI: 10.1109/JSAC.2017.2692307

Shahraki, A., Abbasi, M., Piran, M.J., Chen, M., & Cui, S. (2021). A Comprehensive Survey on 6G Networks: Applications, Core Services, Enabling Technologies, and Future Challenges. *ArXiv*, abs/2101.12475.

Sharma, S. K., & Sharma, N. (2018). An Overview of 5G Technology. *International Journal of Engineering Research & Technology (Ahmedabad)*, 6(17), 1–3. https://www.ijert.org/research/an-overview-of-5g-technology-IJERTCONV6IS17001.pdf

Silva, M. M., & Guerreiro, J. (2020). On the 5G and beyond. *Applied Sciences (Basel, Switzerland)*, 10(20), 7091. DOI: 10.3390/app10207091

Sujood, S., Bano, N., & Siddiqui, S. (2022). Consumers' intention towards the use of smart technologies in tourism and hospitality industry: A deeper insight into the integration of TAM, TPB and trust. *Journal of Hospitality and Tourism Insights*, 7(3), 1412–1434. DOI: 10.1108/JHTI-06-2022-0267

Sulaiman, R. B., & Kareem, A. (2020). A review on concepts and technologies of 6G cellular network and future scope. SSRN *Electronic Journal*. https://doi.org/ DOI: 10.2139/ssrn.3731869

Thantharate, A., Tondwalkar, A., Beard, C., & Kwasinski, A. (2022). Eco6G: Energy and cost analysis for network slicing deployment in beyond 5G networks. *Sensors (Basel)*, 22(22), 8614. DOI: 10.3390/s22228614 PMID: 36433208

Thüemmler, C., Paulin, A., Jell, T., & Lim, A. K. (2017). Information technology – next generation: the impact of 5G on the evolution of health and care services. *Advances in Intelligent Systems and Computing*, 811-817. https://doi.org/DOI: 10.1007/978-3-319-54978-1_100

Uusitalo, M. A., Rugeland, P., Boldi, M., Strinati, E. C., Demestichas, P., Ericson, M., & Zou, Y. (2021). 6G vision, value, use cases and technologies from European 6G flagship project hexa-x. *IEEE Access : Practical Innovations, Open Solutions*, 9, 160004–160020. DOI: 10.1109/ACCESS.2021.3130030

Valcic, M., & Domsic, L. (2012). Information technology for management and promotion of sustainable cultural tourism. *Informatica (Slovenia)*, 36(2), 131–136. Retrieved from https://dblp.uni-trier.de/db/journals/informaticaSI/informaticaSI36.html#ValcicD12

Velásquez, K., Abreu, D. P., Curado, M., & Monteiro, E. (2022). Resource orchestration in 5G and beyond: Challenges and opportunities. *Computer Communications*, 192, 311–315. DOI: 10.1016/j.comcom.2022.06.019

Vignaroli, L., Gramaglia, M., Fuentes, M., Casella, A., Одарченко, P., Natale, L., & D'Andria, F. (2020). *The touristic sector in the 5G technology era: the 5G-tours project approach. 2020 IEEE Globecom Workshops*. GC WKSHPS., DOI: 10.1109/GCWkshps50303.2020.9367418

Vu, D., Dao, N., Won, D., & Cho, S. (2023). Potential enabling technologies for 6g mobile communication networks: a recent review. 2023 *Fourteenth International Conference on Ubiquitous and Future Networks (ICUFN)*.https://doi.org/DOI: 10.1109/ICUFN57995.2023.10199909

Wang, H., Jiang, Z. H., Yu, C., Hou, D., Wang, H., Guo, C., & Zhou, J. (2021). The role of millimeter-wave technologies in 5G/6G wireless communications. *IEEE Journal of Microwaves*, 1(1), 101–122. DOI: 10.1109/JMW.2020.3035541

Wang, W., Kumar, N., Chen, J., Gong, Z., Kong, X., Wei, W., & Gao, H. (2020). Realising the potential of the Internet of Things for smart tourism with 5G and AI. *IEEE Network*, 34(6), 295–301. DOI: 10.1109/MNET.011.2000250

Wei, W. (2019). Research progress on virtual reality (VR) and augmented reality (AR) in tourism and hospitality. *Journal of Hospitality and Tourism Technology*, 10(4), 539–570. DOI: 10.1108/JHTT-04-2018-0030

Xu, H., Klaine, P. V., Onireti, O., Cao, B., Imran, M. A., & Zhang, L. (2020). Blockchain-enabled resource management and sharing for 6G communications. *Digital Communications and Networks*, 6(3), 261–269. DOI: 10.1016/j.dcan.2020.06.002

You, X., Wang, C., Huang, J., Gao, X., Zhang, Z., Wang, M., & Liang, Y. (2020). Towards 6G wireless communication networks: Vision, enabling technologies, and new paradigm shifts. *Science China. Information Sciences*, 64(1), 110301. Advance online publication. DOI: 10.1007/s11432-020-2955-6

Yuen, S. C., Yaoyuneyong, G., & Johnson, E. (2011). Augmented reality: An overview and five directions for AR in education. *Journal of Educational Technology Development and Exchange*, 4(1). Advance online publication. DOI: 10.18785/jetde.0401.10

Chapter 12
Assessing the Economic Impact of 5G/6G Integration on Tea Tourism Sector in Barak Valley, Assam

Jayeeta Bhattacharjee
https://orcid.org/0009-0009-4505-9413
Assam University, India

K. N. Lokesh Kumar
https://orcid.org/0000-0002-0757-466X
Assam University, India

ABSTRACT

The paper focuses on the 5G and 6G integration on the technological advancements in the digital era. It tries to analyze the economic impacts of 5G/6G integration on the tea-estates of Barak Valley in the state of Assam leading to tourism development in the region. It tries to promote sustainable practices along with the operational efficiency of the local communities with the applications of artificial intelligence through virtual and augmented reality. It would attract tourists and enable virtual tours before being physically accessible to the tea-estates. This would develop infra-structural facilities along with good connectivity. For the current study, secondary sources of data are collected. It tries to have a good impact on the local communities of tea estates leading to economic development in the region.

DOI: 10.4018/979-8-3693-6725-4.ch012

1. INTRODUCTION

Technological advancement is the most significant factor in impacting the economy of any country. It creates opportunities for diverse industries to enhance the workforce, existing infrastructure, the operational efficiency and as well as the services. Being in the digital transformation era, we all need faster networking services for convenient purpose. Now the companies and organizations are using the 5G technology which has a speed of 20 Gbps that takes connectivity to the next level with seamless roaming capabilities. It is driven and assured that by 2030, 6G network to be implemented, which would be sharing traits with that of 5G. Now since in the era of 5G technology, it would advance throughout its network architecture in virtualization, cloud-based technologies, and automation with software-defined platforms. With this technology, augmented reality is made possible.

The tourism industry contributed about INR 16.5 trillion in 2023 to the Gross Domestic Product of India, enabling the growth of economy of the entire nation. This sector aims to improve and expand tourism products to ensure employment generation and economic growth. In order to thrive and exist in the global travel market, technology has a significant impact in the travel experiences of both the international and domestic tourists. With the emergence of digital transformation, the tourists can obtain information regarding the tourism services they would like to attain during their tourism purpose.

Tea industry is one of the emerging sectors in the tourism industry. Assam is one of the tea producing regions in India. It is situated in the north-eastern part of the country. It has been divided into Brahmaputra Valley and Barak Valley. The Cachar district is located in the Barak valley area of Assam with 2 more districts, respectively Hailakandi and Karimganj. Cachar district has 56 tea estates in total and also have the largest organic tea estate in the state, Hailakandi district having 19 tea estates and Karimganj district having 23 tea estates. Even there is a branch of Tocklai Tea Research Institute at Silchar, the hub of the district of Cachar. The nature enthusiasts can now plan a proper itinerary and can enjoy by having a proper accommodation in the tea estates. They can have a virtual tour, before arriving at the destination with the help of automation in artificial intelligence.

With the integration of 5G and 6G networks, it would be beneficial for the tea gardens which would attract the tourists and promote the development of tourism in the region. The 5G technology would be able to enable smart farming practices, as devices would be connected that could monitor soil conditions and help the tea crops grow in real time. This would help in sustainable practices and reduce carbon footprints. It would also be helpful in skills development which would boost in employment generation in the local communities of tea gardens. It would also lead to economic diversification with new business models and partnership opportunities.

Despite of the challenges that include lack of good infrastructural facilities, even there are no proper guides, connectivity issue, the tea estates have the potential to attract the tourists. With the proper strategic implementation of technology, the tea tourism would boost in transforming the Barak Valley.

2. STATEMENT OF THE PROBLEM

The integration of 5G and 6G technology would lead to a growth in the tea estates leading to employment generation, skilled workforce, and economic development. But due to the poor infrastructural facility and improper road connectivity, it becomes challenging for the tea regions. Even the local communities are not able to properly connect with the industrial workforce. Now being in the digital era, technology is required in daily basis for work efficiency. With the implementation of technology, virtual tours can be made possible, where augmented reality would be possible. This would help stakeholders, investors to invest in the tourism sectors in the tea-estates. The local communities could arrange homestays on accommodation purpose for the tourists who would love to explore the tea regions. With strategic implementations, technological advancements would enable the locals to initiate their own startups, leading to employment generation and pave the way for economic growth, flourishing the tourism sector in the region.

3. LITERATURE REVIEW

Abhik Gupta (2009) stated the factors that promote the development of Barak Valley. It emphasized on both the historical and geographical prospects. It focused on the resources and its biodiversity that includes the forest products, its agriculture, and wildlife that would promote the tourism in the valley.

Horen Goowalla and Dhurbojyoti Neog (2011) emphasized on the present scenario of tea tourism by focusing on the strength, weakness, opportunities and threats within the tea tourism sector in Assam. There is a lot of potentiality in the tea estates of Assam for creating it as a lucrative destination.

P K Mishra, Himanshu B. Rout and Smita S. Mohapatra (2011) implicated tourism as an efficient tool for promoting economic growth of the host country. It generates employment and foreign exchange by promoting the overall economic and social development. It also focused that government organizations along with central, private and voluntary organizations can collaborate and invest together to attain sustainable growth in the economy as a whole.

Dr. Dinesh Das (2012) explored the potentiality of Assam to attract both the international and domestic tourists. As Assam is rich in natural resources and also promotes cultural and ethnic tourism. The state is also famous for tea tourism as well and can generate employment opportunities in the region.

Dr S.K.Dey and Dr R.Gupta (2012) stated the quantitative relation between safety and productivity developed for the tea sector of Barak Valley. Emphasis was given on safety and health conditions of the locals of the tea estates. Data that was collected was regressed using regression software to find the correlation between the factors.

Dr.R.Sivanesan (2013) analysed the tea industry in India. It focused on the statistical tools to examine the production and sales of tea, its average selling price, along with the exports and imports of tea.

Dr. T. Subash (2015) emphasized on the Indian tourism industry as the pivotal aspect of service sector in India. It acts as an employment generator for the local and host communities and enhances the economic activity in the region. It is also facing obstacle that hinders the growth and involvement by the stakeholders and the community is pertinent for sustainable development of the sector.

Mrs.Mahalakshmi Venkatesh and Dr.P. Stanley Joseph Michal Raj (2016) stated that liberal policies, relaxation in taxes and other statutory rules influence the tourists and generate foreign investment. Intervention by government is necessary to make tourism flourish in India.

Sarvesh Kumar Shah and Vipal A. Pate (2016) emphasized on the key aspects that would enhance growth in the tea industry of India. Though there are challenges co-existing together as well, the tea industry is prospering with government intervention.

Chandan Kumar Sharma and Prarthana Barua (2017) highlighted on the tea plantations in the state of Assam. It focused on the socio-economic profile of the state and how it impacted on the economy of the local communities of the tea estates. It shredded light on the matter that in the past how a dramatic growth in the small planters has brought about a transformative change in the rural landscape of Assam.

Martin Zsarnoczky (2017) explained the importance of artificial intelligence in the prevailing digitalization period. Since, it is used in every fields, tourism industry is highly affected as well. Different smart systems and chatbots are used in travel agencies and air transport companies. The process of continuous progress seems to be unstoppable which will lead to further re-arrangements in the technological revolution.

Biggsa et al. (2018) explored the socio-environmental factors that impose a threat in the tea productivity and the workers in the tea garden regions of Assam. They highlighted on the necessity for a versatile and climate-smart landscape activities with the suggestions from the tea advisory expert committee to promote sustainability for long term socio-environmental cultivation of tea under changing climatic conditions.

Mark Anthony Camilleri (2018) explained the tourist's motivation to travel to a destination. It focuses on the inter-relationships between different sectors of travel, tourism and hospitality. The study also gives light to the essentials of tourism that includes accessibility, accommodation, activities, attractions and amenities. It also tries to elucidate on the destinations that offers different products to satisfy the expectations of the tourists.

Rua-Huan Tsaih and Chih Chun Hsu (2018) proposed a conceptual framework for artificial intelligence in smart tourism for maintaining big data for business innovations. It integrates machine learning, institution and business processes to assist smart tourism stake holders to build a business level IT strategy.

Jayasooriya S. S. W (2019) assessed the potential of tea tourism products that could be developed in the community of Hanthana and neighboring countries. It focused on the consumer behavior and their preferences towards the specific product. It tried to provide justifications for future researchers.

Luca Vignaroli, Marco Gramaglia, Manuel Fuentes, Antonino Casella, Roman Odarchenko, Lorenzo Natalek, Baruch Altman and Francesco D'Andria (2020) discussed about the business potential of 5G network services. They emphasized on the importance of utilizing 5G technology to promote rural tourism through augmented and virtual reality to bridge the gap between physical and virtual world for better tourist experience.

Rimi Rekha Phukon, Papul Das and Golap Borah (2020) explored the potentialities and barriers of the growth of tourism sector in Assam. The state has immense possibilities that are yet to be explored and utilized as commercial resources. It has the potential to grow foreign exchange earnings, create employment and bring development and economic stability to the state.

Suneel Kumar and Shekhar (2020) highlighted the involvement of the strategy makers to identify the factors that propel tourism industry and focus on digitalization. Also, addressing the shortcomings and working on the process. They used the interpretive structural model to develop the strategic framework related to technological advancements.

Lidia Aguiar Castillo, Victor Guerra, Julio Rufo, Jose Rabada and Rafael Perez Jimenez (2021) focused on the applications of optical wireless communication in the tourism sector. It also focused on the implementation of technologies along with the traditional services to face the obstacles that the tourists bear during their travel to their preferred destinations.

Jacques Bulchand-Gidumal (2022) examined the relevance of artificial intelligence in tourism and hospitality industry. Though there are challenges, with the help of artificial intelligence applications such as personalization, robots, smart travel agents, prediction and forecasting systems, smart travel agents have enabled necessary improvements.

Muhammad Sajjad Akbar, Zawar Hussain, Quan Z. Sheng and Subhas Mukhopadhyay (2022) explored on the various aspects of 6G networks for better economic prospects in the future. It would bring high data rates and reliability in the personal mobile networks for convenient use. The study highlighted on the integration of 6G with advanced architectures along with artificial intelligence with its potential futuristic cases of virtual reality and Industry 5.0.

Chandan Jyoti Bezborah (2023) focused on the potential of tea tourism in the state of Assam. It discussed about the challenges and various approaches that could uplift the economy. It also emphasized on the aspect of community based tourism and agri tourism.

Miguel-Angel Garcia-Madurga and Ana-Julia Grillo-Mendez (2023) aimed to identify the emerging areas of artificial intelligence in tourism sector. It emphasizes on the sustainability initiatives and illuminating pathways for future research by augmenting customer service, improving the operational efficiency along with personalizing travel experiences.

Suman Kalyan Chaudhury, Sukanta Sarkar and Chandra Sekhar Pattnaik (2023) explored the prospects and the challenges of sustainable tourism in the state of Assam. It suggested looking after the obstacles through a strategic and sustainable approach, prospering the growth and development in the state.

4. SIGNIFICANCE OF THE STUDY

The 5G/6G integration would play a major role in uplifting the economy of the tea estates in the Barak valley region of Assam. It would enhance tourist experience through virtual and augmented reality and improved digital services. It would also foster regional economic growth by transforming it into a technologically advanced and high appealing tea tourism destination. The local residents would also benefit by enhancing their livelihood having access to digital services for convenient purpose. The infrastructure would also be developed through public-private partnerships. Implementing sustainable practices along with smart technologies can aid in monitoring and maintaining ecological balance. Government policies if structured in proper ways can accelerate the growth of tea tourism sector in the valley.

5. LIMITATIONS OF THE STUDY

1. There is a lack of previous studies done in tea tourism in Barak valley.
2. The technologies are rapidly changing. With passage of time, these technologies maybe outdated.

3. The tourism industry is dynamic and certain findings could be less relevant over time.

6. OBJECTIVES OF THE STUDY

1. To understand the growth and scenario about the tea industry in Barak Valley.
2. To explore the potential of 5G/6G technologies for enhancing tourism in tea estates of Barak Valley
3. To analyze the economic impact of 5G/6G integration on the tea tourism sector in Barak Valley.
4. To provide recommendations for development of tea tourism in the region with special emphasis on 5G/6G integration.

7. RESEARCH METHODOLOGY

The present study is confined to the tea estates in Barak Valley of Assam, comprising of Cachar, Hailakandi and Karimganj district. For the current study, secondary sources of data have been taken into account.

8. TEA INDUSTRY IN BARAK VALLEY

Tea is one of the important beverages in the world. India is one of the major producers of tea around the world. In 1823, the origin of tea industry started in the state of Assam, the north eastern part of the country. In the recent years, after the outbreak of covid-19, there is a decline in the production of tea. The Government of Assam has taking initiatives to revive the situation. As per reports from government sources, the labourers were not receiving their wages proportionately. To improve the situation, schemes related to labour wages were implemented to boost the tea production.

The state of Assam, being geographically divided into two parts, namely, Brahmaputra Valley and Barak Valley. Barak Valley being comprised of three districts-Cachar, Hailakandi and Karimganj. The Valley is bordered by Indian States, Mizoram and Tripura to the South, one country Bangladesh and other Indian states Meghalaya to the West and Manipur to the East. The region is popularly known as Barak Valley for the river Barak. The valley is known for its festivals, fairs and other events. It also plays a good role in the agricultural production of the state. Tea crops are the most influential agricultural crop in the valley. They contribute around 48% of tea production to the state of Assam. To ensure sustainability, the tea estates of Barak Valley produce the organic tea crops for the consumption purpose. There is a tea research centre even, named as Tocklai in the Cachar District of Barak Valley. The centre gives a detailed analysis for researchers, how the tea leaves are being processed to tea for consumption, how termites can be reduced or if any leaves get affected then how can the leaves be used. Even the tourists from all over the world can come and visit the research centre and the other tea estates in the valley.

The local communities in the tea estates are highly impacted by the tourists visit to the tea estates. Here the local people can present their products and commodities for the tourists who come and explore the tea estates. The local communities can also show their culture, ethnicity, handicrafts, traditional folklores and many other that promote their community. The locals can earn a livelihood from such purpose. Their economy would also increase and make them aware of the current happenings around them.

The tea estates have been a natural asset for the state of Assam. Various plans have been incorporated by the state government to promote tea estates in the Barak valley, to attract tourist and prosper the tourism as well as other sectors important for the growth of tea industry in the Valley. Even in many tea estates, where factories aren't used, but only tea leaves are processed and sold and collaborated with other tea estates for further consumption of tea and which can be used by the tourists or even the tea lovers all around the world. One of them is kailashpur Tea Estate, where the tea crops are being collected and sold to other tea estates factories for consumption purpose.

The Board of Tea Association of India, Barak Valley Branch, that handles and takes important decisions regarding all the tea estates in the Valley, conduct an annual meeting every year, to discuss about the activities, finances that took place in a particular financial year. They also prepare budget for the further financial year, take decisions related to the development of tea industry.

Every year, new policies and plans are implemented to develop the tea estates, to improve the road conditions of the valley. The scenic natural beauty of the tea estates and the cultural experiences captivates the mind of all the tourists across the world. However, this 5G connectivity can be established through digital infrastructure, complementing the traditional ways of connectivity.

Figure 1. WEST JALINGA TEA FACTORY IN THE TEA-ESTATE, CACHAR DISTRICT, ASSAM

Figure 2. THE ROAD INFRASTRUCTURE OF THE TEA-ESTATE

9. POTENTIAL OF 5G/6G TECHNOLOGIES IN ENHANCING TOURIST EXPERIENCES

Integration of 5G/6G technologies has a great potential to develop the region and attract tourists at mass level. Artificial intelligence along with traditional ways of prospering and communicating with one another can enhance the tourist experiences in a more beautiful and charismatic way, appealing to the tourist. There are certain factors that attract tourists to the lush green landscapes of natural beauty, more prominently the Barak Valley.

5G/6G can enable high quality virtual and augmented reality. It will help tourists to have virtual tours of tea plantations, how tea is being processed in factories and can enjoy cultural presentations. The tourists before going in the real world of tea gardens can have an artificial demonstration of the tea estates for better tourist experience in real. Customization during the homestays of the tourists in their tour visits in the tea estates can also be done through smart applications that include the seamless navigations, personalized recommendations, and cardless payments. The network towers in every tea estates would lead to improved digitalization for proper connectivity and applications would be used soothingly without any disturbance.

The adoption of 5G technologies can enable smart farming practices. The smart devices connected through 5G can monitor soil conditions, weather patterns, crop health that determines how much it can yield quality crops with less cost and minimum time through automated systems. This would develop the operational sector of the tea factories. It will enhance efficient tracking and logistics for tea distribution and focuses on reducing cost and good quality of the consumable product.

Technology would help in the marketing and promotion of the tea estates and their products. With the use of augmented reality, the target market audience can be identified through personalized marketing campaigns. It can be done through data analytics. Even social media integration can be used as a tool for advertising the potential of the tea-estates by engaging meaningful content by the media influencers and providing content to attract tourists from global audience.

It would create and foster good relationship patterns between the local people and all the people around the globe through virtual interaction. It would also help to sustain the growth of the Barak Valley. It would evolve a sense of social interaction leading to development of the tea-estate regions, improving road connectivity, good social living and livelihood.

10. ECONOMIC IMPACT OF 5G/6G INTEGRATION

5G/6G integration is going to be a dynamic strategy towards the development of tea industry in Barak Valley. This would lead to economic, social, environmental changes in every aspect. Concentrating on the economic aspect of development, it would create a huge impact on the Valley. The local communities will be developed; they would be accessible to literacy, good educational background, improving their standard of living. It would bring transformation in the Valley creating a lot of technology experts.

The capabilities of 5G will enhance innovation and development of new business models within the tea tourism sector. This will lead to economic diversification by providing online platforms for direct tea sales and digital marketing strategies to reach global audiences. There would be technological advancements in the region, leading to mutual partnerships and collaboration among technology providers, tea estates and tourism operators promoting tourism in the Valley. It will empower startups, employment, thus creating jobs and encourage skill development. In this process, technology based jobs will get started related to smart agriculture and supply chain management. Workshops will be collaborated with the government organizations for skilled based awareness programs. This will generate skilled workers. Locals would also be benefitted from the workshops, which would reduce economic inequality.

The locals would be aware and create an environment for tourism purpose. This would create employment opportunities of the local communities, where they will be engaged in showcasing their local culture, handicrafts, and ethnicity and make it known to the tourists, who come and visit the tea estates. There would be stalls, souvenirs shops, restaurants, transportation services linked to the tourist attraction sites. Even with this, the hospitality sector also accounts to the development of tea estate regions of Barak Valley. This would add to the revenue of the state government, contributing to the overall economic growth of Barak Valley. Even with this, the hospitality sector also accounts to the development of tea estate regions of Barak Valley.

The 5G technology which is the current technology network is soon going to be integrated with 6G network by 2030. Good marketing strategies and good experiences of tourist would attract more & would lead to increase in tourist footfall. This will stimulate further technological advancements. The infrastructure will be developed and the connectivity would increase and the local communities would avail services that were not earlier available, that also at minimum cost. Automated and smart logistics would help to reduce the carbon footprint of tea production and distribution. This helps to align with the global sustainable goals for environmental protection. The tea estate owners, stakeholders, investors, traders, the government organizations, the non-governmental organizations can collaborate together and create new projects based on 5G technology. This would foster economic growth in the region implying jobs and the business of the local communities can prosper with strategic implementations.

The tourists who would come to visit the tea-estates could also have a tour of the whole valley including the rest of the places in the state of Assam. For that, the tourists can have their tour planning from the nearest travel agencies, where they can have their tour packages at their preferences. This is where the technology would help them and ease their travel experiences in the beautiful landscapes of tea estates of Barak Valley. The tourists here can opt for virtual tour so to have a demonstration of the tea estates filled with ethnicity and ecstasy. This would boost the revenue of the tea estates pertaining to the overall economic growth and development.

11. RECOMMENDATION FOR DEVELOPMENT OF TEA TOURISM IN BARAK VALLEY THROUGH 5G/6G INTEGRATION

The integration of 5G with 6G technology is going to bring a significant change in the digital transformation of the tea estates in Barak Valley. However with the potentials, there are certain obstacles in the way of its progress. There are certain recommendations for the development of the tea regions of Barak Valley through the intervention of technology.

a. Investments in 5G/6G integration would be pivotal for widespread coverage and reliability. It would create good infrastructural technological equipments leading to mass employment of technology skilled workers. It would enable for economic diversification heading towards economic growth in the Valley.
b. Partnerships and collaboration among government, technology providers and local business to facilitate integration and innovation to boost up the economy of the tea estate regions, would be a good economic strategy.
c. Training programs and workshops on skill awareness program would enhance the local workforce to learn about new technologies and ease the service delivery.
d. It should be recommended that technological advancements align with the sustainable tourism principles to preserve the natural and cultural heritage of Barak Valley.
e. Local communities should be given priority and funds should be provided by the state of Assam to promote the cultural ethnicity and beautiful lush green landscapes of Barak Valley to make it a tourism attraction sites for the tourists across the world.
f. It is highly recommended that infrastructure and connectivity should be given utmost importance. Roads should be maintained and transportation should be made available aptly whenever required.
g. Homestays should be promoted, along with that proper training and awareness must be made about this concept for the locals residing in those tea estate regions. This would lift the economy of the regions.
h. Old bungalows can be converted to heritage hotels for the tourists who prefer to enjoy the natural landscapes and explore the tea-estate regions.

i. Tea estates also could be a wedding destinations as according to recent trends and reports from the ministry, people are very much fascinated to this kind of wedding destinations in the lap of nature.

Barak Valley is a land of natural resources with its sustainable tourism approach along with the technological integration of 5G with 6G network in the tea estate region is going to be a paradigm shift in the tourism industry.

12. CONCLUSION

The integration of 5G and future 6G technologies holds a transformative opportunity for the tea tourism industry in Barak Valley, Assam. By enhancing operational efficiency and improving the tourism experience through technological advancements can significantly influence the economic development in the region. It also enables sustainable practices through strategic approaches involving collaboration between technology providers, local business and government agencies in the tea estate regions.

REFERENCES

Aguiar-Castillo, L., Guerra, V., Rufo, J., Rabadan, J., & Perez-Jimenez, R. (2021). Survey on optical wireless communications-based services applied to the tourism industry: Potentials and challenges. *Sensors (Basel)*, 21(18), 6282. DOI: 10.3390/s21186282 PMID: 34577489

Akbar, M. S., Hussain, Z., Sheng, Q. Z., & Mukhopadhyay, S. (2022). 6G survey on challenges, requirements, applications, key enabling technologies, use cases, AI integration issues and security aspects. *arXiv preprint arXiv:2206.00868*.

Bezborah, C. J. (2023). Tea tourism, the way forward to promote agritourism in the state of Assam: Prospects And Challenges. *International Journal of Multidisciplinary Trends*, 5(9), 1–4. DOI: 10.22271/multi.2023.v5.i9a.330

Biggsa, E. M., Guptac, N., Saikiad, S. D., & Duncana, J. M. (2018). The tea landscape of Assam: Multi-stakeholder insights into sustainable livelihoods under a changing climate. *Environmental Science & Policy*, 82, 9–18. DOI: 10.1016/j.envsci.2018.01.003

Bulchand-Gidumal, J. (2022). Impact of artificial intelligence in travel, tourism, and hospitality. In *Handbook of e-Tourism* (pp. 1943–1962). Springer International Publishing.

Camilleri, M. A., & Camilleri, M. A. (2018). *The tourism industry: An overview*. Springer International Publishing.

Chaudhury, S. K., Sarkar, S., & Pattnaik, C. S. (2023). Assam Tourism-Scenario: The Path Trodden and the Way Forward. *Journal of Global Economics*, 19(4).

Das, D. (2012). Prospects and problems of tourism industry in Assam. *Global Research Methodology Journal*, 2(7), 1–6.

Dey, S. K., & Gupta, R. (2012). Development of Safety and Productivity Correlation Model for Tea Industries of Barak Valley, Assam. *Development*, 2(12).

García-Madurga, M. Á., & Grilló-Méndez, A. J. (2023). Artificial Intelligence in the tourism industry: An overview of reviews. *Administrative Sciences*, 13(8), 172. DOI: 10.3390/admsci13080172

Goonwalla, H., & Neog, D. (2011). Problem and prospect of tea tourism in Assam-a SWOT analysis. In *2011 International conference on advancements in information technology with workshop of ICBMG* (Vol. 20, pp. 243-248).

Gupta, A. (2009). *Development of Barak valley: the Question of Sustainability. Development strategies of Barak valley (Assam).* Akansha Publishing House.

Jayasooriya, S. S. W. (2019). Exploring the potentials, issues, and challenges for community-based tea tourism development (with reference to Hanthana Mountains). *International Journal of Advance Research. Ideas and Innovations in Technology*, 5(2), 475–480.

Kumar, S., & Shekhar, . (2020). Digitalization: A strategic approach for development of tourism industry in India. *Paradigm*, 24(1), 93–108. DOI: 10.1177/0971890720914111

Mishra, P. K., Rout, H. B., & Mohapatra, S. S. (2011). Causality between tourism and economic growth: Empirical evidence from India. *European Journal of Soil Science*, 18(4), 518–527.

Phukon, R., Das, P., & Borah, G. (2020). Prospects and barriers of tourism industry in Assam. *International Journal of Advanced Research in Engineering and Technology*, 11(12), 3367.

Shah, S. K., & Pate, V. A. (2016). Tea production in India: challenges and opportunities. *Journal of Tea Science Research, 6.*

Sharma, C. K., & Barua, P. (2017). Small tea plantation and its impact on the rural landscape of contemporary Assam. *International Journal of Rural Management*, 13(2), 140–161. DOI: 10.1177/0973005217725454

Sivanesan, R. (2013). Tea industry in India–Analysis of import and export of tea. *International Journal of Business and Management Invention*, 2(8), 9–15.

Subash, T. (2015). Tourism in India: Potentials, Challenges and Opportunities. *IJRAR-International Journal of Research and Analytical Reviews*, 2(4), 8–18.

Tsaih, R. H., & Hsu, C. C. (2018). Artificial intelligence in smart tourism: A conceptual framework.

Venkatesh, M., & Raj, D. J. (2016). Impact of tourism in India. *International Journal of Scientific Engineering and Applied Science*, 2(1), 167–184.

Vignaroli, L., Gramaglia, M., Fuentes, M., Casella, A., Odarchenko, R., Natale, L., & D'Andria, F. (2020, December). The touristic sector in the 5G technology era: The 5G-TOURS project approach. In *2020 IEEE Globecom Workshops (GC Wkshps* (pp. 1-6). IEEE.

Zsarnoczky, M. (2017). How does artificial intelligence affect the tourism industry? *Vadyba*, 31(2), 85–90.

Chapter 13

Cooperative Relaying for Hyper Reliable Low Latency Communications in 6G Radio Networks and Its Potential Research Directions

Rajeev Kumar
Indian Institute of Information Technology, Sri City, India

Sunandita Debnath
Indian Institute of Information Technology, Vadodara, India

Ashraf Hossain
National Institute of Technology, Silchar, India

ABSTRACT

This chapter addresses the cooperative relay communications in 6G radio networks for high reliability and low latency communications. In this regard, we consider two-hop buffered relay network. To achieve these goals, advanced protocols need to be developed that are state-of-art in the cooperative relaying networks. To mitigate the delay caused by data packets arrival and scheduling at the relay, the advanced random-access protocol and non-random-access protocol are investigated for buffer state information (BSI). Following this, an average queueing delay (AQD) is formulated in closed-form which is minimized by satisfying certain queueing stability of buffer and maintain the constant power level by exploiting the Markov

DOI: 10.4018/979-8-3693-6725-4.ch013

reward process (MRP) at the relay for RA protocol. Furthermore, the closed-form expression of the AQD is also, derived for NRA protocol. Our proposed numerical results perform better as compared to existing schemes available in the literature. In addition, this chapter provides the research directions in the high-speed mobility scenarios for 6G radio system.

INTRODUCTION

The cooperative relaying in cellular system is a global communication approach in practice to provide reliability to the network, and enhance coverage area, performance gains, and prolong the lifetime of the 6G radio networks. A simple two-hop relay fading channel system has been studied by (Nabar et al., 2004) to execute message in two communication blocks of time-slots: in the first time slot, it receives a packet from the source and forwards the packet to the destination in the second slot. It is noticed that the bufferless relay has been used which bounds the time slot flexibility as the channel between the relay and destination is highly faded i.e., if source transmits packets continuously to relay, these packets are dropped at the relay. Since, the relay does not equip with buffer. To overcome this limitation, the buffer at the relay has been added by Xia et al. in 2008 that stores the received data packets from source in fixed numbers of the time slots and thereafter, it transmits modified data packets to destination. The ultra-reliable and low-latency communications (URLLC) play key role in existing cellular networks (i.e., 5G radio networks) to enable wide range of applications such as industrial automation, intelligent transport system, telemedicine, and tactile internet (Airod et al., 2021). The term '*reliability*' refers to measure the how "*a communication system delivers data accurately and reliably*" from one terminal to another terminal. It is a ratio of number of successful bits received at the receiver to total number of bits transmitted from the source. However, the term '*latency*' is related to measure of how much time takes to transmit a data packet from one terminal to another terminal. In the 5G radio networks, latency has been targeted and set to be to 1 millisecond. It is achieved through physical layer technology is called as transmission delay in the radio access networks. It is noticed that the transmission delay is small fraction in practices. In upper layer networking, a stochastic delay so-called as queuing delay, processing delay, and access delay are key factors to affect the performance gains that realizes the URLLC in the 5G radio system.

In previous literatures, it is also found that the policies of the relay reception and the relay transmission of the information are not dynamic in the time slots which affects the performances of wireless system. Further, Zlatanov et al. (2013) have exploited channel state information (CSI) in the buffered relay network to provide dynamic time slot to relay for transmitting and receiving of the data packets. Without buffer at the relay, and without knowing CSI reduce the performances of wireless systems. Thus, we need buffer at the relay and channel state information to improve the system performances. It has been observed that Islam et al. (2014) proposed link selection protocol to investigate the queuing delay at the relay determine the average end-to-end (E2E) delay for both perfect and outdated CSI. However, the relay selection (RS) policy remains unspecified. It is powerful policy to select suitable relay from array of relays for communication. The relay selection process depends on either quality of channel, i.e., strongest channel from one terminal to another terminals, or buffer state, or both states. The buffer state-based RS scheme at the relay has been applied to investigate average packet processing delay (Luo and The, 2015). The statistical delay guarantees the quality of service (QoS) in presence of random channels. Qiao and Gursoy (2016) introduced statistical delay constraints under delay violation probability to determine the effective capacity of the two-hop buffered relay network (THBRN). In previous works, it has been found that the average delay for RS policies can be achieved within a few milliseconds across time slots. However, ultra-low latency for 5G radio networks has been set to be ~ 1ms by IMT2020. The buffered RS policy has been proposed (Tian et al., 2017) to reduce average delay, where higher priority is given to the relay-to-destination link instead of the source-to-relay link. The full-duplex (FD) operation at the wireless node is a technique by which it can receive and transmit the signal simultaneously over channel in same frequency band. In this operation, the pre-log one-half factor from channel capacity is removed to enhance the performances of wireless system. An adaptive reception/transmission at the FD relay has been employed with self-interference cancellation (SIC) to achieve high data rate and low latency by exploring the concept of adaptive power allocation and fixed power allocation (Razlighi and Zlatanov, 2018). However, an operational method has not been implemented to balance the maximum throughput and low performance delay. The data packet scheduling and threshold-based flow control policies, as discussed by Shin and Baek (2019), have been utilized to investigate the optimal throughput and minimal cost of finite delay for a multi-buffered FD relay system. In previous studies, the concept of delay-sensitive path selection has not been explored. In 2020, Alam et al. considered multihop buffered relay networks to achieve low latency by exploring the concept of delay-sensitive path selection (DSPS) policy at each relay to reduce the signal processing delay, where the DSPS policy depends on greedy scheduling algorithm to mitigate the limitation of delay enforced by processing delay. In 2021, Kumar and Hossain have addressed the cooperative relaying communications and its application in Internet-of-Things (IoT) for 5G radio networks.

The physical layer security and URLLC in the wireless networks are greatly immersed research in an academia and an industry in the recent era. There are needed to provide internet connectivity in the homogeneous-and-heterogeneous cellular systems. Therefore, proper selection relay nodes in cooperative communication can improve the energy efficiency, spectral efficiency and offers the low-latency, and robust the transmission with high security in real-time applications. The short-packet transmission in selective DF relay (Farhat et al., 2018), the relay with jammer (Wang et al., 2019), random mobile receivers (Yu et al., 2022) have been studied to secure and insure the URLLC in the wireless networks. Aforementioned works have not considered unmanned aerial vehicle (UAV) which acts cooperative relay to reduce the latency and robust the network in the dynamic wireless environments. The UAV-aided relay with friendly jammer has been presented to protect the communication by exploiting artificial intelligence in the wireless networks from the eavesdropper (Li et al., 2021). Recently, the intelligent reflecting surface (IRS) is promising a new radio technology to reconfigure environment to enhance spectral efficiency of wireless system. Traditionally, it consists of passive reflecting elements to structure signals in specific format to meet the characteristics of wireless propagation channels. The IRS-aided cooperative communication supports the URLLC in the 5G and 6G radio networks. The IRS-aided URLLC is studied by Zhang et al. (2022) to transfer wireless information and power with the assistance of short packets transmission. The mobile devices in wireless networks need to support heterogeneous services to reduce the effect of traffic features with high QoS that can be overcome with the help of IRS.

In the recent era, the hyper reliable and low latency Communications (HLLC) is state-of-art in the 6G radio networks. The cooperative communications and its application, and other new radio access techniques are proposed for 6G radio networks that are shown in Fig. 1. These are promising techniques that meet the demands of the IoT networks, industry IoT 5.0, automatic riving system, healthcare, and smart grids in the real-time applications that have been discussed in details (Chowdhury et al., 2020). The Lin et al. (2021) demonstrated the capability of edge computing with assistance of software-defined vehicular network infrastructure under ultra-low latency communication for shaping autonomous driving system in human society in next five to ten years. In the cellular network, massive access of the static users with ULLC system has been studied by exploring grant free non-orthogonal multiple access (NOMA) (Liu et al., 2022).

In high mobility scenario, orthogonal frequency division multiplexing (OFDM) may not be functioning properly due to effect of high doppler spreads. So that, dynamic topologies are needed to support the aligning the beams in spatial multiplexing system. An orthogonal time frequency space (OTFS) is optimistic technique that overcome the imitation imposed by Doppler spreads at the physical layer and provide the HRLLC. It was first proposed by Hadani and Monk (2017) in cooperation of the two-dimensional (2D) Doppler delay spread policy to multiplex signals at the transmitter. Here, orthogonality between adjacent subcarriers has been diluted. To fulfil the necessity, Hadani and Monk (2018) have studied OTFS technique for vehicular communication with high doppler shift. But, inter symbol interference (ISI) has not been mitigated. Further, Raviteja et al. (2018) have investigated input-output relation by exploiting OTFS modulation and demodulation for delay Doppler channels, where ISI and inter carrier interference have been cancelled with suitable phase shifting by exploiting message passing algorithm. In the previous works, authors have focused on design of transmission over time invariant channels. However, it has not been supported the time-varying channels. To overcome this problem, Liang et al. (2020) employed index modulation in OTFS (OTFS-IM) modem for transmitting the indices information by combination of active and silent grids in the delay-Doppler environment. In previous literatures, it is notice that the OTFS systems suffers from hardware impairment (HI). Since, an in-phase and quadrature phase (IQ) imbalance, and direct current (DC) offset are caused an image Doppler interference between the subcarriers in the high mobility delay Doppler scenarios. To mitigate this, the OTFS transceiver with cyclic prefix (CP) and a zero-padded OTFS system for estimating and compensating IQ hardware impairments have been employed Neelam and Sahu (2022) in the delay time domain environments. In the existing literature, authors have focused on the HI to examine the compensation and estimation parameters. It is not feasible to compensate the HI impairments practically if we do not identify delay-Doppler information. Further, Singh et al. (2023) have employed deep learning (DL) in the OTFS system to train the signals to reduce an effect of the HI. Furthermore, they have introduced DL based orthogonal time-frequency space multiplexing (OTSM) to detect the signal in presence of HIs. Yuan et al. (2023) addressed a survey on the OTFS modulation and provided sensing predictive beamforming technique to compute the URLLC in the vehicular networks. However, the 1D unitary transformation is suitable for reducing the complexity of modulation and demodulation. Since, it converts received signal into delay-time domain. In addition, an OTFS based on discrete cosine transform has been proposed by Kalpage et al. (2024) to investigate peak-to-average power ratio under same bit error rate.

Motivated from above literature and advancement in next-generation radio system, cooperative relay communication is promising technique to enhance coverage area, performances, and lifetime-time of the networks. In this context, one relay or multiple relays are employed to transfer information from source to destination. The delay caused by data packet arrival from source and scheduling it at the relay are very challenging and open research problem. To overcome these challenging, advanced buffered relaying protocols are needed to implement in 6G radio system.

The contributions of the chapter are summarized as follows. A two-hop buffered relay network is introduced, where source sends data packets to destination through relay node. To mitigate delays caused by packet arrival and scheduling at the relay, both random-access (RA) and non-random-access (NRA) protocols for buffer state information (BSI) are examined to derive closed-form expressions for the average queueing delay (AQD). To improve the QoS, the objective function for the RA protocol is initially formulated as a linear-fractional program and subsequently converted into a linear program using the CVX tool. Our proposed protocols show better performances than existing schemes available in the literatures. In addition, it is notifying that the average queuing delay based on the RA protocol is showed superior performance than the NRA protocol.

The rest of the chapter is structured as follows. Next Section, we provide basics concepts of delay caused by radio access networks and background of the related research works in details. Then, system model and their preliminaries are presented subsequent section. Subsequently, the RA and NRA protocols, utilizing buffer state information, are analyzed to obtain the expressions of the average queueing delay in closed-form for Rayleigh fading channel. In the next-Section, numerical results in term of average queueing delay are discussed in brief. Thereafter, we provide open research directions for next-generation radio systems. Finally, we summarize this chapter.

BACKGROUND

With rapid growth of the real-time applications in the modern era, the time sensitive services, and the hyper-low latency requirements are set forth for 6G radio networks.

BASIC IDEA OF THE DELAY/LATENCY FOR THE RADIO ACCESS NETWORKS

Mostly, the E2E delay is caused by radio access network, fronthaul-and-backhaul transmission, and core network. Thus, the radio delay in wireless communication is defined as sum of following delays:

a) ***Transmit delay:*** it is processing delay which is primarily contributed by accessing process (i.e., beam- management and orientation, and authorization), information encoder, channel encoder, and modulator.
b) ***Receive delay:*** it is triggered by receiver receiving technique, channel decoder, information decoding, and demodulation.

c) ***Transmission delay***: it is elapsed time to reach the signal from transmitter to receiver via wireless/wired channel.
d) ***Frame alignment delay:*** it is contributed by different frame structures and channel access techniques such as time division duplexing and frequency duplexing.
e) ***Queueing delay:*** it is waiting time to execute the data packets in the buffer.
f) ***Caching delay:*** It occurs if system makes the requests to retrieve the packet from cache memory.

THE URLLC BASED ON COOPERATIVE RELAYING COMMUNICATION

In cooperative relaying communication, the queueing of the data packets at the buffered relay is state-of-the art by exploiting Markov chain (MC) that is greatly involved to improve the performance of the wireless system. Shafie et al. (2014) have proposed protocols under consideration of buffer state at the relay and buffer with partial transmit channel state to investigate the source throughput. In contrast to deterministic channel, stochastic channel models attempt to capture the effect of the small-scale fading which is imposed by shadowing and multipath received signals that effects the nodes connectivity if the nodes are separated from one to other with certain distance. Further, the stochastic link selection for THBRN has been presented by Zhou et al. (2015) to maximize the throughput of system by exploring features of Markov decision process. Jamali et al. (2015) have considered two-way buffered relay network to enhance the throughput gains for delay-constrained protocols. Furthermore, they have presented multi-hop buffered relay networks and its performance gain is based on transmission mode along with transmission rate selection policies under delay constrained policy. In previous studies, authors have typically considered independent and identically distributed (i.i.d.) fading channels. However, in practice, channels vary over time for low-speed mobile users. Huang and Cai (2016) assumed a time-correlated fading channel in the THBRN system to evaluate performance in terms of throughput, E2E delay, and outage probability. Here, time-correlated fading affects the performance analysis because state transition probabilities of the buffer occupancy is to be time variant. To mitigate this problem, the quasi-birth-death MC is used to integrate both buffer occupancy and the channel fading processes. Impact of outdated CSI for link selection at the buffered relay has not been considered. Islam et al. (2016) implemented link selection policies using known or outdated CSI to reduce average delay and enhance throughput by considering finite size buffer and infinite size buffer. If the size of the networks is increased, the computational complexity also increases that need strong and efficient methods. Manoj et al. (2016) have considered multi-hop relay system to reduce the outage probability and average packet delay by exploiting max-link RS policy under state-cluster.

The guarantees of the QoS are greatly involved in support of the random-access wireless channel. Siddig and Salleh (2017) proposed the THBRN with an RS scheme for delay-sensitive scenarios to reduce outage and packet dropping probabilities, while improving throughput gains. Wicke et al. (2017) proposed a rate selection scheme at the source and relay to improve throughput and minimum delay under delay constrained protocol. In the relay communications, the packet scheduling along with proper channel selection is very challenging. Two-way buffered relay network (TWRNs) has been presented by Kumar and Hossain (2017) to investigate the advanced protocols by exploiting one dimension birth-death MC at the relay to enhance the throughput gains. The energy harvesting (EH) is beneficial for energy constrained relay node/wireless devices. In Wang et al. (2018), the joint link selection and power allocation to relay have been introduced to secure the communication in the THBRN. Further Raza et al. (2018) have investigated performance of the buffered relay networks by exploring the concept of the RS policy which depends on buffer occupancy along with maximum weighted link from either S2R link or R2D link. Furthermore, the buffer state dependent hybrid RS policy has been introduced by Nomikos et al. (2018) to achieve average delay of the THBRNs. In the previous works, the relay has not been acknowledged to receive and transmit data packet simultaneously from both source terminals. Kumar and Hossain (2019) have explored the concept of the signal-to-noise ratio region to transmit and receive the data packets simultaneously at the relay to reduce the outage probability and improve the E2E throughput and throughput of the system. Kumar and Hossain (2019) presented a brief survey on cooperative relaying communication, highlighting the application of MC and suggesting potential research directions for future studies.

In the previous studies, the RS scheme with a latency mechanism to address signal processing delays at the relay has not been assumed. Alam et al. (2020) proposed a delay-sensitive path selection method to minimize delay in multihop buffered relay networks. However, they did not explore the RS scheme based on buffer state. Xu et al. (2020) introduced an RS scheme based on queuing diversity, selecting the S2R/R2D link to achieve outage probability and average delay in cooperative networks. Additionally, Rjeily (2020) examined diversity gain, coding gain, and asymptotic delay, deriving closed-form expressions for outage probability and average packet delay in single relay networks. In contrast to earlier work, Kumar and Hossain (2021) investigated optimal average queueing delay, focusing on QoS aspects like queueing stability and the Markov reward process (MRP) using a birth-death MC model at the relay in THBRN.

The physical layer security in the wireless networks is open issues for both academia and industry. The infrastructure of the cellular system is designed in such a way that supports internet connectivity in the homogeneous-and-heterogeneous networks. Thus, complex network designs, diverse access techniques, various service necessities, network topologies and mobile phones are bringing serious security challenges during information transmission in wireless networks. If we look the existing cellular network, the delay and reliability are not fully understood if we consider wireless transmission via physical-layer channels. Since, it contributes small fraction of the end-to-end delay. The proper selection of the relays in cooperative communication enhances overall energy efficiency, offers low-latency, and robust the transmission of information in high security scenarios. Farhat et al. (2018) presented selective DF relay networks to investigate short-packet transmission to secure the communication under static and mobile conditions. Further, Wang et al. (2019) considered relay with jammer in the networks that offers spatial degree of freedom to secure communication to achieve high performance gains by insuring URLLC. In the aforementioned works, the researchers have used cooperative policies to improve the security and reliability under static and mobile of the user scenarios. The random mobile receivers in the cooperative networks have been considered by Yu et al. (2022) to secure URLLC. The stochastic delays such as queueing delay, processing delay and access delay are involved in networking layer that limit to meet the requirements of the HLLC. In Zhang et al. (2024) have proposed adaptive finite blocklength scheme by exploiting updated queueing state of MC model of grant free access to reduce the over-air delay significantly.

THE URLLC IN COOPERATIVE UNMANNED AERIAL VEHICLE NETWORKS

In cooperative communications, unmanned aerial vehicles (UAVs) acts as a relay to transfer the received signal from source terminal to destination terminal in the diverse locations. The benefit of the UAVs in wireless network is to reduce the latency and robust networks in the dynamic environments. Zhou et al. (2018) have considered UAVs that act as jammers to secure the communication from legitimate source to destination. Friendly jammer is promising physical layer security technique that exploit artificial intelligence to protect the communication in the wireless networks from eavesdropper. With same line of works, Li et al. (2021) have proposed friendly jamming scheme to secure the physical layer communication using UAV relay-aided URLLC signal. Further, it is notifying that the UAVs have limited power that require more power in the wireless networks. Thus, they need efficient energy to process and secure communication parallelly at the physical layer. The security along with energy harvesting has been addressed by Tedeschi et al. (2020) to reinforce communication in the UAVs networks.

Conventionally, the UAV-aided communication in the wireless networks suffers from two malicious attacks named as active and passive eavesdropping. These attackers notify that how attacker interrupts the wireless channels. In the passive eavesdropping, the eavesdroppers capture the confidential information from the legitimate transmitter and transfer to legitimate without corrupting the received signal at the intendent receiver. On contrast to passive eavesdropping, active eavesdroppers receive the confidential information form transmitter and send it to legitimate receiver by reducing the quality of received signal at the intended receiver. Chen et al. (2023) proposed UAV-aided data collection policy to secure the message from sensor nodes to base station by exploiting optimal transmit power and the blocklength of short-packet. Further, the UAV acts as relay to decode the URLLC control signals from the sensors and forward these signals to actuators in presence of imperfect CSI at the ground eavesdropper. In spite of the UAVs assisted advantages, these UAVs can increase the interference among aerial and ground user communication. To mitigate this challenge, the UAVs-aided low-complexity algorithms have been developed by Ranjha et al. (2023) to synchronize the interferences in the intercell in the cellular infrastructure to enable URLLC facilities.

THE URLLC IN THE IRS-AIDED COOPERATIVE COMMUNICATIONS

In the recent era, the intelligent reflecting surface (IRS) is a new transformative technology that is able to realize the concept of smart and reconfigurable environment to improve the performance of wireless communication significantly. It made of electromagnetic material which is controlled by integrated electronic circuits. It enables network operator to control the negative effect of the wireless propagation channels. Generally, the IRS consists of passive reflecting elements that can structure either reflection amplitude or phase or both of each element to meet the specific need of the wireless propagation environment. The IRS-aided wireless communication has been addressed in details and provided more insight and possible applications of the intelligent surfaces in the 6G communication (Basar et al., 2019; Wu and Zhang, 2020). Zhang and Zhang (2021) considered IRS enabled two user wireless system to investigate capacity region by exploiting two IRS deployment strategies. In practice, single IRS cannot handle the line-of-sight (LOS) link between base station and user in the dense obstacles' environment. To mitigate this issue, multiple IRSs can be deployed to server each user to bypass the scattered obstacles by exploiting inter-IRS passive beamforming gain. Thus, a double IRS has been deployed between base station and user to achieve better power gain by useful insight of the passive beamforming design over single IRS (Hang et al., 2020). In the previous works, the researchers have considered cooperative multi-IRS hop communication to improve the performance of long distant user. Furthermore, Mei and Zhang (2021) presented the multiple distributed IRS scenarios between base station and user to improve the performance of wireless system using optimal beam routing. Naeem et al. (2022) addressed an advanced survey on IRS enabled wireless communication in detail and discussed IRS integration with 6G technologies named as NOMA, artificial intelligence, machine learning, THz/mmWave.

The IRS enabled wireless communication are involved significantly to support the URLLC in the 5G and 6G communications. In 2022, Khan et al. studied on the integration of the UAV communication with IRS to secure the physical layer communication to fulfil the specific need of URLLC. It enables the ULLC in the non-LOS environment due to sufficient improvement of spectral efficiency, energy efficiency, and extension of the coverage area (Almekhlaf et al., 2022). As we know IoT devices in the wireless networks are powered by batteries that have inadequate source of energy supply. In 2022, Zhang et al. presented IRS enabled URLLC communication system to support wireless energy and information transfer by exploiting short packets transmission. Furthermore, machine type communication is emerging services in the IoT network that are able to increase the number of users with the assistance of IRS techniques. The IRS enabled IoT users' system has been considered by Jalali et al. in 2023 to improve the energy efficiency by exploiting resource allocation and short-packet transmission over downlink channel. The major components of the industrial IoT (IIoT) are IoT devices, manufacturing machines, and cloud servers that need unified connectivity to provide reliable and secure communications among the industrial units. Thus, the physical layer security techniques play an important role to secure the communication in the IIoT networks. The IRS enabled physical layer security policy in the IIoT has been addressed to improve the secrecy rate at the user under optimal active and passive beamforming (Ali et al., 2023). Further, they have also notified that large-scale low powered IoT devices in the IIoT are relayed on short-packet signals for URLLC. In addition, the need of mobile devices in the wireless networks is to support heterogeneous services with high QoS and better traffic features. These challenges can be resolved with the assistance of IRS through multiplexing schemes. In 2024, the Souza et al. have proposed uplink traffic multiplexing scheme based on RIS to support the enhanced mobile broadband (eMBB) and URLLC. Moreover, they have used phasor rotation algorithms to mitigate eMBB in URLLC traffics.

COOPERATIVE RELAYING COMMUNICATION SYSTEM AND ITS PRELIMINARIES

To achieve high reliability and low latency in wireless system, we present simple cooperative relaying system such as two-hop buffered relay system where source, S televises information to destination, D through buffered relay, R which is shown in the Fig. 2. We assume that the relay works on half-duplex decode-and-forward relaying mode. Due to heavy path loss, we assume that direct link between source to destination is not available. The relay is equipped with buffer B that can be stored upto finite number of packets K. Further, it is assumed that a packet with N bits is framed in time slot, T seconds. It is also, considered that the source is backlogged means new packets are available always at the source to transmit to relay in each time slot.

We assume that complex fading coefficients over source-and-relay channel, and relay-and-destination channel are denoted as $h_{SR}(i)$ and $h_{RD}(i)$ in i^{th} time slot respectively. The network links experience block fading and are affected by additive white Gaussian noise i.e., $\mathcal{N}(0, \sigma_n^2)$. It is also, assumed that fading gains $|h_{SR}(i)|^2$ and $|h_{RD}(i)|^2$ are characterized by ergodic and stationary random process with means value $\Omega_1 = E\{|h_{SR}(i)|^2\}$ and $\Omega_2 = E\{|h_{RD}(i)|^2\}$ respectively, where $E\{.\}$ represents expectation. Thus, the instantaneous SNR of links, source-to-relay and relay-to-destination are expressed as $\gamma_{SR} = \gamma|h_{SR}(i)|^2$ and $\gamma_{RD} = \gamma|h_{RD}(i)|^2$ respectively, where $\gamma = \frac{P}{\sigma_n^2}$ represents the transmit SNR of each node, which transmits with power P, while noise variance occurs at the receiver is σ_n^2. Outage probability occurs when the transmission rate of any node is greater than or equal to channel capacity i.e., unsuccessful transmission between two adjacent nodes. For Rayleigh fading, it is defined as, $P_{sl} = \Pr\{R_{sl} \geq C_{sl} = \log_2(1 + \gamma|h_{sl}(i)|^2)\} = 1 - \exp\left(-\frac{2^{R_0} - 1}{\Omega_j \gamma}\right)$ with unit channel bandwidth, where $s = \{S, R\}$, $l = \{R, D\}$, $s \neq l$, and $j = \{1, 2\}$. Here, fixed transmission rate from s to l is given by $R_{sl} = R_0 = \frac{N}{T}$ bits/sec/Hz.

PERFORMANCE ANALYSIS OF TWO-HOP BUFFERED RELAY NETWORKS

This section introduces the random-access and non-random-access protocols based on channel state and buffer state conditions. Using these protocols, we derive average queueing delay expression in closed-form.

Analysis of Delay Performance Based on Random-Access (RA) Protocol

This section discusses the RA protocol which is helpful to investigate the expression of the average queuing delay of the THBRN for Rayleigh fading channel.

a) ***The link selection policy for transmission of a packet from one to another terminal for RA Protocol:*** The proposed RA protocol is carried out significant set of rules that are summarized as follows.
   - We assume that when the buffer at the relay contains k packets, it receives a packet over channel from source to relay with probability δ_k where, $0 \leq \delta_k \leq 1$.
   - If the buffered relay has at least one packet, and source is in the silent state, then it transfers a packet to destination through access channel with probability $\overline{\delta_k}$, where we assume that $\overline{F} = 1 - F$ throughout this chapter.
   - When the buffered relay is empty, source sends a packet over source to relay channel with probability $\delta_0 = 1$. However, it is full with packets, source transfer over the same channel with probability $\delta_K = 0$.

b) ***Nature of queuing at the buffered relay after receiving the packets:*** The dynamics of queueing nature at the buffered relay can be performed with the help birth-death MC along with Markov reward process (MRP) that is shown in Fig. 3. The fundament principle of the MC at the buffered relay is summarized as follow:
   - Each state of MC at the buffered relay indicates number of packets. It can store upto K packets.
   - It can receive/transmit only one packet at a time.
   - It is also notifying that source can transfer its decision to relay via transmitting control channel at the commencement of the slot.
   - The low-rate control channel is used to transmit all critical information.
   - The source activity is monitored at the relay with the help of carrier sensing.
   - The received packets at the relay (or, destination) can be identified by acknowledgement (ACK)/ negative ACK (NACK) signal at the end of the time slot.
   - When the relay receives data packets successfully from source within fixed time slot, it notifies ACK signal to the source. Otherwise, source receives NACK signal from the relay. Thereafter, source (or, relay) retransmits data packets in succeeding time slot when the channel from source to relay (or, from relay to destination) is available.

c) ***Formulation of the Average Queueing Delay for RA Protocol:*** It is assumed that number of packet, $x(i)$ is transmitted in time slot i, where present state of the MC depends on the next state or previous state i.e., $i \rightarrow (i+1)$ or $i \rightarrow (i-1)$. Therefore, the probability of state transition can be defined as the product of the channel access probability and probability of the successful transmission from one terminal to another terminal. Hence, it is given by

$$\begin{aligned} \boldsymbol{\alpha_k} &= \mathbf{Pr}\{\boldsymbol{x(i)} = \boldsymbol{k}+1 | \boldsymbol{x(i-1)} = \boldsymbol{k}\} = \boldsymbol{\delta_k \overline{P_{SR}}}, \; 0 \leq \boldsymbol{k} \leq (\boldsymbol{K}-1) \\ \beta_k &= \Pr\{x(i) = k-1 | x(i-1) = k\} = \overline{\delta_k}\, \overline{P_{RD}}, \; 1 \leq k \leq K \end{aligned} \tag{1}$$

where outage probability of links $S-R$ and $R-D$ are given by P_{SR} and P_{RD} respectively. The arrival rate of a packet with buffer of sate k at the relay is investigated as $\mathcal{A}_k = R_0 \alpha_k$ while departure rate of a packet with buffer state k at the relay is defined as $\mathcal{D}_k = R_0 \beta_k$ for RA protocol. Further, average queueing delay at the buffered relay is defined as sum of ratio of product of the number of packets at sate, k with their steady-state probability of birth-death MC and arrival rate of packets. It is formulated as,

$$D_Q^{RA} = \sum_{k=0}^{K} \frac{k \pi_k}{\mathcal{A}_k} \tag{2}$$

where steady-state probability of the birth-death MC with k packets, π_k is given by

$$\pi_k = \pi_0 \prod_{n=0}^{k-1} \frac{\alpha_n}{\beta_{n+1}}, \text{ where } \pi_0 = \left(1 + \sum_{k=1}^{K} \prod_{n=0}^{k-1} \frac{\alpha_n}{\beta_{n+1}}\right)^{-1} \tag{3}$$

By computing (1) and (3) into (2), we can write AQD as,

$$\boldsymbol{D_Q^{RA}} = \frac{\boldsymbol{a}}{1+\boldsymbol{\rho}} \left[\frac{2\boldsymbol{\rho} + \sum_{k=1}^{K-1} \{\boldsymbol{k}(1+\boldsymbol{\rho})^2 + 2\boldsymbol{\rho}\} \boldsymbol{\rho}^k \boldsymbol{V_k} + \boldsymbol{K d \rho^K V_K}}{1 + \sum_{k=1}^{K-1} \boldsymbol{\rho}^k \boldsymbol{V_k} + \frac{\boldsymbol{\rho}^K}{1+\boldsymbol{\rho}} \boldsymbol{V_K}} \right] \tag{4}$$

$$D_Q^{RA} = \frac{a}{1+\rho} \left[\frac{\boldsymbol{A}^{\dagger} \boldsymbol{V} + c}{\boldsymbol{B}^{\dagger} \boldsymbol{V} + 1} \right] \tag{5}$$

where $a = \frac{1}{\overline{P_{SR}}}$, $\rho = \frac{\overline{P_{SR}}}{\overline{P_{RD}}}$, $d = (1+\rho^2)$, and set of the variables $V_k = \frac{\delta_k}{\overline{\delta_k}} V_{k-1}$. From the set of variables expression, we can express channel access probability, $\delta_k = \frac{V_k}{V_k + V_{k-1}}$. To support the queueing stability, we keep transition of death rate of MC is always greater than or equal to transitions of birth rate. For more details, we can follow ref. (Kumar and Hossain, 2021). Thus, AQD obtained in (4) is written in vector form as, where, set of variables in vector form is defined as $V = [V_K, V_{K-1}, \ldots V_2, V_1]^\dagger$, superscript † denotes vector transposition; coefficients are in vector form as, $A = [Kd\rho^K, ((K-1)b+c)\rho^{K-1}, ((K-2)b+c)\rho^{K-2}, \ldots, (2b+c)\rho^2, (b+c)\rho]^\dagger$ and $B = \left[\frac{\rho^K}{1+\rho}, \rho^{K-1}, \rho^{K-2}, \ldots \rho^2, \rho\right]^\dagger$, $b = (1+\rho)^2$; $c = 2\rho$. The AQD should be minimized to get low latency. Thus, we need to investigate objective function subjects to constrains.

Establish Boundary Conditions: With assistance of Markov Reward Process (MRP), we obtain the SNR constraint for minimizing the AQD. It is a discrete stochastic process that transitions from one state to another according to a MC. Each transition is a random variable determined by the 'from' and 'to' states. It is assumed that the relay transmits a packet with an SNR of γ. Therefore, each state of the MC at the buffered relay increases by one if the relay accepts a packet from the source, and decreases by one if the relay transmits a packet to the destination. From Fig. 3, it is assumed that each state of the Markov chain decreases when the relay transmits a packet to the destination with a transmit SNR, $s(i)$, in the ith time slot, which is equal to γ, otherwise it remains zero. The conditional probability of the random variable, $s(i)$ depends on the availability of packets at the buffered relay and the state transitions of the MC, and is expressed as follows:

$$\omega_k(\gamma) = \Pr\{s(i) = \gamma | x(i) = k\} = \begin{cases} \alpha_0 \,; & k = 0 \\ |\alpha_k - \beta_k|; & 1 \le k \le K \end{cases} \tag{6}$$

The probability of state increment transition is always greater than or equal to the probability of state decrement transition for stability of the buffered relay. It is possible only when the link from source to relay is robust as compared to link from relay to destination. Average transmit SNR is associated with conditional probability and steady-state probability of the MC is formulated as,

$$\gamma_{\text{avg}} = \gamma \pi_0 \omega_0(\gamma) + \sum_{k=1}^{K} \gamma \omega_k(\gamma) \pi_k \tag{7}$$

By substituting (1), (3) and (6) into (7), we get-

$$\gamma_{\text{avg}} = \frac{\gamma \overline{P_{SR}} \rho^K V_K}{(1+\rho)\left[1 + \sum_{k=1}^{K-1} \rho^k V_k + \frac{\rho^K}{1+\rho} V_K\right]} \tag{8}$$

It is observed that the γ_{avg} of the relay would be always less than or equal to maximum transmit SNR ($\gamma_{\max}$) of the relay. Thus, we get the condition as,

$$1 + \boldsymbol{G}^{\dagger}\boldsymbol{V} \geq 0$$

$$; \qquad (9)$$

Where, coefficient $\boldsymbol{G} = [u_K, \rho^{K-1}, \rho^{K-2}, \ldots. \rho^2, \rho]^{\dagger}$ and $u_K = \frac{\rho^K}{1+\rho}\left(1 - \frac{\gamma \overline{P}_{SR}}{\gamma_{max}}\right)$. From the (5), (9), and queueing stability conditions, we formulate the objective function of the AQD at the buffered relay which is in linear-fractional programming form as,

$$\text{Minimize, } \boldsymbol{D}_Q^{RA} = \frac{\boldsymbol{a}}{1+\rho}\left[\frac{\boldsymbol{A}^{\dagger}\boldsymbol{V} + \boldsymbol{c}}{\boldsymbol{B}^{\dagger}\boldsymbol{V} + 1}\right]$$

$$\textit{Subject to:}\begin{cases} C1: & V_0 = 1 \\ C2: 0 \leq \dfrac{V_k}{V_k + V_{k-1}} \leq 1 \\ C3: \dfrac{V_k}{V_k + V_{k-1}} \leq \dfrac{1}{1+\rho} \\ C4: & 1 + \boldsymbol{G}^{\dagger}\boldsymbol{V} \geq 0 \end{cases} \qquad (10)$$

It is notice that the letters ***V, A, B, G*** represents in bold are in K- dimensional column vectors. The constraint C1 is obtained from initial condition of the buffer at the relay. The constraint C2 is computed from condition of the channel access probability $0 \leq \delta_k \leq 1$. To stabilize the queueing process at buffered relay, the constraint C3 is obtained. The constraint C4 in vector form is investigated in (9).

ANALYSIS OF DELAY PERFORMANCE BASED ON NON-RANDOM-ACCESS (NRA) PROTOCOL

This Section develops the NRA protocol which is used to investigate closed-form expression average queueing delay of the THBRN. For NRA protocol, transition from one state to another state does not depend on channel access probability, δ_k i.e., it is set to be 1.

a) The link selection policy for transmission of a packet from one to another terminal:
   - The state of MC at the buffered relay is incremented by 1 if source transmits a packet to relay with fixed transmission rate, R_0 which is less than channel capacity from source to relay, C_{SR}.
   - If the source is in silent state and buffer at relay has at least one packet, then relay transmits a packet to destination successfully by satisfying condition such as the relay transmits a packet with rate, R_0 which is less

than or equal to channel capacity from relay to destination, C_{RD}. Thus, each state of MC at the buffered relay is decreased by 1.

- If the buffer at the relay is full, then source is in silent state even it has sufficient packets for transmission and channel is available from source to relay.
- If the buffer at the relay has no packet, the relay is in silent state even channel from relay to destination is available.

b) ***Formulation of the AQD for NRA protocol:*** The transition probabilities from present state k to next state $(k + 1)$ and from present state k to previous $(k - 1)$ are given by,

$$\alpha_k = \alpha = \overline{P_{SR}}, \qquad 0 \leq k \leq (K-1)$$

$$\beta_k = \beta = \overline{P_{RD}}, \qquad 1 \leq k \leq K \tag{11}$$

Let $\{Q_{BR}\}$ be the average size of queue in the buffer at the relay and it receives data packet with rate $\mathscr{A}_k = \mathscr{A} = R_0\alpha$ while it transmits with rate $\mathscr{D}_k = \mathscr{D} = R_0\beta$. Thus, the AQD at the relay for THBRN is formulated as.

$$D_Q^{NRA} = \frac{E\{Q_{BR}\}}{\mathscr{A}} = \frac{\sum_{k=0}^{K} k\pi_k}{\alpha} \tag{12}$$

Substitute (3), and (11) into (12), we get-

$$D_Q^{NRA} = \frac{1}{\overline{P_{SR}}}\left[\frac{\rho - \rho^{K+1}(1 + K(1-\rho))}{(1-\rho)(1-\rho^{K+1})}\right] \tag{13}$$

where transmission rate R_0 =1 packets/slot.

Special case: If the factor ρ is fixed and equal to 1 (i.e., queue of the buffered relay is verge of stability that notify that arrival rate of data packet at the relay is equal it's departure rate of data packet. Thus, the AQD of the two-hop buffered relay system is written as,

$$D_Q^{NRA} = \lim_{\rho\to 1} D_Q^{NRA}(\rho) = \lim_{\rho\to 1}\frac{1}{\overline{P_{SR}}}\left[\frac{\rho - \rho^{K+1}(1 + K(1-\rho))}{(1-\rho)(1-\rho^{K+1})}\right] \tag{14}$$

Finally, the AQD of the THBRN is expressed as,

$$D_Q^{NRA} = \frac{1}{\overline{P_{SR}}}\left[\frac{K(K+1)+1}{2(K+1)}\right] \tag{15}$$

In the next-section, we analyse the results numerically and discuss in brief.

NUMERICAL RESULTS AND DISCUSSIONS BASED ON PROPOSED PROTOCOLS

This Section presents performance of average delay of the THBRN for Rayleigh fading channel. Then, we compare our numerical results with adaptive link selection protocol (ALSP) in theorem 3 case 1 (T3C1) and theorem 3 case 2 (T3C2) (Zlatanov and Schober, 2013), and ref. (Islam et al., 2016). For analysis purpose, it is assumed that mean values, $\Omega_1 = E\{|h_{SR}(i)|^2\} = 0.75$ dB and $\Omega_2 = E\{|h_{RD}(i)|^2\} = 1$, and $\gamma_{max} = 20$ dB. In Fig. 4, we plot the AQD along with size of buffer at the relay. It shows that the AQD for RA protocol is achieved lower than the existing protocols in the literature and authenticates the ALSP, T3C2. The AQD of proposed scheme shows better results if the buffer size is small. Furthermore, if we keep size of buffer is large, the AQD for RA protocol is superior to NRA protocol. Moreover, it is found that the average delay is reached to approximately less than 2 time slots by varying buffer size for RA protocol. In addition, we notify that the average queueing delay for RA protocol does not depend on buffer size.

Further, for more analysis, we evaluate average queueing delay by varying the transmit SNR γ (in dB) which is depicted in Fig. 5. From the results, AQD for RA protocol performances better as compared to existing schemes in literature and authenticates the ALSP, T3C2 if the transmit SNR, $\gamma > -1$ dB and NRA protocol. It is found that AQD decreases by increasing the transmit SNR, γ, because outage probability of link is also decreased with γ. The AQD of the NRA protocol shows better performance over asymmetric channels (i.e., $0 < \rho < 1$) as compared to symmetric channel (i.e., $\rho = 1$). It is notice that at the high SNR, average queueing delay for RA protocol is reached to 1 time slot. Furthermore, average delay of the NRA protocol for symmetric channel is very high as compared to symmetric channel for low SNR. It is also observed that average delay of the NRA protocol for both cases is equivalent at high SNR. Because, transmission rate of links is increased at the high SNR γ. In the Fig. 6, average queueing delay is varied with transmission rate, R_0 in bits/sec/Hz. It is notice that the proposed RA protocol and NRA protocol show low latency if transmission rate varies from 1.5 bits/sec/Hz to 3.5 bits/sec/Hz. From the results, we observe that the average delay is reached to 0.6867 time slot as the transmission rate varies from 2.25 to 2.75 bits/sec/Hz. From overall analysis, we notify that THBRN achieves low-latency over moderate transmission range.

POTENTIAL RESEARCH DIRECTIONS

This Section provides potential research directions for URLLC enabled cooperative communication and its applications in next-generation wireless systems.

Future Research Directions for HRLLC enabled Cooperative Relaying Communication: Cooperative communication is one of the potential areas of research to fulfil the requirements of the next-generation wireless communications. The more rooms are still available for delay analysis by exploiting channel state and buffer state with partial transmit channel state to achieve low latency of the 6G radio networks. By exploiting queuing delay theory at the buffered relay, several research opportunities are still remained to reduce average delay of system. With integration of the full-duplex relaying with QoS constraints, we will also, meet the latency requirements. The cross-layer approach is able to reduce the delay performance of the cooperative communications. The hybrid FD/HD relaying depends on switching operation of the relay as per availability of the channels along with buffer state which is employed to achieve low latency.

Potential Research Directions for HRLLC enabled OTFS System: To fulfil the requirements of smart and intelligent next generation wireless communication, low-latency and high reliability are greatly involved and played important role to achieve high transmission rate in the high mobility scenarios. Since, Doppler spreads affects performances of the wireless system. To mitigate these challenges, OTFS is to be implemented to achieve the HRLLC in high-speed mobility at the physical layer. Most of the authors used deep learning approach to reduce the hardware impairments. But we need suitable hardware practically for orthogonal time-frequency multiplexing to detect signal properly. In addition, it is also notice that the OTFS is not fully understood in high-speed mobility scenarios in the real-time applications.

Open Research Challenges for Securing the Communication with HRLLC signals: To secure the data transmission in the wireless networks, we need to address the security threats and traffic issues by exploring new advanced radio technologies. In this context, physical layer uses resource allocation, beamforming, automatic modulation classification, and advanced channel encoding techniques to protect the confidentiality form eavesdropping and jamming outbreaks. Thus, we need to develop dynamic, lightweight, and energy efficient protocols to address trade-off between inherent challenges and performance gains by considering HRLLC services. Other innovative is open radio access network that can used to enhance the securities of physical layer under HRLLC service. Since, it increases reliability and flexibility, and optimal performance in real-time with desired QoS in the wireless networks. Further, there is open research challenges remain for reducing the dependency on the channel state information as we seeking physical layer policies for URLLC signal. By integrating UAV and IRS in the cooperative communication, we can improve reliability and the physical layer security with low latency communication. In the IoT networks, the extra power is required to enhance the energy efficient for communication in wireless networks. In this regard, advanced energy harvesting techniques with rate splitting are needed for energy constraint devices in the future networks for HRLLC services. By exploring the concept of AI/ML techniques, we can protect the IIoT networks from valuables attack and secure the HRLLC service. In the ML techniques. federated learning and transfer learning have potential to preserve the distributed training data set at edge users.

CONCLUDING REMARKS

This chapter presents the two-hop buffered relay network and develop the RA protocol and NRA protocol in presence of the buffer state information by exploiting birth-death MC. Based on these protocols, we derive average queueing delay in the closed-form expression. Furthermore, we formulate objective function in term of average queuing delay by satisfying QoS such as queuing stability and constant power maintain through MRP at the buffered relay. From the results, we found that the average delay for RA protocol performs better as compared to existing schemes available in literature. The average queueing delay is reached to lowest value at high SNR with fixed-transmission rate. Moreover, the minimum average delay is achieved to 1 time slot at high transmit SNR with fixed rate, R_0=1 bits/sec/Hz. In addition, we found that the average delay is less than 1 time slot in the range of moderate transmissions rate for constant SNR. Finally, we provide the potential research direction toward URLLC enabled cooperative communication for next-generation wireless communications.

REFERENCES

Airod, F. E., Chafnaji, H., & Yanikomeroglu, H. (2021). HARQ in full-duplex relay-assisted transmissions for URLLC. *IEEE Open Journal of the Communications Society*, 2, 409–422. DOI: 10.1109/OJCOMS.2021.3055073

Alam, M. Z., Abkenar, F. S., Adhicandra, I., Murali, S., & Jamalipou, A. (2020). Low-delay path selection for cluster-based buffer-aided vehicular communications. *IEEE Transactions on Vehicular Technology*, 69(9), 9356–9363. DOI: 10.1109/TVT.2020.2976926

Ali, B., Mirza, J., Alvi, S. H., Khan, M. Z., Javed, M. A., & Noorwali, A. (2023). IRS-assisted physical layer security for 5G enabled industrial Internet of Things. *IEEE Access : Practical Innovations, Open Solutions*, 11, 21354–21363. DOI: 10.1109/ACCESS.2023.3250251

Almekhlafi, M., Arfaoui, M. A., Assi, C., & Ghrayeb, A. (2022). Enabling URLLC applications through reconfigurable intelligent surfaces: Challenges and potential. *IEEE Internet Things Mag.*, 5(1), 130–135. DOI: 10.1109/IOTM.007.2100124

Basar, E., Renzo, M. D., Rosny, J. D., Debbah, M., Alouini, M. S., & Zhang, R. (2019). Wireless communications through reconfigurable intelligent surfaces. *IEEE Access : Practical Innovations, Open Solutions*, 7, 116753–116773. DOI: 10.1109/ACCESS.2019.2935192

Chen, X., Zhao, N., Chang, Z., Hämäläinen, T., & Wang, X. (2023). UAV-aided secure short-packet data collection and transmission. *IEEE Transactions on Communications*, 71(4), 2475–2486. DOI: 10.1109/TCOMM.2023.3244954

Chowdhury, M. Z., Shahjalal, M., Ahmed, S., & Jang, Y. M. (2020). 6G wireless communication systems: Applications, requirements, technologies, challenges, and research directions. *IEEE Open Journal of the Communications Society*, 1, 957–975. DOI: 10.1109/OJCOMS.2020.3010270

Farhat, J., Brante, G., & Souza, R. D. (2018). Secure throughput optimization of selective decode-and-forward with finite blocklength. *IEEE 87th Veh. Technol. Conf. (VTC Spring)*, 1–5, Porto, Portugal.

Hadani, R., & Monk, A. (2017). OTFS—A novel modulation technique meeting 5G high mobility and massive MIMO challenges. Available: https://www.cohere-technologies.com/wp-content/uploads/2017/10/ OTFS-Physics-White-Paper.pdf

Hadani, R., & Monk, A. (2018). OTFS: A new generation of modulation addressing the challenges of 5G", arXiv preprint. DOI: https://doi.org//arXiv.1802.02623DOI: 10.48550

Han, Y., Zhang, S., Duan, L., & Zhang, R. (2020). Cooperative double-IRS aided communication: Beamforming design and power scaling. *IEEE Wireless Communications Letters*, 9(8), 1206–1210. DOI: 10.1109/LWC.2020.2986290

Huang, S., & Cai, J. (2016). An analysis framework for buffer-aided relaying under time-correlated fading channels. *IEEE Transactions on Vehicular Technology*, 65(9), 6987–6999. DOI: 10.1109/TVT.2015.2497210

Islam, T., Michalopoulos, D. S., Schober, R., & Bhargava, V. (2014). Delay constrained buffer-aided relaying with outdated CSI. In *Proc. IEEE WCNC*, (pp. 875–880). DOI: 10.1109/WCNC.2014.6952203

Islam, T., Michalopoulos, D. S., Schober, R., & Bhargava, V. K. (2016). Buffer-aided relaying with outdated CSI. *IEEE Transactions on Wireless Communications*, 15(3), 1979–1997. DOI: 10.1109/TWC.2015.2497342

Jalali, J., Khalili, A., Rezaei, A., Berkvens, R., Weyn, M., & Famaey, J. (2023). IRS-based energy efficiency and admission control maximization for IoT users with short packet lengths. *IEEE Transactions on Vehicular Technology*, 72(9), 1–6. DOI: 10.1109/TVT.2023.3266424

Jamali, V., Zlatanov, N., & Schober, R. (2015). Bidirectional buffer-aided relay networks with fixed rate transmission—Part II: Delay-constrained case. *IEEE Transactions on Wireless Communications*, 14(3), 1339–1355. DOI: 10.1109/TWC.2014.2365806

Jamali, V., Zlatanov, N., Shoukry, H., & Schober, R. (2015). Achievable rate of the half-duplex multi-hop buffer-aided relay channel with block fading. *IEEE Transactions on Wireless Communications*, 14(11), 6240–6256. DOI: 10.1109/TWC.2015.2451171

Kalpage, N. V., Priya, P., & Hong, Y. (2024). DCT-based OTFS with reduced PAPR. *IEEE Communications Letters*, 28(1), 158–162. DOI: 10.1109/LCOMM.2023.3337778

Khan, W. U., Lagunas, E., Ali, Z., Javed, M. A., Ahmed, M., Chatzinotas, S., Ottersten, B., & Popovski, P. (2022). Opportunities for physical layer security in UAV communication enhanced with intelligent reflective surfaces. *IEEE Wireless Communications*, 29(6), 22–28. DOI: 10.1109/MWC.001.2200125

Kumar, R., & Hossain, A. (2017). Optimization of throughput of two-way buffer-aided relaying networks with wireless assisted links. *IET Communications*, 11(10), 1626–1632. DOI: 10.1049/iet-com.2016.1364

Kumar, R., & Hossain, A. (2019). Performance of random-access Markov modelling for two-way buffer-aided relaying networks with wireless assisted links. *Wireless Personal Commun.*, Springer, 108(3), 1995–2015.

Kumar, R., & Hossain, A. (2019). Survey on half- and full- duplex relay based cooperative communications and its potential challenges and open issues using Markov chains. *IET Communications*, 13(11), 1537–1550. DOI: 10.1049/iet-com.2018.5823

Kumar, R., & Hossain, A. (2021). Cooperative relaying communications in IoT applications for 5G radio networks. In Innovations and Applications of AI, IoT, and Cognitive Technologies, *Eds. Ramona Trestian, Gabriel-Miro Muntean, IGI Global*, Ch. 2,26 – 41.

Kumar, R., & Hossain, A. (2021). Delay performance of two-hop buffer-aided relay network with wireless assisted-links. *Wireless Personal Communications*, 121(3), 1725–1738. DOI: 10.1007/s11277-021-08694-8

Li, X., Dai, H. N., Shukla, M. K., Li, D., Xu, H., & Imran, M. (2021). Friendly-jamming schemes to secure ultra-reliable and low-latency communications in 5G and beyond communications. *Computer Standards & Interfaces*, 78, 1–14. DOI: 10.1016/j.csi.2021.103540

Liang, Y., Li, L., Fan, P., & Guan, Y. (2020). Doppler resilient orthogonal time-frequency space (OTFS) systems based on index modulation. *in proc. IEEE 91st Vehicular Technology Conference (VTC2020-Spring),* 1 – 5, Antwerp, Belgium.

Lin, S. C., Chen, K. C., & Karimoddini, A. (2021). SDVEC: Software-defined vehicular edge computing with ultra-Low latency. *IEEE Communications Magazine*, 59(12), 66–72. DOI: 10.1109/MCOM.004.2001124

Liu, Y., Deng, Y., Elkashlan, M., Nallanathan, A., & Karagiannidis, G. K. (2022). Optimization of grant-free NOMA with multiple configured-grants for mURLLC. *IEEE Journal on Selected Areas in Communications*, 40(4), 1222–1236. DOI: 10.1109/JSAC.2022.3143264

Luo, S., & Teh, K. C. (2015). Buffer state-based relay selection for buffer-aided cooperative relaying systems. *IEEE Transactions on Wireless Communications*, 14(10), 5430–5439. DOI: 10.1109/TWC.2015.2438296

Manoj, B. R., Mallik, R. K., & Bhatnagar, M. R. (2016). Buffer-aided multi-hop DF cooperative networks: A state-clustering based approach. *IEEE Transactions on Communications*, 64(12), 4997–5010. DOI: 10.1109/TCOMM.2016.2611504

Mei, W., & Zhang, R. (2021). Cooperative beam routing for multi-IRS aided communication. *IEEE Wireless Communications Letters*, 10(2), 426–430. DOI: 10.1109/LWC.2020.3034370

Nabar, R. U., Bölcskei, H., & Kneubühler, F. W. (2004). Fading relay channels: Performance limits and space-time signal design. *IEEE Journal on Selected Areas in Communications*, 22(6), 1099–1109. DOI: 10.1109/JSAC.2004.830922

Naeem, F., Kaddoum, G., Khan, S., Khan, K. S., & Adam, N. (2022). IRS-empowered 6G networks: Deployment strategies, performance optimization, and future research directions. *IEEE Access : Practical Innovations, Open Solutions*, 10, 118676–118696. DOI: 10.1109/ACCESS.2022.3220682

Neelam, S. G., & Sahu, P. R. (2022). Analysis, estimation and compensation of hardware impairments for CP OTFS systems. *IEEE Wireless Communications Letters*, 11(5), 952–956. DOI: 10.1109/LWC.2022.3151707

Neelam, S. G., & Sahu, P. R. (2022). Digital compensation of IQ imbalance DC offset for zero-padded OTFS systems. *IEEE Communications Letters*, 26(10), 2450–2454. DOI: 10.1109/LCOMM.2022.3190047

Nomikos, N., Poulimeneas, D., Charalambous, T., Krikidis, I., Vouyioukas, D., & Johansson, M. (2018). Delay- and diversity-aware buffer-aided relay selection policies in cooperative networks. *IEEE Access : Practical Innovations, Open Solutions*, 6, 73531–73547. DOI: 10.1109/ACCESS.2018.2883894

Qiao, D., & Gursoy, M. C. (2016). Statistical delay tradeoffs in buffer-aided two-hop wireless communication systems. *IEEE Transactions on Communications*, 64(11), 4563–4577. DOI: 10.1109/TCOMM.2016.2607746

Qiao, D., & Gursoy, M. C. (2017). Buffer-aided relay systems under delay constraints: Potentials and challenges. *IEEE Communications Magazine*, 55(9), 168–174. DOI: 10.1109/MCOM.2017.1600213

Ranjha, A., Javed, M. A., Srivastava, G., & Lin, J. C. (2023). Intercell interference coordination for UAV enabled URLLC with perfect/imperfect CSI using cognitive radio. *IEEE Open Journal of the Communications Society*, 4, 197–208. DOI: 10.1109/OJCOMS.2022.3232888

Raviteja, P., Phan, K. T., Hong, Y., & Viterbo, E. (2018). Interference cancellation and iterative detection for orthogonal time frequency space modulation. *IEEE Transactions on Wireless Communications*, 17(10), 6501–6515. DOI: 10.1109/TWC.2018.2860011

Raza, W., Javaid, N., Nasir, H., Aurangzeb, K., Khan, Z. A., & Haider, S. I. (2018). BTRS: Buffer-threshold based relay selection scheme for cooperative wireless networks. *IEEE Access : Practical Innovations, Open Solutions*, 7, 23089–23099. DOI: 10.1109/ACCESS.2018.2890003

Razlighi, M. M., & Zlatanov, N. (2018). Buffer-aided relaying for the two-hop full-duplex relay channel with self-interference. *IEEE Transactions on Wireless Communications*, 17(1), 477–491. DOI: 10.1109/TWC.2017.2767582

Rjeily, C. A. (2020). Toward a better comprehension of decode-and-forward buffer-aided relaying: Case study of a single relay. *IEEE Communications Letters*, 24(5), 1005–1009. DOI: 10.1109/LCOMM.2020.2971494

Shafie, A. E. I., Khafagy, M. G., & Sultan, A. (2014). Optimization of a relay-assisted link with buffer state information at the source. *IEEE Communications Letters*, 18(12), 2149–2152. DOI: 10.1109/LCOMM.2014.2363115

Sheng, Z., Cui, L., Nasir, A. A., Wang, R., & Fang, Y. (2023). URLLC-oriented secure communication for UAV relay-assisted network. *Physical Communication*, 59, 102063. DOI: 10.1016/j.phycom.2023.102063

Shin, T., & Baek, S. J. (2019). Cooperative buffer-aided relaying using full-duplex relays with flow control. *IEEE Transactions on Vehicular Technology*, 68(2), 1824–1838. DOI: 10.1109/TVT.2018.2886753

Siddig, A. A. M., & Salleh, M. F. M. (2017). Buffer-aided relay selection for cooperative relay networks with certain information rates and delay bounds. *IEEE Transactions on Vehicular Technology*, 66(11), 10499–10514. DOI: 10.1109/TVT.2017.2738036

Singh, A., Sharma, S., Deka, K., & Bhatia, V. (2023). DL-based OTFS signal detection in presence of hardware impairments. *IEEE Wireless Communications Letters*, 12(9), 1533–1537. DOI: 10.1109/LWC.2023.3281790

Singh, A., Sharma, S., Sharma, M., & Deka, K. (2023). Low complexity deep-decoder for OTSM with hardware impairments. *IEEE Communications Letters*, 27(12), 3240–3244. DOI: 10.1109/LCOMM.2023.3327249

Souza, J. H. I. D., Croisfelt, V., Kotaba, R., Abrão, T., & Popovski, P. (2024). Uplink multiplexing of eMBB/URLLC services assisted by reconfigurable intelligent surfaces. *IEEE Communications Letters*, 28(9), 2206–2210. DOI: 10.1109/LCOMM.2024.3426613

Tedeschi, P., Sciancalepore, S., & Pietro, R. D. (2020). Security in energy harvesting networks: A survey of current solutions and research challenges. *IEEE Communications Surveys and Tutorials*, 22(4), 2658–2693. DOI: 10.1109/COMST.2020.3017665

Tian, Z., Gong, Y., Chen, G., & Chambers, J. A. (2017). Buffer-aided relay selection with reduced packet delay in cooperative networks. *IEEE Transactions on Vehicular Technology*, 66(3), 2567–2575. DOI: 10.1109/TVT.2016.2573378

Wang, D., Bai, B., Zhao, W., & Han, Z. (2019). A survey of optimization approaches for wireless physical layer security. *IEEE Communications Surveys and Tutorials*, 21(2), 1878–1911. DOI: 10.1109/COMST.2018.2883144

Wang, D., Ren, P., & Cheng, J. (2018). Cooperative secure communication in two-hop buffer-aided networks. *IEEE Transactions on Communications*, 66(3), 972–985. DOI: 10.1109/TCOMM.2017.2776114

Wicke, W., Zlatanov, N., Jamali, V., & Schober, R. (2017). Buffer-aided relaying with discrete transmission rates for the two-hop half-duplex relay network. *IEEE Transactions on Wireless Communications*, 16(2), 967–981. DOI: 10.1109/TWC.2016.2635142

Wu, Q., & Zhang, R. (2020). Towards smart and reconfigurable environment: Intelligent reflecting surface aided wireless network. *IEEE Communications Magazine*, 58(1), 106–112. DOI: 10.1109/MCOM.001.1900107

Xia, B., Fan, Y., Thompson, J., & Poor, H. (2008). Buffering in a three-node relay network. *IEEE Transactions on Wireless Communications*, 7(11), 4492–4496. DOI: 10.1109/T-WC.2008.070867

Xu, P., Chen, G., Yang, Z., & Lei, H. (2020). Buffer-state-based probabilistic relay selection for cooperative networks with delay constraints. *IEEE Wireless Communications Letters*, 9(11), 1855–1859. DOI: 10.1109/LWC.2020.3006079

Yu, K., Yu, J., & Dong, A. (2022). Cooperative communication and mobility for securing URLLC of future wireless networks. *IEEE Transactions on Vehicular Technology*, 71(5), 5331–5342. DOI: 10.1109/TVT.2022.3151063

Yuan, Y., Zou, J., Cui, Y., Li, X., Mu, J., & Han, K. (2023). Orthogonal time frequency space and predictive beamforming-enabled URLLC in vehicular networks. *IEEE Wireless Communications*, 30(2), 56–62. DOI: 10.1109/MWC.005.2200408

Zhang, B., Wang, K., Yang, K., & Zhang, G. (2022). IRS-assisted short packet wireless energy transfer and communications. *IEEE Wireless Communications Letters*, 11(2), 303–307. DOI: 10.1109/LWC.2021.3126596

Zhang, S., & Zhang, R. (2021). Intelligent reflecting surface aided multi-user communication: Capacity region and deployment strategy. *IEEE Transactions on Communications*, 69(9), 5790–5806. DOI: 10.1109/TCOMM.2021.3079128

Zhang, Y., Cheng, W., & Zhang, W. (2024). Multiple access integrated adaptive finite blocklength for ultra-low delay in 6G wireless networks. *IEEE Transactions on Wireless Communications*, 23(3), 1670–1683. DOI: 10.1109/TWC.2023.3290936

Zhou, B., Cui, Y., & Tao, M. (2015). Stochastic throughput optimization for two-hop systems with finite relay buffers. *IEEE Transactions on Signal Processing*, 63(20), 5546–5560. DOI: 10.1109/TSP.2015.2452225

Zhou, Y., Yeoh, P., Chen, H., Li, Y., Schober, R., Zhuo, L., & Vucetic, B. (2018). Improving physical layer security via a UAV friendly jammer for unknown eavesdropper location. *IEEE Transactions on Vehicular Technology*, 67(11), 11280–11284. DOI: 10.1109/TVT.2018.2868944

Zlatanov, N., Schober, R., & Popovski, P. (2013). Buffer-aided relaying with adaptive link selection. *IEEE Journal on Selected Areas in Communications*, 31(8), 1530–1542. DOI: 10.1109/JSAC.2013.130816

Chapter 14
Smart Cities Practices Applied in the Selected City of Slovak Republic

Marica Mazurek
https://orcid.org/0000-0002-0564-3553
University of Zilina, Slovakia

ABSTRACT

The book chapter is focused on the specific cases of smart city applications in the selected Slovakian city, which were implemented to improve the competitiveness as well as to create better co-operation of the chosen city representatives (municipalities) with the city users (citizens, entrepreneurs) and the main focus will be the principle of co-creation and new forms of communication among these entities, which were influenced by new technologies and smart decisions. The digitalization has enormously influenced the new trends of communication and the process of co-creation has been fully applied in the contemporary municipal managerial practices. The specific case study will be focused on the city of Banska Bystrica and the primary and secondary research in this city will be a base of our research and discussion.

DOI: 10.4018/979-8-3693-6725-4.ch014

INTRODUCTION

A territory is a place where economic and social processes transform. The city applying a modern strategic approach to marketing and planning with an innovative goal should also be concerned by the modern approach to governing a territory (a city).

According to Reinisto (2003), not only economic indicators influence competitiveness in cities but also the satisfaction of their inhabitants and entrepreneurs living in the city and their quality of life.

As a result, digitalization is important not only for the business and technical aspects, but also the quality of the environment, helpfulness and friendliness of people, the quality of common communication between citizens and government, and the type of governance, trust in partnerships, trust in government, as well as mutual human relations and the ability to approach a process of co-creation of service provision with providers of services (private, public, etc.).

One good example could be health care facilities, educational institutions, social services, transportation and so on, as has recently been confirmed during the global pandemic scenario. Mutually effective communication facilitated better completion of urgent duties related to resolving emergent situations, influencing greater health situation stability and, as a result, economic stability, implying a reduced risk of competitiveness loss.

Soft factors are also crucial for image, reputation, and competitiveness and fulfill the criteria of the ethical approach to development. Even partnerships are more open to common collaboration and trust if the partners in the territory achieve a good reputation and image. This has been confirmed by several authors, as for instance Kotler (2002), and this especially underlines the importance of co-operation of businesses and government entities in destinations, where the question of image and reputation is crucial.

In their work on the importance of place branding, which means the application of branding principles in places, but it is crucial to mention that in place branding is added one missing component, and this is the need to create partnerships among the partners in a destination. This idea was also supported by Go and Govers (2009). The growing importance of partnerships improves the competitiveness of destinations.

Ashworth and Kavaratzis (2008) also discussed in their work the importance of collaboration in cities and stressed the importance of co-creation. Similarly, the authors, Kotler (2002), Ashworth and Voogd (1990), Reinisto (2001) and Hankinson (2001, 2004, 2005) agreed with this opinion. Kotler and Gertner (2002), confirmed the importance of cooperation for a community and the meaning of trust and ethics for the partnership forming. Asplund (1993), Crouch and Ritchie (2003) similarly to Kotler supported the idea of growing importance of sustainability, improved life quality, and stronger support for such factors as education and culture in territorial development strategies.

Sundbo (2008) underlined the importance of psychographic factors in marketing and their growing impact. The continual change from marketing management to marketing means also a discourse based on Vargo and Lusch (2004), Lusch (2007) the authors of a New Dominant Logic. These approaches promote such ideas as the value of services, exchange processes, and connections as partnerships.

The authors Vargo and Lusch (2006) discussed in their work that "the client is always a value generator" and these ideas were supported also by the authors Prahalad and Ramaswarny (2000, 2004).

For this reason, it is crucial to take into account governance and its role, partnerships, and the meaning of value creation in cities for competitiveness improvement. According to the above statement, Boisen (2007 discussed in his work that the traditional approach to marketing can be joined with the neoliberal approach, which includes crucial concepts such as governance, partnership creation (collaborative governance), and co-creation. This type of city governance is called participatory governance. One of the trends that have been proposed recently in city development strategies is the creation of smart cities and creative clusters. Some of these concepts have been discussed by academics as Buhalis (2014). In these ideas were presented such concepts as the investments in human and social capital, governance, innovations and smart destination approaches. It means for instance the growth of creativity in cities and forming of clusters among cities.

The authors, such as Carvalho et al. (2015), Hollands (2015), Kitchin (2015), Nam et al. (2011), Shelton (2015), Suzuki (2017), also discussed the idea of smart cities and this idea has been depicted in the following Figure 1:

Figure 1. Components of smart city

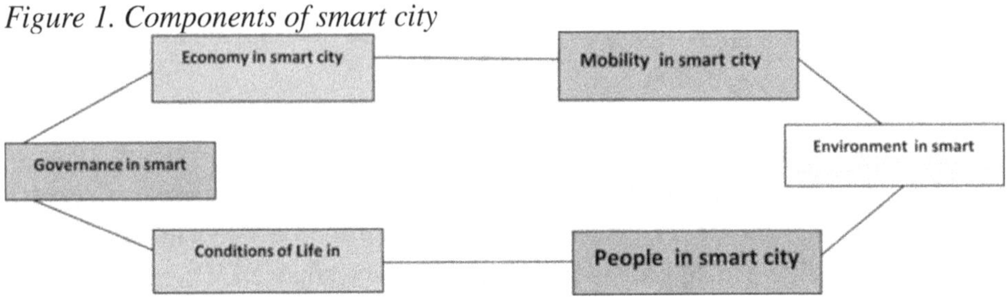

Source: Amended and based on www.smart-cities.eu, 2016.

City innovation and smart cities building means a mutual co-existence of trends in technical development with an ethical approach and social rules. This means that co-creation principles and consumer-centric marketing are good examples of this approach. The following Figure 2, which was based on the work of Lendel (2009) and slightly amended, contains the concept of consumer-centric marketing.

Figure 2. Concept of customer-centric marketing - Stars model amended for the territorial purposes

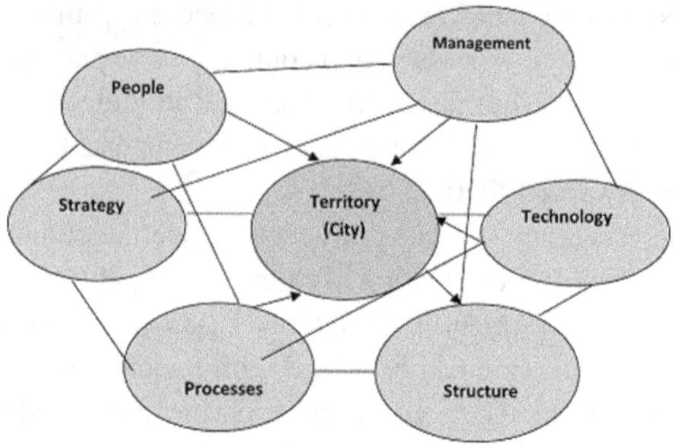

Source: amended upon Lendel, 2009.

METHODOLOGY

The methodological approach to this study will be based on the primary and secondary research amended in the selected Slovak municipality Banska Bystrica.

The following research questions have been created in order to support our research and for this specific study a third research question has been specifically applied:

> Question 1 - Does the innovative marketing of place contribute to the local development of a territory?
> Question 2 - How important is the active co-operation of partners in territory for a successful strategy and marketing?
> Question 3 - How important is for the competitive marketing strategy to respect the needs of the users of territory (the innovative approach to marketing based on co-creation principles and prosumer approach concept)?

In this primary and secondary research has been applied mixed methods approach as has been presented by some authors as Tashakkori and Teddlle, 1998, 2003, Cresswell, 2002. The multilevel model (Tashakkori and Teddlie, 1998) has been applied, which allows to combine quantitative methods (a survey among the users of territory) and qualitative methods (interviews with the representatives of municipalities, private sector and non-governmental sector representatives have been applied). The representatives of municipalities were elected deputies of the local municipal government, which were chosen randomly and the questions were formulated in the structured questionnaires similarly as the citizens of the city asked in the structured questionnaires. Additionally, the city representatives were interviewed by the appointment in personal interviews and this kind of research was less formal and based on the unstructured and taped interviews, coded results and compared with the demand side research results. Some form of triangulation of results has been amended in order to compare the similarity of answers of respondents.

CONCLUSIONS

A primary research focused at the demand side (customers of the city) has been conducted in a city of Banska Bystrica and the used analysis of the importance and performance of city and its potential has been researched by the application of the Importance-Performance Analysis (IPA) created by Martilla and James (1977) as the managerial method used in marketing research.

PERFORMANCE

Table 1. The Importance and performance Matrix

I M P O R T A N C E	**Quadrant 1 (High importance; Lower performance)** This quadrant displays attributes which are considered to be important, yet are not being performed to the same standard, or level. If an attribute falls within this quadrant effort should be made to shift it to quadrant two (upper right).	**Quadrant 2 (High importance; High performance)** This quadrant represents those attributes of VRM which are considered to be important and are also being performed to a higher standard. Attributes that fall into this quadrant should be maintained to ensure they remain at this level of performance.
	Quadrant 3 (Lower importance; Lower performance) This quadrant indicates attributes which are of a lower priority. Resources are not being channeled towards them as much as other attributes. Attributes which fall into this quadrant should remain untouched.	**Quadrant 4 (Lower importance; High performance)** This quadrant indicates a possible overuse or resources. Organizations are allocating resources to, and highly performing these attributes, yet they are not deemed to be exceptionally important to the organization.

Source: Martilla and James, 1977.

The creation of IPA quadrants and the importance of particular factors have been factors placed into four quadrants as has been recommended by Martilla and James (1997). The following table explains the reorganization of quadrants and factors in the more visual form as has been later used for the performance of results of research (Figure 3).

Figure 3. Importance of Performance Analysis

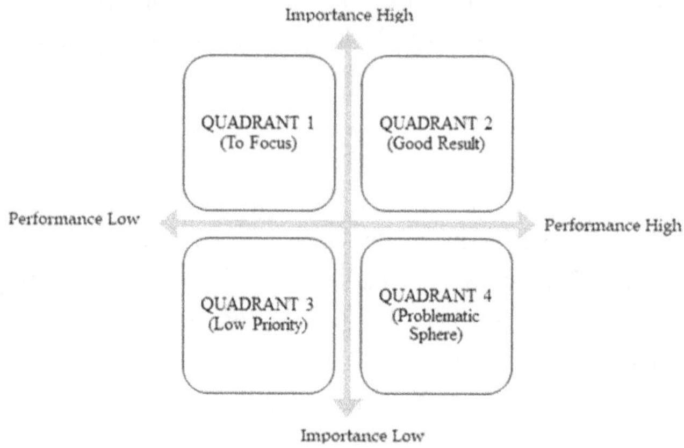

Source: own analysis based on Martilla and James Importance and Performance Matrix, 1977.

A sample of 134 respondents (citizens, visitors and entrepreneurs/developers) was used in Banska Bystrica. The answers were compared with the interview results of the municipality representatives. The evaluation has been performed with the use of non-parametric methods, Friedman and Wilcox test. The importance of factors has been confirmed by the application of the function Monte Carlo Sig. In Banska Bystrica, the following chart (Figure 4) of importance and satisfaction with the place and the factors has been created and analysed.

Figure 4. in the form of Graph: The Results of research of importance and performance of services in Banska Bystrica

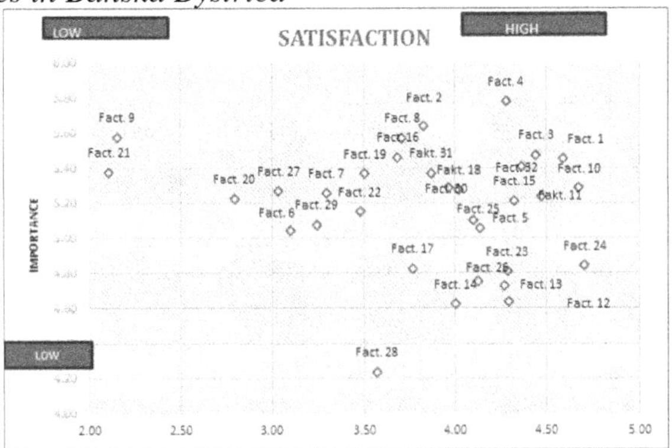

Source: own research in the city of Banska Bystrica

The most critical factors in the city from the side of citizens, visitors and entrepreneurs were chosen the opportunities to work, the quality of infrastructure (also transportation), entrepreneurship opportunities, innovative approach to governance and city competitiveness.

The following table consists of data of the factors of importance and satisfaction, which have been used in the study.

Table 2. Factors of importance and satisfaction in the city Banska Bystrica

Factor	D	S	Description of the factor
Fact 1	5,45	4,59	Attractiveness
Fact 2	5,64	3,83	Environment
Fact 3	5,47	4,44	Atmosphere
Fact 4	5,78	4.28	Security
Fact 5	5,05	4,14	Peacefulness
Fact 6	5,04	3,10	Innovativeness
Fact 7	5,26	3,30	Competitiveness
Fact 8	5,57	3,71	Accessibility
Fact 9	5,58	2,15	Employment
Fact 10	5,28	4,68	Dwelling (apartments, houses)
Fact 11	5,23	4,47	Network of stores
Fact 12	4,64	4.29	Culture
Fact 13	4.73	4,26	Sport facilities

continued on following page

Table 2. Continued

Factor	D	S	Description of the factor
Fact 14	4,62	4.00	Leisure facilities
Fact 15	5.21	4.32	Quality of the educational facilities
Fact 16	5,46	3,68	Quality of health care facilities
Fact 17	4,82	3,77	Quality of social facilities
Fact 18	5,27	4,02	Satisfaction with services
Fact 19	5,37	3,51	Transportation services
Fact 20	5,22	2,79	Parking facilities
Fact 21	5,38	2,11	Road maintenance
Fact 22	5,15	3,48	Transportation signs in the city
Fact 23	4,81	4,28	Accommodation opportunities
Fact 24	4,84	4,71	The amount of restaurants
Fact 25	5,10	4,10	The infrastructure for citizens
Fact 26	4,75	4,12	The tourism attractions
Fact 27	5,27	3,04	Entrepreneurship opportunities
Fact 28	4,24	3,57	Not-for Profit Organizations and their activities in the city
Fact 29	5,08	3,24	Partnership of the city representatives with citizens
Fact 30	5,29	3,96	Communication of the city representatives with citizens
Fact 31	5,37	3,87	Satisfaction with services in the city
Fact 32	5,41	4,36	Qualified labour

Source: own preparation of the evaluation

CONCLUSION

The most critical factors in the city from the side of citizens, visitors and entrepreneurs were chosen the opportunities to work, the quality of infrastructure, entrepreneurship opportunities, innovative approach to governance and city competitiveness. Based on the research by the application of IPA analysis (analysis of importance and performance – satisfaction) in the city of Banska Bystrica, Slovakia the results indicate that the public sector representatives have to concentrate on the improvement of employment possibilities, improvement of the infrastructure (quality of roads, enlargement of parking places).

Deficit could be found in a support of new business opportunities, growth of competitiveness, which is also a reason of lack of employment opportunities for young people or seniors after the age of 50. Generally speaking, employment opportunities and attracting of investors are two major weaknesses of the city of Banska Bystrica. Despite of the existence of the industrial park in Salkova and the efforts of local public sector representatives, the city is still behind with smaller number of citizens or the area, for instance Zvolen, which is situated only 20 km from Banska Bystrica. Problematic seems to be also the partnerships and the co-operation of public, private and not-for profit sector, which might be a reason of lack of confidence to co-operate or a deficit of existing co-ordination of common activities. Marketing would be one of possible tools to improve this situation, especially the co-ordination of the co-operation, which could be performed by the Department of Development and Marketing.

Satisfactory results have been achieved in the quality of environment, accessibility, quality of health care facilities and services, educational institutions and quality of educational services. Communication with citizens is also at very sound level and the quality of services generally is also very satisfactory. A paradox is that lack of the employment opportunities is in a city with citizens who are highly educated and skilled, which means that city has to focus more on the innovation and new technologies, which means to attract companies and employers offering such type of employment. The city has a strong deficit in the innovation strategies, weaknesses are also in the capacity of accommodation facilities and better infrastructure. Building of a smart city would be an excellent opportunity for this city to improve its competitiveness.

This research might be helpful for the public sector planners and politicians to recognize a city and especially desires of its users from the demand side (side of territory users and partners of the process of co-creation with the public sector). In the end of this discussion, we conclude with the words of Douglas North, a Nobel Prize winner, who mentioned in the concept of the „adaptive effectiveness "a necessity of the effective change. For this reason, cities have to be open to continual changes and improvements and marketing as well as new communication trends as the use of 5G/6G networks and city representatives need to understand the meaning of place marketing, branding and new communication technologies in their activities and be able to use it for their effective and adaptive governance with a co-operation of a community.

REFERENCES

Ashworth, G. J., & Vooght, T. H. (1990). *Selling the City: Marketing Approaches in Public Sector Urban Planning*. Belhaven Press.

Asplund, C. (1993). Placehunting International. Om konsten att gora sig mer attraktiv for investeringar. Eurofutures, Stockholm. In REINISTO, S. K. 2003. *Success Factors of Place Marketing: A Study of Place Marketing Practices in Northern Europe and the United States.* Helsinki University of Technology, Institute of Strategy and International Business, Doctoral Dissertations 2003/4.

Boisen, M. (2007). Strategic *Marketing for middle-sized cities in the Netherlands.* Master Thesis, University of Utrecht, Utrecht.

Buhalis, D. (2014). *Contribution on smart tourism to the UNWTO 3rd Global Summit on city tourism New paradigms in City Tourism Developemt", in Barcelona, December 9th – 10th..*

Carvalho, L. (2015). Smart cities from scratch? A socio-technical perspective. *Cambridge Journal of Regions, Economy and Society*, 8(1), 43–60. DOI: 10.1093/cjres/rsu010

Cresswell, J. W. (2003). *Research Design. Qualitative, Quantitative and Mixed Method Approaches*. Sage Publications.

Crouch, G. I., & Ritchie, J. R. B. (2003) The *Competitive Destination: A Sustainable Tourism Perspective*, Cambridge: CABI Publishing. 272 s.

Go, F., & Govers, R. (2009). *Place Branding*. Palgrave MacMillan.

Hankinson, G. (2001). Location branding: A study of twelve English cities. *Journal of Brand Management*, 9(2), 127–142. DOI: 10.1057/palgrave.bm.2540060

Hankinson, G. (2004). Relational network brands: Towards a conceptual model of place brands. *Journal of Vacation Marketing*, 10(2), 109–121. DOI: 10.1177/135676670401000202

Hankinson, G. (2005). Destination Brand Images: A Business Tourism Pespective. *Journal of Services Marketing*, 19(1), 24–32. DOI: 10.1108/08876040510579361

Hollands, R. (2015). Critical interventions into the corporate smart city. *Cambridge Journal of Regions, Economy and Society*, 8(1), 61–77. DOI: 10.1093/cjres/rsu011

Kavaratzis, M., & Ashworth, G. (2008). Place Marketing: How Did we Get there and Where we Going? *Journal of Place Management and Development*, 1(2), 150–165. DOI: 10.1108/17538330810889989

Kitchin, R. (2015). Making sense of smart cities: Addressing present shortcomings. *Cambridge Journal of Regions, Economy and Society*, 8(1), 131–136. DOI: 10.1093/cjres/rsu027

Kotler, P., ... (2002). *Marketing Asian Places.*John Wiley & Sons (Asia), Singapore, 443 p.

Kotler, P, Gertner, D (2002) Country as brand, product and beyond: A place marketing and brand management perspective. *Journal of Brand Management*, April, 9(4/5).

Lendel, V. (2009). *Riadenie vzťahov so zákazníkmi v podniku.* Dizertačná práca. Žilinská univerzita, Fakulta riadenia a informatiky.

Lusch, R. F., Vargo, S., O'Brien, M. (2007). Competing through service: Insights from service- dominant logic. *Journal of Retailing* 83 (1, 2007), pp. 5-18.

Martilla, J. A., & James, J. C. (1977). Importance-Performance Analysis. *Journal of Marketing*, 41(1), 77–79. DOI: 10.1177/002224297704100112

Nam, T., & Pardo, T. A. (2011). Smart City as Urban Innovation: Focusing on Management, Policy and Context. In Estevez, E., & Janssen, M. (Eds.), *Proceedings of the 5th International Conference on Theory and Practice of Electronic Governance (ICEGOV2011).* Tallinn, Estonia: ACM Press. DOI: 10.1145/2072069.2072100

Prahalad, C. K., & Ramaswarny, V. (2000). Co-opting customer competence of the corporation. *Harvard Business Review*, 78, 79–87.

Prahalad, C. K., & Ramaswarny, V. (2004). Co-creation Experiences: The Next Practice in Value Creation. *Journal of Interactive Marketing*, 18(3), 5–14. DOI: 10.1002/dir.20015

Reinisto, S. K. (2001). *Lisensiaatintutkimus:Kaupungin kehittaminen merkkituotteena. Lahden ja Helsingin tapaustutkimukset. Licenciate Study: City Branding – Case Studies Lahti and Helsinki.* Helsinki University of Technology.

Reinisto, S. K. (2003). Success Factors of Place Marketing: A Study of Place Marketing Practices in Northern Europe and the United States. Helsinki University of Technology, Institute of Strategy and International Business, Doctoral Dissertations 2003/4.

https://blogs.bournemouth.ac.uk/ekotourismlab/research-projects/phd-projects/smart-tourism-destinations-explore-how-smartness-increases-competitiveness-in-the-context-of-tourism-destinations, (Retrieved 15.12. 2014).

Shelton, T., Zook, M., & Wiig, A. (2015). The 'actually existing smart city'. *Cambridge Journal of Regions, Economy and Society*, 8(1), 13–25. DOI: 10.1093/cjres/rsu026

Sundbo, J. (2008). *Creating Experiences In the Experience Economy*. Edward Elgar Publishing. DOI: 10.4337/9781848444003

Suzuki, L. R. (2017). Smart Cities IoT: Enablers and Technology Road Map. In Rassia, S., & Pardalos, P. (Eds.), *Smart City Networks. Springer Optimization and Its Applications* (Vol. 125). Springer. DOI: 10.1007/978-3-319-61313-0_10

Tashakkori, A., & Teddlie, C. (1998). *Mixed Methodology: Combining Qualitative and Quantitative Approaches*. SAGE.

Tashakkori, A., & Teddlie, C. (2003). *Handbook of mixed methods in the social and behavioural sciences*. Sage.

Vargo, S. L., & Lusch, R. F. (2004). Evolving to a new dominant logic for marketing. *Journal of Marketing*, 68(1), 1–17. DOI: 10.1509/jmkg.68.1.1.24036

Vargo, S. L., & Lusch, R. F. (2006). Service-dominant logic: What it is, what it is not, what it might be. In Lusch, R. F., & Vargo, S. L. (Eds.), *The service-dominant logic of marketing: Dialog, debate, and directions* (pp. 43–56). ME Sharpe.

Compilation of References

Abdellatif, A. A., Mohamed, A., Chiasserini, C. F., Tlili, M., & Erbad, A. (2019). Edge computing for smart health: Context-aware approaches, opportunities, and challenges. *IEEE Network*, 33(3), 196–203. DOI: 10.1109/MNET.2019.1800083

Abdulla, F. Y., & Mapjabil, J. (2021). Review of theories and model of research on liminality in tourism. *Proceeding for Global Tourism Conference (GTC) 2021*, 175-177. https://doi.org/DOI: 10.46754/gtc.2021.11.048

Acker, A. (2014). The Short Message Service: Standards, infrastructure and innovation. *Telematics and Informatics*, 31(4), 559–568. DOI: 10.1016/j.tele.2014.01.004

Agarwal, A., & Agarwal, K. (2014). The Next Generation Mobile Wireless Cellular Networks – 4G and Beyond. *American Journal of Electrical and Electronic Engineering*, 2(3), 92–97. DOI: 10.12691/ajeee-2-3-6

Aguiar-Castillo, L., Guerra, V., Rufo, J., Rabadan, J., & Perez-Jimenez, R. (2021). Survey on optical wireless communications-based services applied to the tourism industry: Potentials and challenges. *Sensors (Basel)*, 21(18), 6282. DOI: 10.3390/s21186282 PMID: 34577489

Ahire, M., Patil, V., & Kolapkar, P.A., Student, & Wagh, K.K. (2021). A Survey on Mobile Communications and Network Technologies. [OAIJSE]. *Open Access International Journal of Science& Engineering*, 6(6), 5–9.

Ahmadi, H., Nag, A., Khan, Z., Sayrafian, K., & Rahadrja, S. (2021). *Networked Twins and Twins of Networks: an Overview on the Relationship Between Digital Twins and 6G.*

Ahmed, R., & Matin, M. A. (2020). Towards 6g wireless networks-challenges and potential technologies. *Journal of Electrical Engineering*, 71(4), 290–297. DOI: 10.2478/jee-2020-0040

Airod, F. E., Chafnaji, H., & Yanikomeroglu, H. (2021). HARQ in full-duplex relay-assisted transmissions for URLLC. *IEEE Open Journal of the Communications Society*, 2, 409–422. DOI: 10.1109/OJCOMS.2021.3055073

Aissaoui, N. (2022). The Digital Divide: A Literature Review and Some Directions for Future Research in Light of COVID-19. *Global Knowledge, Memory and Communication*, 71(8/9), 686–708. DOI: 10.1108/GKMC-06-2020-0075

Akbar, M. S., Hussain, Z., Sheng, Q. Z., & Mukhopadhyay, S. (2022). 6G survey on challenges, requirements, applications, key enabling technologies, use cases, AI integration issues and security aspects. *arXiv preprint arXiv:2206.00868.*

Akyildiz, I. F., Kak, A., & Nie, S. (2020). 6G and Beyond: The Future of Wireless Communications Systems. *IEEE Access : Practical Innovations, Open Solutions*, 8, 133995–134030. DOI: 10.1109/ACCESS.2020.3010896

Akyildiz, I., & Jornet, J. (2010). The Internet of nano-things. *IEEE Wireless Communications*, 17(6), 58–63. DOI: 10.1109/MWC.2010.5675779

Alahi, E. E., Sukkuea, A., Tina, F. W., Nag, A., Kurdthongmee, W., Suwannarat, K., & Mukhopadhyay, S. (2023). Integration of IoT-Enabled Technologies and Artificial Intelligence (AI) for Smart City Scenario: Recent Advancements and Future Trends. *Sensors (Basel)*, 23(11), 5206. DOI: 10.3390/s23115206 PMID: 37299934

Alam, M. Z., Abkenar, F. S., Adhicandra, I., Murali, S., & Jamalipou, A. (2020). Low-delay path selection for cluster-based buffer-aided vehicular communications. *IEEE Transactions on Vehicular Technology*, 69(9), 9356–9363. DOI: 10.1109/TVT.2020.2976926

Al-Emran, M., Mezhuyev, V., & Kamaludin, A. (2019). PLS-SEM in information systems research: a comprehensive methodological reference. In *Proceedings of the International Conference on Advanced Intelligent Systems and Informatics 20184* (pp. 644-653). Springer International Publishing. DOI: 10.1007/978-3-319-99010-1_59

Al-Fuqaha, A., Guizani, M., Mohammadi, M., Aledhari, M., & Ayyash, M. (2015). Internet of Things: A Survey on Enabling Technologies, Protocols, and Applications [Fourthquarter.]. *IEEE Communications Surveys and Tutorials*, 17(4), 2347–2376. DOI: 10.1109/COMST.2015.2444095

Algarni, S., & Sheldon, F. (2023). Systematic Review of Recommendation Systems for Course Selection. *Mach. Learn. Knowl. Extr.*, 5(2), 560–596. DOI: 10.3390/make5020033

Alhassan, A. H., Rahman, T. A., & Abd-Alhameed, R. A.. (2021). Comprehensive Survey on 6G Wireless Networks: Key Technologies, Challenges, Standardization, and Future Directions. *IEEE Access : Practical Innovations, Open Solutions*, 9, 148191–148230.

Al-Hyari, H. S. A., Al-Smadi, H. M., & Weshah, S. R. (2023). The impact of artificial intelligence (AI) on guest satisfaction in hotel management: An empirical study of luxury hotels. *Geo Journal of Tourism and Geosites*, 48(2 supplement), 810–819. DOI: 10.30892/gtg.482spl15-1081

Ali, A., & Frew, A. J. (2014). Technology innovation and applications in sustainable destination development. *Information Technology & Tourism*, 14(4), 265–290. DOI: 10.1007/s40558-014-0015-7

Ali, B., Mirza, J., Alvi, S. H., Khan, M. Z., Javed, M. A., & Noorwali, A. (2023). IRS-assisted physical layer security for 5G enabled industrial Internet of Things. *IEEE Access : Practical Innovations, Open Solutions*, 11, 21354–21363. DOI: 10.1109/ACCESS.2023.3250251

Alkandari, A., & AlAhmad, M. A. (2022). Overview On 5G and 6G Wireless Communication with IoT Technology. *Journal of Engineering Science and Technology*, 17(1), 95–105.

Allam, Z., & Jones, D. (2021). Future (post-COVID) digital, smart and sustainable cities in the wake of 6G: Digital twins, immersive realities and new urban economies. *Land Use Policy*, 101, 105201. DOI: 10.1016/j.landusepol.2020.105201

Almekhlafi, M., Arfaoui, M. A., Assi, C., & Ghrayeb, A. (2022). Enabling URLLC applications through reconfigurable intelligent surfaces: Challenges and potential. *IEEE Internet Things Mag.*, 5(1), 130–135. DOI: 10.1109/IOTM.007.2100124

Alraih, S., Shayea, I., Behjati, M., Nordin, R., Abdullah, N. F., Abu-Samah, A., & Nandi, D. (2022). Revolution or Evolution? Technical Requirements and Considerations towards 6G Mobile Communications. *Sensors (Basel)*, 22(3), 762. DOI: 10.3390/s22030762 PMID: 35161509

Alreshidi, E. (2019). Smart sustainable agriculture (SSA) solution underpinned by internet of things (IoT) and artificial intelligence (AI). arXiv preprint arXiv:1906.03106.

Alsharif, M. H., Kelechi, A. H., Albreem, M. A., Chaudhry, S. A., Zia, M., & Kim, S. H. (2020). Sixth generation (6G) wireless networks: Vision, research activities, challenges and potential solutions. *Symmetry*, 12(4), 676. DOI: 10.3390/sym12040676

Alsulami, H., Serbaya, S. H., Abualsauod, E. H., Othman, A. M., Rizwan, A., & Jalali, A. (2022). A federated deep learning empowered resource management method to optimise 5G and 6G quality of services (QoS). *Wireless Communications and Mobile Computing*, 2022, 1–9. DOI: 10.1155/2022/1352985

Aluvala, R., & Suryanarayana, A. (2024). Monumental Transition From 5G To 6G: Skill Development Needs And Workforce Evolution Imperative. *Educational Administration: Theory and Practice*, 13699–13706. DOI: 10.53555/kuey.v30i5.5986

Alwis, C., Kalla, A., Pham, Q., Kumar, P., Dev, K., Hwang, W., & Liyanage, M. (2021). Survey on 6G frontiers: Trends, applications, requirements, technologies and future research. *IEEE Open Journal of the Communications Society*, 2, 836–886. DOI: 10.1109/OJCOMS.2021.3071496

Andrews, J. G., Buzzi, S., Choi, W., Hanly, S. V., Lozano, A., Soong, A. C., & Zhang, J. C. (2014). What will 5G be? *IEEE Journal on Selected Areas in Communications*, 32(6), 1065–1082. DOI: 10.1109/JSAC.2014.2328098

Ankarali, Z. E., Peköz, B., & Arslan, H. (2017). Flexible radio access beyond 5G: A future projection on waveform, numerology, and frame design principles. *IEEE Access : Practical Innovations, Open Solutions*, 5, 18295–18309. DOI: 10.1109/ACCESS.2017.2684783

Ansari, J., Christian, A., de Bruin, P., Farkas, J., Grosjean, L., Sachs, J., Torsner, J., Varga, B., Harutyunyan, D., König, N., & Schmitt, R. (2022). Performance of 5G Trials for Industrial Automation. *Electronics (Basel)*, 11(3), 412. DOI: 10.3390/electronics11030412

Anvigh, A. A., Khavan, Y., & Pourghebleh, B. (2024). Transforming vehicular networks: How 6G can revolutionize intelligent transportation? Science. *Engineering and Technology*, 4(1), 80–93.

Apeh, E. T., Gabrys, B., & Schierz, A. (2011), Customer profile classification using transactional data. *Nature and Biologically Inspired Computing (NaBIC), 2011 Third World Congress on*. IEEE.

Apruzzese, M., Bruni, M. E., Musso, S., & Perboli, G. (2023). 5G and Companion Technologies as a Boost in New Business Models for Logistics and Supply Chain. *Sustainability (Basel)*, 15(15), 11846. DOI: 10.3390/su151511846

Aranda, J., Sacoto-Cabrera, E. J., Haro-Mendoza, D., & Astudillo-Salinas, F. (2021). 5G networks: A review from the perspectives of architecture, business models, cybersecurity, and research developments. *Novasinergia*, 4(1), 6–41. DOI: 10.37135/ns.01.07.01

Arastouei, N. (2023). 6G technologies: Key features, challenges, security and privacy issues. *Communications in Computer and Information Science*, 1807, 94–109. DOI: 10.1007/978-3-031-36096-1_7

Arena, F., Collotta, M., Pau, G., & Termine, F. (2022). An overview of augmented reality. *Computers*, 11(2), 28. DOI: 10.3390/computers11020028

Arulsamy, A. S., Singh, I., Kumar, S., Panchal, J. J., & Bajaj, K. K. (2023). Employee Training and Development Enhancing Employee Performance -A Study. *Samdarshi, 16*(3). https://www.researchgate.net/publication/373775939_Employee_Training_and_Development_Enhancing_Employee_Performance_-A_Study#fullTextFileContent

Asghar, M. Z., Memon, S. A., & Hämäläinen, J. (2022). Evolution of wireless communication to 6G: Potential applications and research directions. *Sustainability (Basel)*, 14(10), 6356. DOI: 10.3390/su14106356

Ashworth, G. J., & Vooght, T. H. (1990). *Selling the City: Marketing Approaches in Public Sector Urban Planning*. Belhaven Press.

Aspin, D. N., & Chapman, J. D. (2007). Lifelong Learning: Concepts and Contexts. *International Journal of Lifelong Education.*

Asplund, C. (1993). Placehunting International. Om konsten att gora sig mer attraktiv for investeringar. Eurofutures, Stockholm. In REINISTO, S. K. 2003. *Success Factors of Place Marketing: A Study of Place Marketing Practices in Northern Europe and the United States.* Helsinki University of Technology, Institute of Strategy and International Business, Doctoral Dissertations 2003/4.

Babin, B. J., Darden, W. R., & Griffin, M. (1994). Work and/or fun: Measuring hedonic and utilitarian shopping value. *The Journal of Consumer Research*, 20(4), 644–656. DOI: 10.1086/209376

Barakabitze, A. A., & Walshe, R. (2022). SDN and NFV for QOE-driven multimedia services delivery: The road towards 6G and beyond networks. *Computer Networks*, 214, 109133. DOI: 10.1016/j.comnet.2022.109133

Basar, E., Di Renzo, M., de Rosny, J., Debbah, M., Alouini, M.-S., & Zhang, R. (2019). Wireless Communications Through Reconfigurable Intelligent Surfaces. *IEEE Access : Practical Innovations, Open Solutions*, 7, 116753–116773. DOI: 10.1109/ACCESS.2019.2935192

Batala, B., & Slevitch, L. (2024). Keeping two balls in the air: The bleisure travel experience. *Annals of Tourism Research Empirical Insights*, 5(1), 100115. DOI: 10.1016/j.annale.2023.100115

Battarbee, K., & Koskinen, I. (2005). The Importance of Empathy in Design: Perspectives from the Field. *Design Issues*.

Bennis, M., Debbah, M., & Poor, V. (2018). Ultra-Reliable and Low-Latency Wireless Communication: Tail, Risk, and Scale. *Proceedings of the IEEE*, 106(10), 1834–1853. DOI: 10.1109/JPROC.2018.2867029

Berardinelli, G., Mahmood, N. H., Rodriguez, I., & Mogensen, P. (2018). *Beyond 5G wireless IRT for Industry 4.0: Design principles and spectrum aspects. In 2018 IEEE Globecom Workshops (GC Wkshps)*. IEEE.

Berry, M., & Linoff, G. (2000). *Mastering Data Mining*. John Wiley & Sons, Inc.

Berryman, D. (2012). Augmented reality: A review. *Medical Reference Services Quarterly*, 31(2), 212–218. DOI: 10.1080/02763869.2012.670604 PMID: 22559183

Beshley, M., Oda, Y., & Vasile, S. (2022). Emerging network technologies for digital transformation: 5G/6G, IoT, SDN/IBN, cloud computing, and blockchain. In Proceedings of the IEEE International Conference on Advanced Trends in Radioelectronics, Telecommunications and Computer Engineering. Cham: Springer Nature Switzerland.

Bessen, J. E. (2019). Emerging Technical Skills for the Future of Work: Trends and Implications. *The Journal of Economic Perspectives*.

Bezborah, C. J. (2023). Tea tourism, the way forward to promote agritourism in the state of Assam: Prospects And Challenges. *International Journal of Multidisciplinary Trends*, 5(9), 1–4. DOI: 10.22271/multi.2023.v5.i9a.330

Bhandari, N., Devra, S., & Singh, K. (2017). Evolution of Cellular Network: From 1G to 5G. *International Journal of Engineering and Techniques*, 3(5), 98–105.

Bharwani, S., & Mathews, D. (2021). Techno-business strategies for enhancing guest experience in luxury hotels: A managerial perspective. *Worldwide Hospitality and Tourism Themes*, 13(2), 168–185. DOI: 10.1108/WHATT-09-2020-0121

Bhatt, D., Sharma, S., & Rana, V. (2023). A study of the remote work-family balance of female hotel managers. *African Journal of Hospitality, Tourism and Leisure*, 12(3), 989–1010.

Biggsa, E. M., Guptac, N., Saikiad, S. D., & Duncana, J. M. (2018). The tea landscape of Assam: Multi-stakeholder insights into sustainable livelihoods under a changing climate. *Environmental Science & Policy*, 82, 9–18. DOI: 10.1016/j.envsci.2018.01.003

Biswas, R., Jana, S., & Pal, K. (2024). Vision and Research Directions of 6G Technologies and Applications. *International Journal of Advanced Research in Science. Tongxin Jishu*, 4(1), 19–23. DOI: 10.48175/IJARSCT-15304

BMW. (2020). *BMW Group uses 5G for flexible, efficient production.* Retrieved from https://www.bmwgroup.com/en/news/general/2020/bmw-group-5g-production.html

Bochechka, G., & Tikhvinskiy, V. (2014). *Proceedings of the 2014 ITU Kaleidoscope Academic Conference: Living in a Converged World-Impossible Without Standards?* Geneva, Switzerland: ITU.

Böhme, C. (2023). Shaping the Guest Experience: How Personalized Services Improve Customer Satisfaction in Luxury Hotels and Create a Competitive Advantage. *Signature (Ramsey, N.J.).*

Boisen, M. (2007). Strategic *Marketing for middle-sized cities in the Netherlands.* Master Thesis, University of Utrecht, Utrecht.

Brown, S., & Black, P. (2022). Professional Development in the Age of 6G. *Journal of Continuing Education*, 28(4), 150–165.

Brown, T. (2008). Design Thinking and Human-centric Innovation. *Harvard Business Review.*

BT. (2021). BT and King's College London to deploy 5G at King's College Hospital. Retrieved from https://www.bt.com/news/2021/may/bt-and-kings-college-london-to-deploy-5g-at-kings-college-hospital

Buchanan, R., & Layton, N. (2019). Innovation in Assistive Technology: Voice of the User. *Societies (Basel, Switzerland)*, 9(2), 1–10. DOI: 10.3390/soc9020048

Budimir, M. (2014). An overview of research on technological and telecommunication development. *Ekonomski Vjesnik*, (1), 171–182. https://doaj.org/article/e129e602b44a479e9e39870a77bb5323

Buhalis, D. (2014). *Contribution on smart tourism to the UNWTO 3rd Global Summit on city tourism New paradigms in City Tourism Developemt", in Barcelona, December 9th – 10th..*

Buhalis, D., & Moldavska, I. (2021). In-room voice-based AI digital assistants transforming on-site hotel services and guests' experiences. In *Information and Communication Technologies in Tourism 2021:Proceedings of the ENTER 2021 eTourism Conference,January 19–22, 2021* (pp. 30-44). Springer International Publishing.

Buhalis, D., Harwood, T., Bogicevic, V., Viglia, G., Beldona, S., & Hofacker, C. F. (2019). Technological disruptions in services: Lessons from tourism and hospitality. *Journal of Service Management*, 30(4), 484–506. DOI: 10.1108/JOSM-12-2018-0398

Buhalis, D., & Moldavska, I. (2022). Voice assistants in hospitality: Using artificial intelligence for customer service. *Journal of Hospitality and Tourism Technology*, 13(3), 386–403. DOI: 10.1108/JHTT-03-2021-0104

Bulchand-Gidumal, J. (2020). Impact of artificial intelligence in travel, tourism, and hospitality. *Handbook of E-Tourism*, 1-20. https://doi.org/DOI: 10.1007/978-3-030-05324-6_110-1

Bulchand-Gidumal, J. (2022). Impact of artificial intelligence in travel, tourism, and hospitality. In *Handbook of e-Tourism* (pp. 1943–1962). Springer International Publishing.

Caicedo-Barreth, A., Santos Pavón, E., & Lima Santos, L. (2020). Competitiveness of Guayaquil towards bleisure tourism. *EJTHR*, 10(2), 118–133.

Calisto, M. L., & Sarkar, S. (2024). A systematic review of virtual reality in tourism and hospitality: The known and the paths to follow. *International Journal of Hospitality Management*, 116, 103623. DOI: 10.1016/j.ijhm.2023.103623

Calvanese Strinati, E., Barbarossa, S., Cassiau, N., Maret, L., Dehos, C., Ktenas, D., & Gonzalez-Jimenez, J. L. (2019). 6G: The Next Frontier: From Holographic Messaging to Artificial Intelligence Using Subterahertz and Visible Light Communication. *IEEE Vehicular Technology Magazine*, 14(3), 42–50. DOI: 10.1109/MVT.2019.2921162

Cambria, E. (2016). Affective computing and sentiment analysis. *IEEE Intelligent Systems*, 31(2), 102–107. DOI: 10.1109/MIS.2016.31

Camilleri, M. A., & Camilleri, M. A. (2018). *The tourism industry: An overview*. Springer International Publishing.

Carlson Wagonlit Travel. (2016). Combing business and leisure trips: A quantitative look at the bleisure phenomenon. In CWT Solution Group White Paper. Available at https://www.mycwt.com/news/pr/carlson-wagonlit-travel-research-business-travelers-feel-50-percent-safer-at-airports-than-at-train-or-subway-stations/ [Retrieved 22 May 2020].

Carnevale, A. P., & Desrochers, D. M. **(2001),** "*Technical Skills and Competencies in the Modern Workplace*", Workforce Education Forum.

Car, T., Stifanich, L. P., & Šimunić, M. (2019). *Internet of Things (IoT) in tourism and hospitality: opportunities and challenges*. Tourism in Southern and Eastern Europe., DOI: 10.20867/tosee.05.42

Carvalho, L. (2015). Smart cities from scratch? A socio-technical perspective. *Cambridge Journal of Regions, Economy and Society*, 8(1), 43–60. DOI: 10.1093/cjres/rsu010

Caytiles, R. D., & Park, B. (2013). Future Directions of Information and Telecommunication Systems Through the Technological Advancement Convergence. *International Journal of Multimedia and Ubiquitous Engineering*, 8(1), 101–106. https://gvpress.com/journals/IJMUE/vol8_no1/9.pdf

Celik, N., Darema, F., Runsewe, T., Saad, W., & Yavuz, A. (2024). Dddas for optimised design and management of 5G and beyond 5G (6G) networks. *Lecture Notes in Computer Science*, 13984, 123–132. DOI: 10.1007/978-3-031-52670-1_12

Chae, B., Sheu, C., Yang, C., & Olson, D. (2014). The impact of advanced analytics and data accuracy on operational performance: A contingent resource based theory (RBT) perspective. *Decision Support Systems*, 59, 119–126. DOI: 10.1016/j.dss.2013.10.012

Chang, J., Miska, R. A., & Shober, R. A. (1993). Wireless systems and technologies: An overview. *AT & T Technical Journal*, 72(4), 11–18. DOI: 10.1002/j.1538-7305.1993.tb00546.x

Chaoub, A., Giordani, M., Lall, B., Bhatia, V., Kliks, A., Mendes, L., Rabie, K., Saarnisaari, H., Singhal, A., Zhang, N., Dixit, S., & Zorzi, M. (2022). 6G for Bridging the Digital Divide: Wireless Connectivity to Remote Areas. *IEEE Wireless Communications*, 29(1), 160–168. DOI: 10.1109/MWC.001.2100137

Chataut, R., Nankya, M., & Akl, R. (2024). 6G Networks and the AI Revolution—Exploring Technologies, Applications, and Emerging Challenges. *Sensors (Basel)*, 24(6), 1888. DOI: 10.3390/s24061888 PMID: 38544151

Chaudhury, S. K., Sarkar, S., & Pattnaik, C. S. (2023). Assam Tourism-Scenario: The Path Trodden and the Way Forward. *Journal of Global Economics*, 19(4).

Chauhan, D., Hadia, S. K., Desai, D., Parekh, D., & Patel, K. (2019). Evolution of mobile communication network: From 0G to 5G. *International Journal of Research and Analytical Reviews*, 6(2), 278–281. https://ijrar.com/upload_issue/ijrar_issue_20543738.pdf

Chauhan, R., Chirputkar, A., & Pathak, P. (2022, March 23). Blockchain and IoT in developing Fintech Ecosystem- An assistance to Insurance Industry. *2022 International Conference on Decision Aid Sciences and Applications (DASA)*. https://doi.org/DOI: 10.1109/DASA54658.2022.9765026

Cheng, J., Chen, W., Tao, F., & Lin, C.-L. (2018). Industrial IoT in 5G environment towards smart manufacturing. *Journal of Industrial Information Integration*, 10, 10–19. DOI: 10.1016/j.jii.2018.04.001

Chen, L., & Wang, H. (2022). Quantum Communication in 6G: Opportunities and Challenges. *IEEE Communications Magazine*, 60(8), 45–51. DOI: 10.1109/MCOM.2022.9755279

Chen, P., Wang, P., & Sun, J. (2011). Design and implement of the OFDM communication system. In *Proceedings of the 2011 IEEE International Workshop on Open-source Software for Scientific Computation* (pp. 59-63). Hangzhou, China: IEEE. DOI: 10.1109/OSSC.2011.6184695

Chen, S., Law, R., Zhang, M., & Si, Y. (2021). Mobile communications for tourism and hospitality: A review of historical evolution, present status, and future trends. *Electronics (Basel)*, 10(15), 1804. DOI: 10.3390/electronics10151804

Chen, X., Zhao, N., Chang, Z., Hämäläinen, T., & Wang, X. (2023). UAV-aided secure short-packet data collection and transmission. *IEEE Transactions on Communications*, 71(4), 2475–2486. DOI: 10.1109/TCOMM.2023.3244954

Chen, Y. L., Kuo, M. H., Wu, S. Y., & Tang, K. (2009). Discovering recency, frequency, and monetary (RFM) sequential patterns from customers' purchasing data. *Electronic Commerce Research and Applications*, 8(5), 241–251.

Chen, Y. L., Tang, K., Shen, R. J., & Hu, Y. H. (2005). Market basket analysis in a multiple store environment. *Decision Support Systems*, 40(2), 339–354.

Chen, Y., Qing, W., Chen, H., Song, X., Tang, H., & Tian, M. (2019). An overview of augmented reality technology. *Journal of Physics: Conference Series*, 1237(2), 022082. DOI: 10.1088/1742-6596/1237/2/022082

Cherkasova, O. P., Serdyukov, D. S., Nemova, E. F., Ratushnyak, A. S., Kucheryavenko, A. S., Dolganova, I. N., Xu, G., Skorobogatiy, M., Reshetov, I. V., Timashev, P. S., Spektor, I. E., Zaytsev, K. I., & Tuchin, V. V. (2021). Cellular effects of terahertz waves. *Journal of Biomedical Optics*, 26(9), 090902. DOI: 10.1117/1.JBO.26.9.090902 PMID: 34595886

Choi, J., Lee, B., & Jindal, N. (2021). 6G Era: Challenges and Solutions for Future Communications and Networking. *IEEE Transactions on Wireless Communications*, 20(2), 1160–1170.

Chowdhury, M. Z., Shahjalal, M., Ahmed, S., & Jang, Y. M. (2020). 6G Wireless Communication Systems: Applications, Requirements, Technologies, Challenges, and Research Directions. *IEEE Open Journal of the Communications Society*, 1, 957–975. DOI: 10.1109/OJCOMS.2020.3010270

Chowdhury, M. Z., Shahjalal, M., Hasan, M. K., & Jang, Y. M. (2019). The role of optical wireless communication technologies in 5G/6G and IoT solutions: Prospects, directions, and challenges. *Applied Sciences (Basel, Switzerland)*, 9(20), 4367. DOI: 10.3390/app9204367

Cimini, C., Lagorio, A., Pinto, R., Pezzotta, G., Adrodegari, F., & Cavalieri, S. (2023). 5G supporting digital servitisation in manufacturing: An exploratory survey. *IET Collaborative Intelligent Manufacturing*, 5(3), e12083. Advance online publication. DOI: 10.1049/cim2.12083

Clancy, J., Mullins, D., Deegan, B., Horgan, J., Ward, E., Denny, P., ... & Glavin, M. (2023). Feasibility study of V2X communications in initial 5G NR deployments. IEEE Access.

Clark, A., & Adams, P. (2023). Government Initiatives for 6G Skill Development. *Public Policy and Technology Journal*, 21(4), 145–160.

Clemente, D., Cabral, T., Rosa-Santos, P., & Taveira-Pinto, F. (2023). Blue seaports: The smart, sustainable and electrified ports of the future. *Smart Cities*, 6(3), 1560–1588. DOI: 10.3390/smartcities6030074

Cotton, S. L., Bultitude, R. J. C., & Scanlon, W. G. (2011). Millimeter-Wave Technology in Wireless PAN, LAN, and MAN: From IEEE 802.15.3c to IEEE 802.11ad and Beyond. *IEEE Communications Magazine*, 49(4), 68–74.

Cousik, T., Kleine, K., Shafin, R., Zhou, Z., Reed, J., & Liu, L. (2019). *CogRF: A New Frontier for Machine Learning and Artificial Intelligence for 6G RF Systems*. cornell university. https://doi.org//arxiv.1909.06862DOI: 10.48550

Cresswell, J. W. (2003). *Research Design. Qualitative, Quantitative and Mixed Method Approaches*. Sage Publications.

Crispin P. Noguerra, Jr. (2023). "User-Centered Design Principles for Effective Information System Development." International Journal of Advanced Research in Science, Communication and Technology: 826–30.

Crouch, G. I., & Ritchie, J. R. B. (2003) The *Competitive Destination: A Sustainable Tourism Perspective*, Cambridge: CABI Publishing. 272 s.

D. S. P. (2024). New technologies in the tourism and hospitality service sector. International. *Journal of Multidisciplinary Research*, 6(1), 12513. Advance online publication. DOI: 10.36948/ijfmr.2024.v06i01.12513

Dangi, R., Lalwani, P., Choudhary, G., You, I., & Pau, G. (2021). Study and Investigation on 5G Technology: A Systematic Review. *Sensors (Basel)*, 22(1), 26. DOI: 10.3390/s22010026 PMID: 35009569

Dangwal, A., Kukreti, M., Angurala, M., Sarangal, R., Mehta, M., & Chauhan, P. (2023). A Review on the Role of Artificial Intelligence in Tourism. In *2023 10th International Conference on Computing for Sustainable Global Development (INDIACom)* (pp. 164-168). IEEE.

Das, D. (2012). Prospects and problems of tourism industry in Assam. *Global Research Methodology Journal*, 2(7), 1–6.

Davis, M., Clark, J., & Lewis, R. (2022). Enhancing Cybersecurity Skills for 6G. *Cybersecurity Education Journal*, 13(5), 210–225.

De Guzman, J. A., Thilakarathna, K., & Seneviratne, A. (2019). A first look into privacy leakage in 3d mixed reality data. In *Computer Security–ESORICS 2019: 24th European Symposium on Research in Computer Security, Luxembourg, September 23–27, 2019* [Springer International Publishing.]. *Proceedings*, 24(Part I), 149–169.

De', R., Pandey, N., & Pal, A. (2020). Impact of Digital Surge during Covid-19 Pandemic: A Viewpoint on Research and Practice. *International Journal of Information Management*, 55(June), 102171. DOI: 10.1016/j.ijinfomgt.2020.102171 PMID: 32836633

Devarasetty, P., Balasai, A., & Brox, M. (2022). The evolution of virtual and augmented reality and its various practical applications. *Journal of Student Research*, 11(3). Advance online publication. DOI: 10.47611/jsrhs.v11i3.3509

Dey, S. K., & Gupta, R. (2012). Development of Safety and Productivity Correlation Model for Tea Industries of Barak Valley, Assam. *Development*, 2(12).

DIGITAL 2020: JULY GLOBAL STATSHOT. (2020). Retrieved March 11, 2022, from https://www.bonfire.com.au/wp-content/uploads/2021/08/Hootsuite-digital2021julyglobalstatshotreportv02-210721025325.pdf

Doborjeh, Z. G., Hemmington, N., Doborjeh, M. G., & Kasabov, N. (2021). Artificial intelligence: A systematic review of methods and applications in hospitality and tourism. *International Journal of Contemporary Hospitality Management*, 34(3), 1154–1176. DOI: 10.1108/IJCHM-06-2021-0767

Dogra, A., Jha, R. K., & Jain, S. (2021). A survey on beyond 5G network with the advent of 6G: Architecture and emerging technologies. *IEEE Access : Practical Innovations, Open Solutions*, 9, 67512–67547. DOI: 10.1109/ACCESS.2020.3031234

Dou, Z., Song, R., & Wen, J. R. (2007), A large-scale evaluation and analysis of personalized search strategies, in: *WWW '07: Proceedings of the 16th international conference on World Wide Web*, ACM Press, pp. 581–590. DOI: 10.1145/1242572.1242651

Dutta, A., & Hammad, E. (2020). 5G Security Challenges and Opportunities: A System Approach. *2020 IEEE 3rd 5G World Forum (5GWF)*.https://doi.org/DOI: 10.1109/5GWF49715.2020.9221122

Dutton, W. H., & Reisdorf, B. C. (2019). Cultural Divides and Digital Inequalities: Attitudes Shaping Internet and Social Media Divides*. *Information Communication and Society*, 22(1), 18–38. DOI: 10.1080/1369118X.2017.1353640

Dwarkadas, A. L., Krishna Challa, R., & Talasila, V., & K G, S. (2023). Augmented/ virtual reality: technological advancement with use cases. *2023 Global Conference on Information Technologies and Communications (GCITC)*. https://doi.org/DOI: 10.1109/GCITC60406.2023.10425860

Eickhoff, C., Collins-Thompson, K., Bennett, P. N., & Dumais, S. (2013), Personalizing atypical web search sessions, in: *Proceedings of the Sixth ACM International Conference on Web Search and Data Mining*, ACM, New York, NY, USA, pp. 285–294. DOI: 10.1145/2433396.2433434

Eissa, H. A., Bozed, K. A., & Younis, H. (2019). "Software Defined Networking." 19th International Conference on Sciences and Techniques of Automatic Control and Computer Engineering, STA 2019 (November): 620–25.

Elsen, I., Hartung, F., Horn, U., Kampmann, M., & Peters, L. (2001). Streaming technology in 3G mobile communication systems. *Computer*, 34(9), 46–52. DOI: 10.1109/2.947089

Ericson. (n.d.). 6G – Preparing for the next decade of connectivity. Retrieved July 12, 2024, from https://www.ericsson.com/en/6g

Ericsson. (2021). *Ericsson smart factory: 5G-enabled manufacturing*. Retrieved from https://www.ericsson.com/en/news/2021/5/ericsson-smart-factory-5g-enabled -manufacturing

Ezhilarasan, E., & Dinakaran, M. (2017). A Review on Mobile Technologies: 3G, 4G and 5G. *2017 Second International Conference on Recent Trends and Challenges in Computational Models (ICRTCCM)*, 369–373. DOI: 10.1109/ICRTCCM.2017.90

Fang, L., Zhao, B., Li, Y., Liu, Z., Ge, C., & Meng, W. (2020). Countermeasure based on smart contracts and AI against DoS/DDoS attack in 5G circumstances. *IEEE Network*, 34(6), 54–61. DOI: 10.1109/MNET.021.1900614

Farhadi Zavleh, A., & Bakhshi, H. (2021). Resource Allocation in Sparse Code Multiple Access-based Systems for Cloud-radio Access Network in 5G Networks. *Transactions on Emerging Telecommunications Technologies*, 32(1), e4153. DOI: 10.1002/ett.4153

Farhat, J., Brante, G., & Souza, R. D. (2018). Secure throughput optimization of selective decode-and-forward with finite blocklength. *IEEE 87th Veh. Technol. Conf. (VTC Spring)*, 1–5, Porto, Portugal.

Farooq, M., Ahmed, M. I., & Al, U. M. (2013). Future generations of mobile communication networks. *Academy of Contemporary Research Journal*, 2(1), 24–30.

Farris, I., Taleb, T., Khettab, Y., & Song, J. (2018). A survey on emerging SDN and NFV security mechanisms for IoT systems. *IEEE Communications Surveys and Tutorials*, 21(1), 812–837. DOI: 10.1109/COMST.2018.2862350

Farsad, N., & Goldsmith, A. (2017). Detection algorithms for communication systems using deep learning. *arXiv preprint arXiv:1705.08044*.

Fayyad, U., Piatetsky-Shapiro, G., & Smyth, P. (1996), From data mining to knowledge discovery: An overview. In *Advances in Knowledge Discovery and Data Mining,* chapter. 1, U. Fayyad, G. Piatesky-Shapiro, P. Smyth, and R. Uthurusamy, Eds. AAAI Press, Menlo Park, CA, and the MIT Press, Cambridge, MA, pp.1–34.

Fereidouni, M. A. and Kawa, A. (2019). Dark side of digital transformation in tourism. *Intelligent Information and Database Systems*, 510-518. https://doi.org/ DOI: 10.1007/978-3-030-14802-7_44

Fernandes, T., & Oliveira, E. (2021). Understanding consumers' acceptance of automated technologies in service encounters: Drivers of digital voice assistants' adoption. *Journal of Business Research*, 122, 180–191. DOI: 10.1016/j.jbusres.2020.08.058

Field, J. (2006). Lifelong Learning Policies and Strategies in Europe: Trends and Issues. *Comparative Education*.

Fowdur, T. P., & Doorgakant, B. (2023). A review of machine learning techniques for enhanced energy efficient 5G and 6G communications. *Engineering Applications of Artificial Intelligence*, 122, 106032. DOI: 10.1016/j.engappai.2023.106032

Foxconn. (2021). *Foxconn integrates 5G for smart manufacturing*. Retrieved from https://www.foxconn.com/news/2021/foxconn-integrates-5g-for-smart-manufacturing

Gade, J., Johnpaul, M., Agnihotri, S., & Miryala, R. K. (2024). The role of artificial intelligence in enhancing the experience of solo travelers from Europe to India: A literature-based study. In Talukder, M. B., Kumar, S., & Tyagi, P. K. (Eds.), *Hotel and travel management in the AI era* (pp. 551–567). IGI Global., DOI: 10.4018/979-8-3693-7898-4.ch026

Gangakhedkar, S., Cao, H., Ali, A. R., Ganesan, K., Gharba, M., & Eichinger, J. (2018). Use cases, requirements and challenges of 5G communication for industrial automation. In 2018 IEEE International Conference on Communications Workshops (ICC Workshops) (pp. 1-6). Kansas City, MO, USA: IEEE.

Gantumur, Z. (2024). Secure Network Slicing in 5G.

Garcia, L., Brown, M., & Nguyen, T. (2022). Addressing Resource Constraints in Skill Development. *Resource Management Journal*, 29(4), 180–195.

Garcia, L., Nguyen, T., & Brown, M. (2022). Critical Thinking in Technical Professions. *Journal of Professional Development*, 20(3), 88–99.

García-Madurga, M. Á., & Grilló-Méndez, A. J. (2023). Artificial Intelligence in the tourism industry: An overview of reviews. *Administrative Sciences*, 13(8), 172. DOI: 10.3390/admsci13080172

Garcia-Martin, M., Gramaglia, M., & Serrano, P. (2024). Network automation and data analytics in 3Gpp 5G systems. *IEEE Network*, 38(4), 182–189. DOI: 10.1109/MNET.2023.3321524

Garg, A., Rawat, S., & Kumar, P. (2018). Analysis and evolution of dynamic and modern techniques of internet generation. *2018 International Conference on Computational Techniques, Electronics and Mechanical Systems (CTEMS)*. https://doi.org/DOI: 10.1109/CTEMS.2018.8769325

Gaur, L., Afaq, A., Solanki, A., Singh, G., Sharma, S., Jhanjhi, N. Z., & Le, D. (2021). Capitalising on big data and revolutionary 5G technology: Extracting and visualising ratings and reviews of global chain hotels. *Computers & Electrical Engineering*, 95, 107374. DOI: 10.1016/j.compeleceng.2021.107374

Gera, B., Sharma, A., & Gupta, S. (2023). Leveraging AI-enabled 6G-driven IoT for sustainable smart cities. *International Journal of Communication Systems*, 36(16), e5588. DOI: 10.1002/dac.5588

Ghosh, A., Ratasuk, R., Mondal, B., Mangalvedhe, N., & Thomas, T. (2010). LTE-advanced: Next-generation wireless broadband technology [Invited Paper]. *IEEE Wireless Communications*, 17(3), 10–22. DOI: 10.1109/MWC.2010.5490974

Giordani, M., Mezzavilla, M., Polese, M., Zorzi, M., & Rangan, S. (2020). Toward 6G Networks: Use Cases and Technologies. *IEEE Communications Magazine*, 58(3), 55–61. DOI: 10.1109/MCOM.001.1900411

Gladson Robin, L. (2023). Chapter 1 – exploring the advancements, benefits and challenges of emerging technologies. https://doi.org/DOI: 10.59646/advcivilmechc1/026

Go, F., & Govers, R. (2009). *Place Branding*. Palgrave MacMillan.

Gohar, A., & Nencioni, G. (2021). The role of 5G technologies in a smart city: The case for intelligent transportation system. *Sustainability (Basel)*, 13(9), 5188. DOI: 10.3390/su13095188

Goleman, D. (1998). Emotional Intelligence and its Impact on Leadership Effectiveness. *Harvard Business Review*.

Goonwalla, H., & Neog, D. (2011). Problem and prospect of tea tourism in Assam-a SWOT analysis. In *2011 International conference on advancements in information technology with workshop of ICBMG* (Vol. 20, pp. 243-248).

Gormley, C., & Tong, Z. (2015). *Elastic search: The Definitive Guide* (1st ed.). O'Reilly Media, Inc.

Govardhana Reddy, G., & Raghavendra, K. (2022). Vector space modelling-based intelligent binary image encryption for secure communication. *Journal of Discrete Mathematical Sciences and Cryptography*, 25(4), 1157–1171. DOI: 10.1080/09720529.2022.2075090

Gretzel, U., & Fesenmaier, D. R. (2006). Persuasion in recommender systems. *International Journal of Electronic Commerce*, 11(2), 81–100.

Gretzel, U., Σιγάλα, M., Xiang, Z., & Koo, C. (2015). Smart tourism: Foundations and developments. *Electronic Markets*, 25(3), 179–188. DOI: 10.1007/s12525-015-0196-8

Guger, C., Ince, N. F., Korostenskaja, M., & Allison, B. Z. (2024). Brain-Computer Interface Research: A State-of-the-Art Summary 11. In Guger, C., Allison, B., Rutkowski, T. M., & Korostenskaja, M. (Eds.), *Brain-Computer Interface Research* (pp. 1–11)., DOI: 10.1007/978-3-031-49457-4_1

Gui, G., Liu, M., Tang, F., Kato, N., & Adachi, F. (2020). 6G: Opening New Horizons for Integration of Comfort, Security, and Intelligence. *IEEE Wireless Communications*, 27(5), 126–132. DOI: 10.1109/MWC.001.1900516

Guo, W. (2020). Explainable Artificial Intelligence for 6G: Improving Trust between Human and Machine. *IEEE Communications Magazine*, 58(6), 39–45. DOI: 10.1109/MCOM.001.2000050

Gupta, A. (2009). *Development of Barak valley: the Question of Sustainability. Development strategies of Barak valley (Assam)*. Akansha Publishing House.

Gupta, A., & Jha, R. K. (2015). A Survey of 5G Network: Architecture and Emerging Technologies. *IEEE Access : Practical Innovations, Open Solutions*, 3(3), 1206–1232. DOI: 10.1109/ACCESS.2015.2461602

Gürsoy, D., Malodia, S., & Dhir, A. (2022). The Metaverse in the hospitality and tourism industry: An overview of current trends and future research directions. *Journal of Hospitality Marketing & Management*, 31(5), 527–534. DOI: 10.1080/19368623.2022.2072504

Guterres, A. (2020). *Digital Divide 'a Matter of Life and Death'amid COVID-19 Crisis*. United Nations.

Hadani, R., & Monk, A. (2017). OTFS—A novel modulation technique meeting 5G high mobility and massive MIMO challenges. Available: https://www.cohere-technologies.com/wp-content/uploads/2017/10/ OTFS-Physics-White-Paper.pdf

Haensge, K., Trossen, D., Robitzsch, S., Boniface, M., & Phillips, S. (2019). Cloud-native 5G service delivery platform. *2019 IEEE Conference on Network Function Virtualization and Software Defined Networks (NFV-SDN)*. https://doi.org/DOI: 10.1109/NFV-SDN47374.2019.9040042

Hager, P., & Halliday, J. (2009). Learning Throughout Life: An *Interdisciplinary Perspective. Studies in Continuing Education*.

Hair, J. F.Jr, Sarstedt, M., Ringle, C. M., & Gudergan, S. P. (2017). *Advanced Issues in Partial Least Squares Structural Equation Modelling*. Sage Publications.

Hair, J., & Alamer, A. (2022). Partial Least Squares Structural Equation Modeling (PLS-SEM) in second language and education research: Guidelines using an applied example. *Research Methods in Applied Linguistics*, 1(3), 100027. DOI: 10.1016/j.rmal.2022.100027

Hall, T., & Brown, J. (2022). Corporate Approaches to 6G Workforce Training. *Corporate Training Journal*, 25(3), 110–125.

Hamidova, M. and Egamberdiyev, S. (2023). Analysis of cases of use of smart technologies in tourism and hotel enterprises. Yashil Iqtisodiyot Va Taraqqiyot, 1(mxs). https://doi.org/DOI: 10.55439/GED/vol1_issmaxsusson/a653

Hankinson, G. (2001). Location branding: A study of twelve English cities. *Journal of Brand Management*, 9(2), 127–142. DOI: 10.1057/palgrave.bm.2540060

Hankinson, G. (2004). Relational network brands: Towards a conceptual model of place brands. *Journal of Vacation Marketing*, 10(2), 109–121. DOI: 10.1177/135676670401000202

Hankinson, G. (2005). Destination Brand Images: A Business Tourism Pespective. *Journal of Services Marketing*, 19(1), 24–32. DOI: 10.1108/08876040510579361

Han, S., i, C., Xu, Z., & Rowell, C. (2015). Large-Scale Antenna Systems with Hybrid Analog and Digital Beamforming for Millimeter Wave 5G. *IEEE Communications Magazine*, 53(1), 186–194. DOI: 10.1109/MCOM.2015.7010533

Hansen, E. B., & Bøgh, S. (2021). Artificial Intelligence and Internet of Things in Small and Medium-Sized Enterprises: A Survey. *Journal of Manufacturing Systems*, 58, 362–372. DOI: 10.1016/j.jmsy.2020.08.009

Hänsge, K., Trossen, D., Robitzsch, S., Boniface, M., & Phillips, S. (2020). 5G service delivery platform: 360 VR vertical as an example. *Internet Technology Letters*, 3(6), e186. Advance online publication. DOI: 10.1002/itl2.186

Han, Y., Zhang, S., Duan, L., & Zhang, R. (2020). Cooperative double-IRS aided communication: Beamforming design and power scaling. *IEEE Wireless Communications Letters*, 9(8), 1206–1210. DOI: 10.1109/LWC.2020.2986290

Harris, P., & Clark, R. (2022). Nokia's 6G Training Program. *Journal of CorporateTraining Initiatives*, 17(4), 140–155.

Harte, L., Kitka, R., & Levine, R. (2002). *3G Wireless Demystified*. McGraw-Hill.

Hassan, R., Ahmed, A. S., Qamar, F., Aman, A. H. M., & Hasan, M. K. (2020). Internet of Things and Its Applications: A Comprehensive Survey. *Symmetry*, 12(10), 1674. DOI: 10.3390/sym12101674

Henseler, J., Ringle, C. M., & Sarstedt, M. (2015). A New Criterion for Assessing Discriminant Validity in Variance-Based Structural Equation Modelling. *Journal of the Academy of Marketing Science*, 43(1), 115–135. DOI: 10.1007/s11747-014-0403-8

Hewa, T., Gur, G., Kalla, A., Ylianttila, M., Bracken, A., & Liyanage, M. (2020). The Role of Blockchain in 6G: Challenges, Opportunities and Research Directions. *2020 2nd 6G Wireless Summit (6G SUMMIT)*, 1–5. DOI: 10.1109/6GSUMMIT49458.2020.9083784

Hidalgo, A., Gabaly, S., Morales-Alonso, G., & Urueña, A. (2020). The Digital Divide in Light of Sustainable Development: An Approach through Advanced Machine Learning Techniques. *Technological Forecasting and Social Change*, 150, 119754. https://www.sciencedirect.com/science/article/pii/S0040162519310637. DOI: 10.1016/j.techfore.2019.119754

Hodara, H., & Skaljo, E. (2021). From 1G to 5G. *Fiber and Integrated Optics*, 40(2–3), 85–183. DOI: 10.1080/01468030.2021.1919358

Hoffman, D. L., & Novak, T. P. (1996). Marketing in hypermedia computer-mediated environments: Conceptual foundations. *Journal of Marketing*, 60(3), 50–68. DOI: 10.1177/002224299606000304

Hollands, R. (2015). Critical interventions into the corporate smart city. *Cambridge Journal of Regions, Economy and Society*, 8(1), 61–77. DOI: 10.1093/cjres/rsu011

Hoshina, H., Yamazaki, S., Tsubouchi, M., & Harata, M. (2021). Terahertz irradiation effects on the morphology and dynamics of actin biopolymer. *JPhys Photonics*, 3(3), 034015. DOI: 10.1088/2515-7647/ac0958

https://blogs.bournemouth.ac.uk/ekotourismlab/research-projects/phd-projects/smart-tourism-destinations-explore-how-smartness-increases-competitiveness-in-the-context-of-tourism-destinations, (Retrieved 15.12. 2014).

Huang, M., Xu, Z., & Liu, Y. (2023). Terahertz Technology for 6G Wireless Communications. *Wireless Networks Journal*, 38(2), 98–112.

Huang, S., & Cai, J. (2016). An analysis framework for buffer-aided relaying under time-correlated fading channels. *IEEE Transactions on Vehicular Technology*, 65(9), 6987–6999. DOI: 10.1109/TVT.2015.2497210

Huq, K. M. S., Busari, S. A., Rodriguez, J., Frascolla, V., Bazzi, W., & Sicker, D. C. (2019). Terahertz-Enabled Wireless System for beyond-5G Ultra-Fast Networks: A Brief Survey. *IEEE Network*, 33(4), 89–95. DOI: 10.1109/MNET.2019.1800430

Huynh-The, T., Hua, C. H., Pham, Q. V., & Kim, D. S. (2020). MCNet: An efficient CNN architecture for robust automatic modulation classification. *IEEE Communications Letters*, 24(4), 811–815. DOI: 10.1109/LCOMM.2020.2968030

Ibrahim, J. (2002). 4G Features. *Bechtel Telecommunications Technical Journal*, 1(1), 11–14.

Iivari, N., Sharma, S., & Ventä-Olkkonen, L. (2020). Digital Transformation of Everyday Life – How COVID-19 Pandemic Transformed the Basic Education of the Young Generation and Why Information Management Research Should Care? *International Journal of Information Management*, 55(June), 102183. DOI: 10.1016/j.ijinfomgt.2020.102183 PMID: 32836640

Imoize, A. L., Adedeji, O., Tandiya, N., & Shetty, S. (2021). 6G Enabled Smart Infrastructure for Sustainable Society: Opportunities, Challenges, and Research Roadmap. *Sensors (Basel)*, 21(5), 1709. DOI: 10.3390/s21051709 PMID: 33801302

Islam, T., Michalopoulos, D. S., Schober, R., & Bhargava, V. (2014). Delay constrained buffer-aided relaying with outdated CSI. In *Proc. IEEE WCNC*, (pp. 875–880). DOI: 10.1109/WCNC.2014.6952203

Islam, T., Michalopoulos, D. S., Schober, R., & Bhargava, V. K. (2016). Buffer-aided relaying with outdated CSI. *IEEE Transactions on Wireless Communications*, 15(3), 1979–1997. DOI: 10.1109/TWC.2015.2497342

ITU-R M.2370-0. (2015). IMT Traffic Estimates for the Years 2020 to 2030. Geneva, Switzerland: ITU Publications. https://www.itu.int/dms_pub/itu-r/opb/rep/R-REP-M.2370-2015-PDF-E.pdf

Jakob, N. (2011). Extracting Opinion Targets from User-Generated Discourse with an Application to Recommendation Systems.

Jalali, J., Khalili, A., Rezaei, A., Berkvens, R., Weyn, M., & Famaey, J. (2023). IRS-based energy efficiency and admission control maximization for IoT users with short packet lengths. *IEEE Transactions on Vehicular Technology*, 72(9), 1–6. DOI: 10.1109/TVT.2023.3266424

Jamali, V., Zlatanov, N., & Schober, R. (2015). Bidirectional buffer-aided relay networks with fixed rate transmission—Part II: Delay-constrained case. *IEEE Transactions on Wireless Communications*, 14(3), 1339–1355. DOI: 10.1109/TWC.2014.2365806

Jamali, V., Zlatanov, N., Shoukry, H., & Schober, R. (2015). Achievable rate of the half-duplex multi-hop buffer-aided relay channel with block fading. *IEEE Transactions on Wireless Communications*, 14(11), 6240–6256. DOI: 10.1109/TWC.2015.2451171

Jannach, D., & Ludewig, M. (2017), Investigating personalized search in e-commerce, in: *FLAIRS Conference*, AAAI Press, pp. 645–650.

Jayanthi, J., & Rabara, S. A. (2010). Next-generation internet protocol -technical realms. *2010 3rd International Conference on Computer Science and Information Technology*. https://doi.org/DOI: 10.1109/ICCSIT.2010.5564916

Jayasooriya, S. S. W. (2019). Exploring the potentials, issues, and challenges for community-based tea tourism development (with reference to Hanthana Mountains). *International Journal of Advance Research. Ideas and Innovations in Technology*, 5(2), 475–480.

Jha, A. V., Sharma, P., & Kumar, R. (2024). 6G for intelligent transportation systems: Standards, technologies, and challenges. *Telecommunication Systems*, 86(2), 1–28. DOI: 10.1007/s11235-024-01126-5

Jiang, W., Han, B., Habibi, M. A., & Schotten, H. D. (2021a). The Road Towards 6G: A Comprehensive Survey. *IEEE Open Journal of the Communications Society*, 2, 334–366. DOI: 10.1109/OJCOMS.2021.3057679

Jiang, X., Shokri-Ghadikolaei, H., Fodor, G., Modiano, E., Pang, Z., Zorzi, M., & Fischione, C. (2018). Low-latency networking: Where latency lurks and how to tame it. *Proceedings of the IEEE*, 107(2), 280–306. DOI: 10.1109/JPROC.2018.2863960

Ji, B., Wen, H., Menon, V. G., Wang, Y., Song, K., Li, C., & Mumtaz, S. (2021). A Survey of Computational Intelligence for 6G: Key Technologies, Applications and Trends. *IEEE Transactions on Industrial Informatics*, 17(10), 7145–7154. DOI: 10.1109/TII.2021.3052531

Johnpaul, M., Miryala, R. S. B., Mazurek, M., Jayaprakashnarayana, G., & Miryala, R. K. (2025). Artificial intelligence and machine learning in eCommerce. In Kaur, G., Arora, J., Jain, V., & Shaikh, A. (Eds.), *Strategic innovations of AI and ML for eCommerce data security* (pp. 31–50). IGI Global., DOI: 10.4018/979-8-3693-5718-7.ch002

Johnson, P., Smith, J., & Davis, M. (2022). Economic Impacts of 6G Innovation. *Journal of Economic Development and Innovation*, 26(3), 220–235.

Johnson, P., Taylor, A., & White, G. (2022). Bridging the AI Skills Gap in Telecommunications. *Journal of AI Research and Development*, 18(2), 200–215.

Johri, S., Singh, S. K., Reddy, C. V., Nijhawan, G., Alawadi, A. H., & Reddy, U. (2023). Improving security and effectiveness in banking operations with IoT integration. *2023 10th IEEE Uttar Pradesh Section International Conference on Electrical, Electronics and Computer Engineering (UPCON)*. https://doi.org/DOI: 10.1109/UPCON59197.2023.10434723

Jones, T., & Brown, R. **(2023).** Global Connectivity in the 6G Era. Journal ofGlobal Communications, 14(2), 80-95.

Jones, T., & Brown, R. (2023). Security Challenges in 6G Networks. *Cybersecurity and Privacy Journal*, 29(7), 90–105.

Juneja, S., Gahlan, M., Dhiman, G., & Kautish, S. (2021). *Futuristic Cyber-Twin Architecture for 6G Technology to Support Internet of Everything*. Hindawi Scientific Programming. DOI: 10.1155/2021/9101782

Kadir, E. A., Shubair, R. M., Rahim, S. K. A., Himdi, M., Kamarudin, M. R., & Rosa, S. L. (2021). B5G and 6G: next generation wireless communications technologies, demand and challenges. *2021 International Congress of Advanced Technology and Engineering (ICOTEN)*.https://doi.org/DOI: 10.1109/ICOTEN52080.2021.9493470

Kalpage, N. V., Priya, P., & Hong, Y. (2024). DCT-based OTFS with reduced PAPR. *IEEE Communications Letters*, 28(1), 158–162. DOI: 10.1109/LCOMM.2023.3337778

Kalra, B., & Chauhan, D. K. (2014). A comparative study of mobile wireless communication network: 1G to 5G. *International Journal of Computer Science and Information Technology Research*, 2(3), 430–433.

Kamruzzaman, M. M. (2022). Key technologies, applications and trends of Internet of Things for energy-efficient 6G wireless communication in smart cities. *Energies*, 15(15), 5608. DOI: 10.3390/en15155608

Kansakar, P., Munir, A., & Shabani, N. (2019). Technology in the hospitality industry: Prospects and challenges. *IEEE Consumer Electronics Magazine*, 8(3), 60–65. DOI: 10.1109/MCE.2019.2892245

Karimzadeh, A., & Shoghli, O. (2020). Predictive Analytics for Roadway Maintenance: A Review of Current Models, Challenges, and Opportunities. *Civil Engineering Journal*, 6(3), 602–625. DOI: 10.28991/cej-2020-03091495

Karuppusamy, M. (2023). *6G: The Future of Wireless*. Bannari Amman Institute of Technology. https://www.bitsathy.ac.in/6g-the-future-of-wireless/#:~:text=Here%20are%20some%20of%20the%20potential%20use%20cases%20for%206G,are%20not%20at%20the%20venue

Katz, M., Pirinen, P., & Posti, H. (2019). Towards 6G: getting ready for the next decade. 2019 *16th International Symposium on Wireless Communication Systems (ISWCS)*.https://doi.org/DOI: 10.1109/ISWCS.2019.8877155

Kaul, A., & Gupta, J. P. (2021). Revolutionary 6G: technologies, architecture, coverage, and performance. *2021 12th International Conference on Computing Communication and Networking Technologies (ICCCNT)*.https://doi.org/DOI: 10.1109/ICCCNT51525.2021.9579602

Kaur, J., Khan, M. A., Iftikhar, M., Imran, M., & Emad Ul Haq, Q. (2021). Machine Learning Techniques for 5G and Beyond. *IEEE Access : Practical Innovations, Open Solutions*, 9, 23472–23488. DOI: 10.1109/ACCESS.2021.3051557

Kaur, P., Sharma, T., & Singh, J. P. (2020). Evaluation of internet generations on wireless network: A survey. *Journal of Technology Management for Growing Economies*, 11(2), 55–61. DOI: 10.15415/jtmge.2020.112007

Kavaratzis, M., & Ashworth, G. (2008). Place Marketing: How Did we Get there and Where we Going? *Journal of Place Management and Development*, 1(2), 150–165. DOI: 10.1108/17538330810889989

Khalid, R., & Naqi Raza, M. (2024). Analysing the impact of optical wireless communication technologies on 5g/6g and IoT solutions: Prospects, developments, and challenges. *Journal of Engineering Research and Sciences*, 3(5), 23–36. DOI: 10.55708/js0305003

Khanna, A., & Kaur, S. (2019). Evolution of Internet of Things (IoT) and its significant impact in the field of Precision Agriculture. *Computers and Electronics in Agriculture*, 157, 218–231. https://doi.org/. DOI: 10.1016/j.compag.2018.12.039

Khan, W. U., Lagunas, E., Ali, Z., Javed, M. A., Ahmed, M., Chatzinotas, S., Ottersten, B., & Popovski, P. (2022). Opportunities for physical layer security in UAV communication enhanced with intelligent reflective surfaces. *IEEE Wireless Communications*, 29(6), 22–28. DOI: 10.1109/MWC.001.2200125

Khatib, E. J., & Barco, R. (2021). Optimization of 5G Networks for Smart Logistics. *Energies*, 14(6), 1758. DOI: 10.3390/en14061758

Kimball, R., & Merz, R. (2000). *The Data Webhouse Toolkit: Building the Web-Enabled Data Warehouse*. John Wiley & Sons, Inc.

Kim, J., Park, S., & Lee, H. (2023). IoT Integration in 6G Networks. *Sensors and Actuators Journal*, 45(9), 321–334.

Kimura, S., Aburakawa, Y., Watanabe, F., Torashima, S., Igarashi, S., Nakamura, T., & Yamaguchi, M. (2021). Holographic Video Communication System Realizing Virtual Image Projection and Frontal Image Capture. *ITE Transactions on Media Technology and Applications*, 9(1), 105–112. DOI: 10.3169/mta.9.105

Kitanov, S., Nikolić, V., & Janevski, T. (2022). Quality of artificially intelligently orchestrated services in 6G mobile networks. *Journal of Electrical Engineering and Information Technologies*, 7(1), 43–52. DOI: 10.51466/JEEIT2271195043k

Kitchin, R. (2015). Making sense of smart cities: Addressing present shortcomings. *Cambridge Journal of Regions, Economy and Society*, 8(1), 131–136. DOI: 10.1093/cjres/rsu027

Kohavi, R., Brodley, C., Frasca, B., Mason, L., & Zheng, Z. (2000). KDD-Cup 2000 organizers' report: Peeling the onion. *SIGKDD Explorations*, 2(2), 86–98. www.ecn.purdue.edu/KDDCUP. DOI: 10.1145/380995.381033

Kolios, P., Panayiotou, C. G., Ellinas, G., & Polycarpou, M. M. (2015). Event-based communication for IoT networking. *2015 IEEE 2nd World Forum on Internet of Things (WF-IoT)*. https://doi.org/DOI: 10.1109/WF-IoT.2015.7389076

Kolluru, D. S., & Reddy, P. B. (2021). Review on communication technologies in telecommunications from conventional telephones to smartphones. *AIP Conference Proceedings*, 020003. Advance online publication. DOI: 10.1063/5.0074088

Kong, D. (2016). *Personalized Feature based Re-Ranking Method for Ecommerce Search at CIKM Cup 2016. Technical Report*. CIKM Cup.

Kota, S. L. (2011). Hybrid/integrated networking for ngn services. In 2011 2nd International Conference on Wireless Communication, Vehicular Technology, Information Theory and Aerospace &Electronic Systems Technology (Wireless VITAE) (pp. 1-6). Chennai, India: IEEE.

Kotler, P, Gertner, D (2002) Country as brand, product and beyond: A place marketing and brand management perspective. *Journal of Brand Management*, April, 9(4/5).

Kotler, P., ... (2002). *Marketing Asian Places*.John Wiley & Sons (Asia), Singapore, 443 p.

Kuan, F. (2022). *What is IPS? – An Introduction to Indoor Positioning System*. Moko Smart. https://www.mokosmart.com/introduction-to-indoor-positioning-system/#:~:text=What%20is%20an%20IPS?as%20a%20smartphone%20or%20tablet

Kulkarni, R., & Kulkarni, S. (2021). How 6G Has an Influence on Smart Cities: An Overview. *International Journal of Engineering Research & Technology (Ahmedabad)*, 10(5), 832–838. https://www.ijert.org/research/how-6g-has-an-influence-on-smart-cities-an-overview-IJERTV10IS050447.pdf

Kumar, A., Liu, D. Y., & Sengupta, D. J. (2010). Divya4,"Evolution of Mobile Wireless Communication Networks: 1G to 4G", Vol. 1, Issue 1.

Kumar, B. A., & Rao, P. T. (2015). Overview of advances in communication technologies. *2015 13th International Conference on Electromagnetic Interference and Compatibility (INCEMIC)*.https://doi.org/DOI: 10.1109/INCEMIC.2015.8055856

Kumar, R., & Hossain, A. (2019). Performance of random-access Markov modelling for two-way buffer-aided relaying networks with wireless assisted links. *Wireless Personal Commun.*, Springer, 108(3), 1995–2015.

Kumar, R., & Hossain, A. (2021). Cooperative relaying communications in IoT applications for 5G radio networks. In Innovations and Applications of AI, IoT, and Cognitive Technologies, *Eds. Ramona Trestian, Gabriel-Miro Muntean, IGI Global*, Ch. 2,26 – 41.

Kumari, A., & Thakur, P. (2023). Sustainable tourism: A review of current practices and future directions-coalescence of social media and e-commerce. *International Journal of Research Publication and Reviews*, 4(5), 1655–1662. DOI: 10.55248/gengpi.234.5.40417

Kumar, M. (2023). 5G technology is revolutionising the wireless industry with unparalleled efficiency. *Sci Wave Bulletin*, 01(03), 21–28. DOI: 10.61925/SWB.2023.1303

Kumar, R., & Hossain, A. (2017). Optimization of throughput of two-way buffer-aided relaying networks with wireless assisted links. *IET Communications*, 11(10), 1626–1632. DOI: 10.1049/iet-com.2016.1364

Kumar, R., & Hossain, A. (2019). Survey on half- and full- duplex relay based cooperative communications and its potential challenges and open issues using Markov chains. *IET Communications*, 13(11), 1537–1550. DOI: 10.1049/iet-com.2018.5823

Kumar, R., & Hossain, A. (2021). Delay performance of two-hop buffer-aided relay network with wireless assisted-links. *Wireless Personal Communications*, 121(3), 1725–1738. DOI: 10.1007/s11277-021-08694-8

Kumar, S., & Shekhar, . (2020). Digitalization: A strategic approach for development of tourism industry in India. *Paradigm*, 24(1), 93–108. DOI: 10.1177/0971890720914111

Kürner, T., & Priebe, S. (2014). Towards THz communications-status in research, standardization and regulation. *Journal of Infrared, Millimeter and Terahertz Waves*, 35(1), 53–62. DOI: 10.1007/s10762-013-0014-3

Kuzlu, M., Fair, C., & Guler, O. (2021). "Role of Artificial Intelligence in the Internet of Things (IoT) Cybersecurity." Discover. *Internet of Things : Engineering Cyber Physical Human Systems*, 1(1), 7.

Kwon, T., Kim, S., Lee, K., & Chung, J. (2022). Special issue on 6G and satellite communications. *ETRI Journal*, 44(6), 881–884. DOI: 10.4218/etr2.12551

Κατσαρός, K., Gkounis, D., Kaleshi, D., Thomas, B., Harris, J. W., Falaki, H. R., & Simeonidou, D. (2019). Enhancing tourist experiences through 5G - the 5G smart tourism case study. *2019 IEEE 2nd 5G World Forum (5GWF)*. https://doi.org/DOI: 10.1109/5GWF.2019.8911700

Laricchia, F. (2023, March 10). *Number of mobile devices worldwide 2020-2025*. Statista. https://www.statista.com/statistics/245501/multiple-mobile-device-ownership-worldwide/

Law, R., Leung, D., & Chan, I. C. C. (2019). Progression and development of information and communication technology research in hospitality and tourism. *International Journal of Contemporary Hospitality Management*, 32(2), 511–534. DOI: 10.1108/IJCHM-07-2018-0586

Law, R., Leung, R., & Buhalis, D. (2009). Information technology applications in hospitality and tourism: A review of publications from 2005 to 2007. *Journal of Travel & Tourism Marketing*, 26(5-6), 599–623. DOI: 10.1080/10548400903163160

Lee, H., & Kim, S. (2022). Adapting to Technological Change in Telecom. *Technology and Innovation Journal*, 33(4), 150–165.

Lee, H., & Kim, S. (2022). Overcoming Workforce Resistance to Technological Change. *Journal of Change Management*, 34(2), 100–115.

Lee, P. (2021). Progress in smart tourism 2010-2017: A systematic literature review. *Journal of Smart Tourism*, 1(1), 19–30. DOI: 10.52255/smarttourism.2021.1.1.4

Lendel, V. (2009). *Riadenie vzťahov so zákazníkmi v podniku*. Dizertačná práca. Žilinská univerzita, Fakulta riadenia a informatiky.

Levy, J. (2005). *The Great Pyramid of Giza: Measuring Length, Area, Volume, and Angles*. The Rosen Publishing Group, Inc.

Liang, Y., Li, L., Fan, P., & Guan, Y. (2020). Doppler resilient orthogonal time-frequency space (OTFS) systems based on index modulation. *in proc. IEEE 91st Vehicular Technology Conference (VTC2020-Spring),* 1 – 5, Antwerp, Belgium.

Liang, F., Mu, L., Wang, D., & Kim, B. (2021). A new model path for the development of smart leisure sports tourism industry based on 5G technology. *IET Communications*, 16(5), 485–496. DOI: 10.1049/cmu2.12271

Li, C., Guo, W., Sun, S. C., Al-Rubaye, S., & Tsourdos, A. (2020). Trustworthy in 6G-enabled mass autonomy: From concept to quality-of-trust key performance indicators. *IEEE Vehicular Technology Magazine*, 15(4), 112–121. DOI: 10.1109/MVT.2020.3017181

Li, H., La Guardia, M. J., Liu, H., Hale, R. C., Mainor, T. M., Harvey, E., Sheng, G., Fu, J., & Peng, P. (2019). Brominated and organophosphate flame retardants along a sediment transect encompassing the Guiyu, China e-waste recycling zone. *The Science of the Total Environment*, 646, 58–67. DOI: 10.1016/j.scitotenv.2018.07.276 PMID: 30048869

Li, L. (2022). Reskilling and Upskilling the Future-ready Workforce for Industry 4.0 and Beyond. *Information Systems Frontiers*. Advance online publication. DOI: 10.1007/s10796-022-10308-y PMID: 35855776

Lin, M., & Zhao, Y. (2020). Artificial intelligence-empowered resource management for future wireless communications: A survey. *China Communications*, 17(3), 58–77. DOI: 10.23919/JCC.2020.03.006

Lin, S. C., Chen, K. C., & Karimoddini, A. (2021). SDVEC: Software-defined vehicular edge computing with ultra-Low latency. *IEEE Communications Magazine*, 59(12), 66–72. DOI: 10.1109/MCOM.004.2001124

Li, S., Yang, T., & Chen, Q. (2022). Machine Learning for 6G: An Overview. *Future Internet Journal*, 14(5), 200–215.

Liu, D. (2019). 5G and intelligence medicine—How the next generation of wireless technology will reconstruct healthcare? *Precision Clinical Medicine*, 2(4), 205–208. DOI: 10.1093/pcmedi/pbz020 PMID: 31886033

Liu, J., Wan, J., Jia, D., Zeng, B., Li, D., Hsu, C., & Chen, H. (2017). High-Efficiency Urban Traffic Management in Context-Aware Computing and 5G Communication. *IEEE Communications Magazine*, 55(1), 34–40. DOI: 10.1109/MCOM.2017.1600371CM

Liu, S., Xiao, F., Ou, W., & Si, L. (2017), Cascade ranking for operational e-commerce search, in: *Proceedings of the 23rd ACM SIGKDD International Conference on Knowledge Discovery and Data Mining*, ACM, New York, NY, USA, pp. 1557–1565.

Liu, Y., Deng, Y., Elkashlan, M., Nallanathan, A., & Karagiannidis, G. K. (2022). Optimization of grant-free NOMA with multiple configured-grants for mURLLC. *IEEE Journal on Selected Areas in Communications*, 40(4), 1222–1236. DOI: 10.1109/JSAC.2022.3143264

Liu, Y., Liu, X., Mu, X., Hou, T., Xu, J., Di Renzo, M., & Al-Dhahir, N. (2021). Reconfigurable Intelligent Surfaces: Principles and Opportunities [Thirdquarter.]. *IEEE Communications Surveys and Tutorials*, 23(3), 1546–1577. DOI: 10.1109/COMST.2021.3077737

Li, W., Logenthiran, T., Phan, V.-T., & Woo, W. L. (2018). Implemented IoT-Based Self-Learning Home Management System (SHMS) for Singapore. *IEEE Internet of Things Journal*, 5(3), 2212–2219. DOI: 10.1109/JIOT.2018.2828144

Li, X., Dai, H. N., Shukla, M. K., Li, D., Xu, H., & Imran, M. (2021). Friendly-jamming schemes to secure ultra-reliable and low-latency communications in 5G and beyond communications. *Computer Standards & Interfaces*, 78, 1–14. DOI: 10.1016/j.csi.2021.103540

Lodhi, R. N., Del Gesso, C., Asif, M., & Cobanoglu, C.LODHI. (2024). Exploring virtual and augmented reality in the hospitality industry: A bibliometric analysis. *Tourism and Hospitality Management*, 30(1), 67–84. DOI: 10.20867/thm.30.1.6

Loghin, D., Cai, S., Chen, G., Dinh, T. T. A., Fan, F., Lin, Q., Ng, J., Ooi, B. C., Sun, X., Ta, Q.-T., Wang, W., Xiao, X., Yang, Y., Zhang, M., & Zhang, Z. (2020). The Disruptions of 5G on Data-Driven Technologies and Applications. *IEEE Transactions on Knowledge and Data Engineering*, 32(6), 1179–1198. DOI: 10.1109/TKDE.2020.2967670

Loureiro, S. M. C. (2022). Technology and luxury in tourism and hospitality. *The Emerald Handbook of Luxury Management for Hospitality and Tourism*, 273-284. https://doi.org/DOI: 10.1108/978-1-83982-900-020211013

Loureiro, S. M. C., & Nascimento, J. (2021). Shaping a view on the influence of technologies on sustainable tourism. *Sustainability (Basel)*, 13(22), 12691. DOI: 10.3390/su132212691

Lovley, D. R. (2017). e-Biologics: Fabrication of sustainable electronics with “green” biological materials. *mBio*, 8(3), e00695–e17. DOI: 10.1128/mBio.00695-17 PMID: 28655820

Lovreglio, R. (2020). Virtual and Augmented Reality for Human Behaviour in Disasters: A Review. *Conference Proceedings, Fire and Evacuation Modeling Technical Conference (FEMTC),* September 9-11, 2020,1-14.

Lu, H., Chen, M., Kim, H., Li, Y., & Serikawa, S. (2017). Brain Intelligence: Go beyond Artificial Intelligence. *Mobile Networks and Applications*, 23(2), 368–375. DOI: 10.1007/s11036-017-0932-8

Lukanova, G., & Ilieva, G. (2019). Robots, artificial intelligence, and service automation in hotels. In *Robots, artificial intelligence, and service automation in travel, tourism and hospitality* (pp. 157–183). Emerald Publishing Limited. DOI: 10.1108/978-1-78756-687-320191009

Luo, S., & Teh, K. C. (2015). Buffer state-based relay selection for buffer-aided cooperative relaying systems. *IEEE Transactions on Wireless Communications*, 14(10), 5430–5439. DOI: 10.1109/TWC.2015.2438296

Lusch, R. F., Vargo, S., O'Brien, M. (2007). Competing through service: Insights from service- dominant logic. *Journal of Retailing* 83 (1, 2007), pp. 5-18.

Madrigal, A. C. (2011, September 16). *The 1947 Paper That First Described a Cell-Phone Network*. The Atlantic. https://www.theatlantic.com/technology/archive/2011/09/the-1947-paper-that-first-described-a-cell-phone-network/245222/

Mahmood, N. H., Böcker, S., Munari, A., Clazzer, F., Moerman, I., Mikhaylov, K., . . . Seppänen, P. (2020). White paper on critical and massive machine type communication towards 6G. arXiv preprint arXiv:2004.14146.

Mambretti, J., & Schmidt, A. (1999). Next generation Internet: creating advanced networks and services. *John Wiley & Sons, Inc. eBooks*. Retrieved from http://ci.nii.ac.jp/ncid/BA44156623

Manoj, B. R., Mallik, R. K., & Bhatnagar, M. R. (2016). Buffer-aided multi-hop DF cooperative networks: A state-clustering based approach. *IEEE Transactions on Communications*, 64(12), 4997–5010. DOI: 10.1109/TCOMM.2016.2611504

Mansouri, Y., & Ali Babar, M. (2021). A Review of Edge Computing: Features and Resource Virtualization. *Journal of Parallel and Distributed Computing*, 150, 155–183. DOI: 10.1016/j.jpdc.2020.12.015

Mao, B., Tang, F., Kawamoto, Y., & Kato, N. (2021). Optimizing Computation Offloading in Satellite-UAV-Served 6G IoT: A Deep Learning Approach. *IEEE Network*, 35(4), 102–108. DOI: 10.1109/MNET.011.2100097

Mariani, M., & Borghi, M. (2021). Customers' evaluation of mechanical artificial intelligence in hospitality services: A study using online reviews analytics. *International Journal of Contemporary Hospitality Management*, 33(11), 3956–3976. DOI: 10.1108/IJCHM-06-2020-0622

Markowitz, D. M., Hancock, J., & Bailenson, J. (2023). Linguistic markers of AI-Generated text versus human-generated text: Evidence from hotel reviews and news headlines. *PsyArXiv. January, 30.*

Martilla, J. A., & James, J. C. (1977). Importance-Performance Analysis. *Journal of Marketing*, 41(1), 77–79. DOI: 10.1177/002224297704100112

Martin, J., & White, S. (2023). Barriers to 6G Adoption. *Telecommunications Barriers Journal*, 11(2), 77–89.

Martín-Sacristán, D., Monserrat, J. F., Peñuelas, J., Calabuig, D., Garrigas, S., & Cardona, N. (2009). On the Way towards Fourth-Generation Mobile: 3GPP LTE and LTE-Advanced. EURASIP Journal on Wireless Communications and Networking, 2009(1). DOI: 10.1155/2009/354089

Matinmikko-Blue, M., Aalto, S., Asghar, M. I., Berndt, H., Chen, Y., Dixit, S., . . . Ziegler, V. (2020). White paper on 6G drivers and the UN SDGs. arXiv preprint arXiv:2004.14695.

Matinmikko-Blue, M., Yrjölä, S., & Ahokangas, P. (2020, March). Spectrum management in the 6G era: The role of regulation and spectrum sharing. In 2020 2nd 6G Wireless Summit (6G SUMMIT) (pp. 1-5). IEEE.

Matinmikko-Blue, M., Latva-aho, M., Säily, J., & Mustonen, M. (2019). Industry 4.0 and 5G: Where do they meet? *Business Horizons*, 62, 47–56.

Matracia, M., Rahman, A. U., Wang, R., Kishk, M. A., & Alouini, M. S. (2023). Bridging the digital divide. In *Fundamentals of 6G Communications and Networking* (pp. 113–139). Springer International Publishing.

Mazurek, M., Ayoobkhan, M., Kaldeen, M., Mohamed, M., Said, T., Umachandran, K., Arora, S., Sharma, A., Pratap, R., Kaurav, S., Baber, R., Rajput, S., Chilembwe, J. M., Mweiwa, V. R., Ray, S., O'Connor, D. P., Garg, A., Urlu, D. K., Albattat, D. A., . . . Ercan, D. A. tin, İ., Koba, D.Y., & Coghlan, D.A. (2020). The Emerald Handbook of ICT in Tourism and Hospitality. *Emerald Publishing Limited eBooks.* https://doi.org/DOI: 10.1108/9781839826887

McCandless, M., Hatcher, E., & Gospodnetic, O. (2010), Lucene in Action, Second Edition: Covers Apache Lucene 3.0. Manning Publications Co, Greenwich, CT, USA.

McKercher, B. and Prideaux, B. (2020). Tourism theories, concepts and models. https://doi.org/DOI: 10.23912/9781911635352-4280

Meah, J. (2024, May 27). *6G Networks Use Cases, Capabilities, Release Date & What It Means for 5G/4G*. Techopedia.

Meccawy, M. (2022). Creating an Immersive XR Learning Experience: A Roadmap for Educators. *Electronics (Basel)*, 11(21), 3547. DOI: 10.3390/electronics11213547

Meena, P., Pal, M. B., Jain, P. K., & Pamula, R. (2022). 6G Communication Networks: Introduction, Vision, Challenges, and Future Directions. *Wireless Personal Communications*, 125(2), 1097–1123. DOI: 10.1007/s11277-022-09590-5

Megha Mudholkar, D.P. (2024). Impact of Artificial Intelligence and Internet of Things on Performance Management: A Systematic Review. *Journal of Informatics Education and Research*. https://doi.org/DOI: 10.52783/jier.v4i1.569

Mei, J., Wang, X., Zheng, K., Boudreau, G., Sediq, A. B., & Abou-Zeid, H. (2021). Intelligent radio access network slicing for service provisioning in 6G: A hierarchical deep reinforcement learning approach. *IEEE Transactions on Communications*, 69(9), 6063–6078. DOI: 10.1109/TCOMM.2021.3090423

Mei, W., & Zhang, R. (2021). Cooperative beam routing for multi-IRS aided communication. *IEEE Wireless Communications Letters*, 10(2), 426–430. DOI: 10.1109/LWC.2020.3034370

Meril, A. S., Basthikodi, M., & Faizabadi, A. R. (2019). Review: Comprehensive study of 5G and 6G Communication Networks. *Journal of Emerging Technologies and Innovative Research*, 6(5), 715–719. https://www.jetir.org/view?paper=JETIRCU06132

Mihovska, A., & Prasad, R. (2020). "Overview of 5G New Radio and Carrier Aggregation: 5G and beyond Networks." In 2020 23rd International Symposium on Wireless Personal Multimedia Communications (WPMC), IEEE, 1–6.

Mihret, E. T., & Haile, G. (2021). 4G, 5G, 6G, 7G and Future Mobile Technologies. *American Journal of Computer Science and Information Technology*, 9(2), 75.

Millard, S. (2013), Big Data Brewing Value in Human Capital Management – Ventana Research. Retrieved April 2, 2015 from http://stephanmillard.ventanaresearch.com/2013/08/28/big-data-brewing-value-in-humancapital-management

Miller, J., & Davis, L. (2022). R&D for Future Telecommunications. *Research and Development Journal*, 32(6), 90–105.

Mishra, P. K., Rout, H. B., & Mohapatra, S. S. (2011). Causality between tourism and economic growth: Empirical evidence from India. *European Journal of Soil Science*, 18(4), 518–527.

Mitchell, G. W., Skinner, L. B., & White, B. J. (2010). Essential soft skills for success in the twenty-first century workforce as perceived by business educators. *Delta Pi Epsilon Journal*, 52(1).

Mobasher, B. (2007). *Data mining for web personalization. The adaptive web.* Springer Berlin Heidelberg.

Mohsan, S. A. H., Mazinani, A., Malik, W., Younas, I., Qasem, N., Amjad, H., & Mahmood, A. (2020). 6G: Envisioning the Key Technologies, Applications and Challenges. *International Journal of Advanced Computer Science and Applications*, 11(9). Advance online publication. DOI: 10.14569/IJACSA.2020.0110903

Moldovan, A., Karunakaran, P., Akyildiz, I. F., & Gerstacker, W. H. (2017). Coverage and achievable rate analysis for indoor terahertz wireless networks. In *2017 IEEE International Conference on Communications (ICC)* (pp. 1-7). Paris, France: IEEE. DOI: 10.1109/ICC.2017.7996402

Monaco, S., & Sacchi, G. (2023). Travelling the Metaverse: Potential benefits and main challenges for tourism sectors and research applications. *Sustainability (Basel)*, 15(4), 3348. DOI: 10.3390/su15043348

Moon, N. W., Baker, P. M. A., & Goughnour, K. (2019). Designing Wearable Technologies for Users with Disabilities: Accessibility, Usability, and Connectivity Factors. *Journal of Rehabilitation and Assistive Technologies Engineering*, 6, 205566831986213. DOI: 10.1177/2055668319862137 PMID: 35186318

Mosavi, A., & Vaezipour, A. (2013). *Developing Effective Tools for Predictive Analytics and Informed Decisions. Technical Report.* University of Tallinn.

Motorola Solutions. (n.d.). Cell phone development. Motorola Solutions. https://www.motorolasolutions.com/en_us/about/history/explore-motorola-heritage/cell-phone-development.html

Mukherjee, A., Swindlehurst, A. L., Huang, J., & Fakoorian, S. A. A. (2014). Principles of Physical Layer Security in Multiuser Wireless Networks: A Survey. *IEEE Communications Surveys and Tutorials*, 16(3), 1550–1573. DOI: 10.1109/SURV.2014.012314.00178

Murroni, M., Cossu, S., & Toma, L. (2023). 6G—Enabling the New Smart City: A Survey. *Sensors (Basel)*, 23(17), 7528. DOI: 10.3390/s23177528 PMID: 37687986

Nabar, R. U., Bölcskei, H., & Kneubühler, F. W. (2004). Fading relay channels: Performance limits and space-time signal design. *IEEE Journal on Selected Areas in Communications*, 22(6), 1099–1109. DOI: 10.1109/JSAC.2004.830922

Naeem, F., Kaddoum, G., Khan, S., Khan, K. S., & Adam, N. (2022). IRS-empowered 6G networks: Deployment strategies, performance optimization, and future research directions. *IEEE Access : Practical Innovations, Open Solutions*, 10, 118676–118696. DOI: 10.1109/ACCESS.2022.3220682

Nair, P. S. (2018). An Augmentation and Virtuality of Reality: - An Overview. *International Journal of Research*, 5, 250–255.

Nam, K., Dutt, C. S., Chathoth, P. K., & Khan, M. S. (2019). Blockchain technology for smart city and smart tourism: Latest trends and challenges. *Asia Pacific Journal of Tourism Research*, 26(4), 454–468. DOI: 10.1080/10941665.2019.1585376

Nam, T., & Pardo, T. A. (2011). Smart City as Urban Innovation: Focusing on Management, Policy and Context. In Estevez, E., & Janssen, M. (Eds.), *Proceedings of the 5th International Conference on Theory and Practice of Electronic Governance (ICEGOV2011)*. Tallinn, Estonia: ACM Press. DOI: 10.1145/2072069.2072100

Nannelli, M., Capone, F., & Lazzeretti, L. (2023). Artificial intelligence in hospitality and tourism. State-of-the-art and future research avenues. *European Planning Studies*, 31(7), 1325–1344. DOI: 10.1080/09654313.2023.2180321

National Integrated Database of Hospitality Industry. (2022). Available at https://nidhi.nic.in/HotelDivision/ProposalSubmissions/CompleteViewApplication.as px [Retrieved 2 May, 2024].

Nayyar, A., Mahapatra, B., Le, D., & Suseendran, G. (2018). Virtual reality (VR) and Augmented reality (AR) technologies for tourism and hospitality industry. *International Journal of Engineering &Amp; Technology*, 7(2.21), 156. https://doi.org/DOI: 10.14419/ijet.v7i2.21.11858

Neelam, S. G., & Sahu, P. R. (2022). Analysis, estimation and compensation of hardware impairments for CP OTFS systems. *IEEE Wireless Communications Letters*, 11(5), 952–956. DOI: 10.1109/LWC.2022.3151707

Neelam, S. G., & Sahu, P. R. (2022). Digital compensation of IQ imbalance DC offset for zero-padded OTFS systems. *IEEE Communications Letters*, 26(10), 2450–2454. DOI: 10.1109/LCOMM.2022.3190047

Network, I T U. (2019). "2030–A Blueprint of Technology, Applications and Market Drivers Towards the Year 2030 and Beyond."

Neuhofer, B., Buhalis, D., & Ladkin, A. (2015). Smart technologies for personalised experiences: A case study in the hospitality domain. *Electronic Markets*, 25(3), 243–254. DOI: 10.1007/s12525-015-0182-1

Neuhofer, B., Magnus, B., & Celuch, K. (2020). The impact of artificial intelligence on event experiences: A scenario technique approach. *Event Management*, 31(3), 601–617. DOI: 10.1007/s12525-020-00433-4 PMID: 38624486

Nguyen, T., & Smith, K. (2023). Case Study: University of Technology's 6G Curriculum. *Journal of Higher Education in Technology*, 22(1), 55–70.

Nicola-Gavrilă, L. (2023). Empirical approaches regarding interdependency between technology and sustainable tourism. *Journal of Environmental Management and Tourism*, 14(4), 2140. DOI: 10.14505/jemt.v14.4(68).25

Nieto De Santos, F. J., & Villalonga, S. G. (2015). *Exploiting Local Clouds in the Internet of Everything Environment.* institute of electrical electronics engineers. DOI: 10.1109/PDP.2015.117

Nomikos, N., Poulimeneas, D., Charalambous, T., Krikidis, I., Vouyioukas, D., & Johansson, M. (2018). Delay- and diversity-aware buffer-aided relay selection policies in cooperative networks. *IEEE Access : Practical Innovations, Open Solutions*, 6, 73531–73547. DOI: 10.1109/ACCESS.2018.2883894

Norman, D. A. (2002). Human-Centered Systems: Information, Interactivity, and Intelligence. *Interacting with Computers.*

Norman, D., & Euchner, J. (2023). Design for a Better World: A Conversation with Don Norman. *Research Technology Management*, 66(3), 11–18. DOI: 10.1080/08956308.2023.2183015

Nuriev, M., Kalyashina, A., Smirnov, Y., Gumerova, G., & Gadzhieva, G. (2024). The 5G revolution transforming connectivity and powering innovations. *E3S Web of Conferences*, 515, 04008. https://doi.org/DOI: 10.1051/e3sconf/202451504008

NVIDIA MAXINE. (2020). Retrieved March 11, 2022, from https://developer .nvidia.com/maxine

O'neill, J. L. (2021). "Accessibility for All Abilities: How Universal Design, Universal Design for Learning, and Inclusive Design Combat Inaccessibility and Ableism." Journal and Inclusive Design Combat Inaccessibility and Ableism. *Journal of Open Access to Law*, 9, 1–15.

O'Shea, T., & Hoydis, J. (2017). An Introduction to Deep Learning for the Physical Layer. *IEEE Transactions on Cognitive Communications and Networking*, 3(4), 563–575. DOI: 10.1109/TCCN.2017.2758370

Oinas-Kukkonen, H., Karppinen, P., & Kekkonen, M. (2021). 5G and 6G Broadband Cellular Network Technologies as Enablers of New Avenues for Behavioral Influence with Examples from Reduced Rural-Urban Digital Divide. *Urban Science (Basel, Switzerland)*, 5(3), 60. DOI: 10.3390/urbansci5030060

Oughton, E., Frias, Z., Russell, T., Sicker, D., & Cleevely, D. D. (2018). Towards 5G: Scenario-based assessment of the future supply and demand for mobile telecommunications infrastructure. *Technological Forecasting and Social Change*, 133, 141–155. DOI: 10.1016/j.techfore.2018.03.016

Padhi, P. K., & Charrua-Santos, F. (2021). 6G enabled industrial Internet of everything: Towards a theoretical framework. *Applied System Innovation*, 4(1), 11. DOI: 10.3390/asi4010011

Palattella, M. R., Döhler, M., Ruta, M., Rizzo, G., Torsner, J., Engel, T., & Ladid, L. (2016). Internet of Things in the 5G Era: Enablers, Architecture, and Business Models. *IEEE Journal on Selected Areas in Communications*, 34(3), 510–527. DOI: 10.1109/JSAC.2016.2525418

Palotti, J. (2016). *Learning to Rank for Personalized E-Commerce Search at CIKM Cup 2016. Technical Report.* CIKM Cup.

Pan, C., Ren, H., Wang, K., Elkashlan, M., Nallanathan, A., Wang, J., & Hanzo, L. (2019). Intelligent reflecting surface enhanced MIMO broadcasting for simultaneous wireless information and power transfer. *IEEE Journal on Selected Areas in Communications*, 38(8), 1719–1734. DOI: 10.1109/JSAC.2020.3000802

Pandi, S., & Albert, A. J. (2024). A Novel Enhanced Security Architecture for Sixth Generation (6G) Cellular Networks Using Authentication and Acknowledgement (AA) Approach. *Results in Engineering*, 21, 101669. DOI: 10.1016/j.rineng.2023.101669

Pant, B. C. (2019). Status and Potential of India as a MICE Destination: An Informative Study. *Emerging Trends in Indian Tourism and Hospitality: Transformation and Innovation*, 257.

Pan, T., & Fu, R. J. (2024). *Navigating the AI horizon in hospitality: a novel classification and future research agenda.* International Hospitality Review.

Park, J., Park, S., & Ji, H. (2020). Dynamic Spectrum Sharing for 5G and Beyond: A Survey. *IEEE Wireless Communications*, 27(5), 12–18.

Parvez, I., Rahmati, A., Guvenc, I., Sarwat, A. I., & Dai, H. (2018). A survey on low latency towards 5G: RAN, core network and caching solutions. *IEEE Communications Surveys and Tutorials*, 20(4), 3098–3130. DOI: 10.1109/COMST.2018.2841349

Patel, R., Singh, P., & Green, D. (2023). Quantum Technologies for Future Networks. *Journal of Quantum Computing*, 12(1), 67–79.

Patel, S. C. (2013). Incremental Journey for World Wide Web: Introduced with Web 1.0 to Recent Web 5.0 - A Survey Paper. *International Journal of Advanced Research in Computer Science and Software Engineering*, 3, 112–125. https://www.scirp.org/reference/referencespapers?referenceid=2228357

Peng, B., Li, Z., & Li, S. (2021). A Survey on 5G and 6G Technologies: Opportunities, Challenges, and Solutions. *IEEE Communications Magazine*, 59(1), 38–43.

Pérez, P., Corregidor, D., Garrido, E., Benito, I., González-Sosa, E., Cabrera, J., Berjón, D., Díaz, C., Morán, F., García, N., Igual, J., & Ruiz, J. (2022). Live Free-Viewpoint Video in Immersive Media Production Over 5G Networks. *IEEE Transactions on Broadcasting*, 68(2), 439–450. DOI: 10.1109/TBC.2022.3154612

Persad, U., Langdon, P., & John Clarkson, P. (2019). "A Framework for Analytical Inclusive Design Evaluation." Proceedings of ICED 2007, the 16th International Conference on Engineering Design: 1–12.

Peštek, A., & Sarvan, M. (2020). Virtual reality and modern tourism. *Journal of Tourism Futures*, 7(2), 245–250. DOI: 10.1108/JTF-01-2020-0004

Pham, Q.-V., Fang, F., Ha, V. N., Piran, M. J., Le, M., Le, L. B., Hwang, W.-J., & Ding, Z. (2020). A Survey of Multi-Access Edge Computing in 5G and beyond: Fundamentals, Technology Integration, and State-of-the-Art. *IEEE Access : Practical Innovations, Open Solutions*, 8, 116974–17. DOI: 10.1109/ACCESS.2020.3001277

Phukon, R., Das, P., & Borah, G. (2020). Prospects and barriers of tourism industry in Assam. *International Journal of Advanced Research in Engineering and Technology*, 11(12), 3367.

Pinho, M., & Marques, J. (2021, September). The bleisure tourism trend and the potential for this business-leisure symbiosis in Porto. [). Routledge.]. *Journal of Convention & Event Tourism*, 22(4), 346–362. DOI: 10.1080/15470148.2021.1905575

Polese, F. (2009). Local Government and Networking Trends Supporting Sustainable Tourism: Some Empirical Evidence. *Cultural Tourism and Sustainable Local Development*, 131–148.

Popovski, P., Trillingsgaard, K. F., Simeone, O., & Durisi, G. (2018). 5G Wireless Network Slicing for eMBB, URLLC, and mMTC: A Communication-Theoretic View. *IEEE Access : Practical Innovations, Open Solutions*, 6, 55765–55779. DOI: 10.1109/ACCESS.2018.2872781

Porcu, D., Castro, S. M., Otura, B., Encinar, P., Chochliouros, I. P., Ciornei, I., Hadjidemetriou, L., Ellinas, G., Santiago, A. R., Grigoriou, E., Antonopoulos, A., Cadenelli, N., di Pietro, N., Betzler, A., Prieto, I., Battista, F., Brodimas, D., Rumenova, R., & Bachoumis, A. (2022). Demonstration of 5G Solutions for Smart Energy Grids of the Future: A Perspective of the Smart5Grid Project. *Energies*, 15(3), 839–839. DOI: 10.3390/en15030839

Pradini, G., Awaloedin, D. T., Agustiani, F., & Kelly, A. (2024). 5G Technology Transformation In Tourism. *International Journal of Economics, Management, Business, and Social Science*, 4(1), 523–536. DOI: 10.59889/ijembis.v4i1.376

Prahalad, C. K., & Ramaswarny, V. (2000). Co-opting customer competence of the corporation. *Harvard Business Review*, 78, 79–87.

Prahalad, C. K., & Ramaswarny, V. (2004). Co-creation Experiences: The Next Practice in Value Creation. *Journal of Interactive Marketing*, 18(3), 5–14. DOI: 10.1002/dir.20015

Prakasam, P., Shohel Sayeed, M., & Ajayan, J. (2021). Guest Editorials: P2P Computing for 5G, beyond 5G (B5G) Networks and Internet-of-Everything (IoE). *Peer-to-Peer Networking and Applications*, 14(1), 240–242. DOI: 10.1007/s12083-020-01001-5

Prasad, R., & Ojanpera, T. (1998). An overview of CDMA evolution toward wideband CDMA. *IEEE Communications Surveys and Tutorials*, 1(1), 2–29. DOI: 10.1109/COMST.1998.5340404

Psiha, M., & Vlamos, P. (2017). IoT applications with 5G connectivity in medical tourism sector management: Third-party service scenarios. *Advances in Experimental Medicine and Biology*, 989, 141–154. DOI: 10.1007/978-3-319-57348-9_12 PMID: 28971423

Puri, R. (2013), How Online Retailers Use Predictive Analytics To Improve Your Shopping Experience.Retrieved April5, 2015 from https://blogs.sap.com/innovation/analytics/how-online-retailers-use-predictiveanalytics- to-improve-your-shopping-experience-0108060

Qadir, Z., Ali, M., & Ahmed, S. (2023). Towards 6G Internet of Things: Recent advances, use cases, and open challenges. *ICT Express*, 9(3), 296–312. DOI: 10.1016/j.icte.2022.06.006

Qamar, F., Siddiqui, M. U. A., Hindia, M. H. D. N., Hassan, R., & Nguy n, Q. N. (2020). Issues, challenges, and research trends in spectrum management: A comprehensive overview and new vision for designing 6G networks. *Electronics (Basel)*, 9(9), 1416. DOI: 10.3390/electronics9091416

Qiao, D., & Gursoy, M. C. (2016). Statistical delay tradeoffs in buffer-aided two-hop wireless communication systems. *IEEE Transactions on Communications*, 64(11), 4563–4577. DOI: 10.1109/TCOMM.2016.2607746

Qiao, D., & Gursoy, M. C. (2017). Buffer-aided relay systems under delay constraints: Potentials and challenges. *IEEE Communications Magazine*, 55(9), 168–174. DOI: 10.1109/MCOM.2017.1600213

Qin, Y., Gacanin, H., & Saad, W.. (2021). Integrating 6G with AI: Vision, Challenges, and Opportunities. *IEEE Network*, 35(2), 20–26.

Rahman, A., Wadud, M., Islam, M. J., Kundu, D., Bhuiyan, T. M. A.-U.-H., Muhammad, G., & Ali, Z. (2024). Internet of medical things and blockchain-enabled patient-centric agent through SDN for remote patient monitoring in 5G network. *Scientific Reports*, 14(1), 5297. Advance online publication. DOI: 10.1038/s41598-024-55662-w PMID: 38438526

Rajatheva, N., ... (2020). "White Paper on Broadband Connectivity in 6G." arXiv preprint arXiv:2004.14247.

Ramachandran, M., Archana, T., Deepika, V., Kumar, A., & Sivalingam, K. M. (2022). 5G network management system with machine learning-based analytics. *IEEE Access : Practical Innovations, Open Solutions*, 10, 73610–73622. DOI: 10.1109/ACCESS.2022.3190372

Ramezanpour, K., Jagannath, J., & Jagannath, A. (2023). Security and Privacy Vulnerabilities of 5G/6G and WiFi 6: Survey and Research Directions from a Coexistence Perspective. *Computer Networks*, 221, 221. DOI: 10.1016/j.comnet.2022.109515

Rana, D. S., Dhondiyal, S. A., & Chamoli, S. K. (2019). Software Defined Networking (SDN) Challenges, Issues and Solution. *International Journal on Computer Science and Engineering*, 7(1), 884–889.

Rane, N. L. (2023). Enhancing Customer Loyalty through Artificial Intelligence (AI), Internet of Things (IoT), and Big Data Technologies: Improving Customer Satisfaction, Engagement, Relationship, and Experience. *Social Science Research Network*.https://doi.org/DOI: 10.2139/ssrn.4616051

Ranjha, A., Javed, M. A., Srivastava, G., & Lin, J. C. (2023). Intercell interference coordination for UAV enabled URLLC with perfect/imperfect CSI using cognitive radio. *IEEE Open Journal of the Communications Society*, 4, 197–208. DOI: 10.1109/OJCOMS.2022.3232888

Rantcell. (n.d.). Comparison of 2G, 3G, 4G, 5G. Rantcell. Retrieved July 12, 2024, from https://rantcell.com/comparison-of-2g-3g-4g-5g.html

Rappaport, T. S., MacCartney, G. R., Sun, S., Yan, H., & Deng, S. (2017). Small-scale, local area, and transitional millimeter wave propagation for 5G communications. *IEEE Transactions on Antennas and Propagation*, 65(12), 6474–6490. DOI: 10.1109/TAP.2017.2734159

Rappaport, T. S., Xing, Y., Kanhere, O., Ju, S., Madanayake, A., Mandal, S., Alkhateeb, A., & Trichopoulos, G. C. (2019). Wireless Communications and Applications Above 100 GHz: Opportunities and Challenges for 6G and Beyond. *IEEE Access : Practical Innovations, Open Solutions*, 7, 78729–78757. DOI: 10.1109/ACCESS.2019.2921522

Rappaport, T. S., Xing, Y., Kanhere, O., Ju, S., Madanayake, A., Mandal, S., & Trichopoulos, G. C. (2019). Wireless communications and applications above 100 GHz: Opportunities and challenges for 6G and beyond. *IEEE Access : Practical Innovations, Open Solutions*, 7, 78729–78757.

Rappaport, T. S., Xing, Y., MacCartney, G. R., Molisch, A. F., Mellios, E., & Zhang, J. (2017). Overview of Millimeter Wave Communications for Fifth-Generation (5G) Wireless Networks—With a Focus on Propagation Models. *IEEE Transactions on Antennas and Propagation*, 65(12), 6213–6230. DOI: 10.1109/TAP.2017.2734243

Rathore, M. M., Ahmad, A., Paul, A., & Rho, S. (2016). Urban planning and building smart cities based on the Internet of Things using Big Data analytics. *Computer Networks*, 101, 63–80. DOI: 10.1016/j.comnet.2015.12.023

Raviteja, P., Phan, K. T., Hong, Y., & Viterbo, E. (2018). Interference cancellation and iterative detection for orthogonal time frequency space modulation. *IEEE Transactions on Wireless Communications*, 17(10), 6501–6515. DOI: 10.1109/TWC.2018.2860011

Raza, W., Javaid, N., Nasir, H., Aurangzeb, K., Khan, Z. A., & Haider, S. I. (2018). BTRS: Buffer-threshold based relay selection scheme for cooperative wireless networks. *IEEE Access : Practical Innovations, Open Solutions*, 7, 23089–23099. DOI: 10.1109/ACCESS.2018.2890003

Razlighi, M. M., & Zlatanov, N. (2018). Buffer-aided relaying for the two-hop full-duplex relay channel with self-interference. *IEEE Transactions on Wireless Communications*, 17(1), 477–491. DOI: 10.1109/TWC.2017.2767582

Reddick, C. G., Enriquez, R., Harris, R. J., & Sharma, B. (2020). Determinants of broadband access and affordability: An analysis of a community survey on the digital divide. *Cities (London, England)*, 106, 102904.

Reddy, L. S., & Suryanarayana, A. (2024). Migrating From 5g To 6g Technologies: A Paradigm Shift With Profound Concomitant Changes. *Educational Administration Theory and Practice*, 13955–13962. DOI: 10.53555/kuey.v30i5.6165

Reddy, A. M., & Tanguturi, R. (2022). "Improved Secure Communication Mechanism for IoT Platform Devices," *2022 International Conference on Smart Generation Computing, Communication and Networking (SMART GENCON)*, Bangalore, India, pp. 1-6.

Reinisto, S. K. (2003). Success Factors of Place Marketing: A Study of Place Marketing Practices in Northern Europe and the United States. Helsinki University of Technology, Institute of Strategy and International Business, Doctoral Dissertations 2003/4.

Reinisto, S. K. (2001). *Lisensiaatintutkimus: Kaupungin kehittaminen merkkituotteena. Lahden ja Helsingin tapaustutkimukset. Licenciate Study: City Branding – Case Studies Lahti and Helsinki*. Helsinki University of Technology.

Ren, J., & Xia, D. (2023). Introduction to 5G C-V2X. In *Autonomous driving algorithms and Its IC Design* (pp. 283–294). Springer Nature Singapore. DOI: 10.1007/978-981-99-2897-2_13

Riis, M., Skriver, A. J. V., & Lodahl, J. (2004). Deployment of Mobile Switching Centers in a Telecommunications Network: A Stochastic Programming Approach. *Telecommunication Systems*, 26(1), 93–109. DOI: 10.1023/B:TELS.0000029022.63021.09

Rikkeisoft. (2024, April 4). How real-time data analysis empowers your business. Retrieved July 12, 2024, from https://rikkeisoft.com/blog/how-real-time-data-analysis-empowers-your-business/

Rjeily, C. A. (2020). Toward a better comprehension of decode-and-forward buffer-aided relaying: Case study of a single relay. *IEEE Communications Letters*, 24(5), 1005–1009. DOI: 10.1109/LCOMM.2020.2971494

Robertson, S., & Zaragoza, H. (2009). The probabilistic relevance framework: Bm25 and beyond. Found. *Trends Information Retrieval*, 3(4), 333–389. DOI: 10.1561/1500000019

Robles, M. M. (2012). Soft skills and job performance: An empirical study. *Business Communication Quarterly*.

Roth, R. (2017). User interface and user experience (UI/UX) design. Geographic Information Science & Technology Body of Knowledge, 2017(Q2).

Roy, B. K., & Pagaldiviti, S. R. (2023). Advancements in arena technology: Enhancing customer experience and employee adaptation in the tourism and hospitality industry. *Smart Tourism*, 4(1), 2330. DOI: 10.54517/st.v4i2.2330

Roy, P., Ramaprasad, B. S., Chakraborty, M., Prabhu, N., & Rao, S. (2020). CustomeracceptanceofUseofartificialintelligencein hospitality services: An Indian hospitality sector perspective. *Global Business Review*, •••, 0972150920939753.

Rückert, R. (2023). *Disruptions in public transport: from dispatching assistance to improved robustness* (Doctoral dissertation, Dissertation, Halle (Saale), Martin-Luther-Universität Halle-Wittenberg, 2023).

Rupprecht, D., Kohls, K., Holz, T., & Pöpper, C. (2020). "Call Me Maybe: Eavesdropping Encrypted {LTE} Calls with {ReVoLTE}." In 29th USENIX Security Symposium (USENIX Security 20), 73–88.

Rush, K. L., Seaton, C. L., Corman, K., Hawe, N., Li, E. P. H., Dow-Fleisner, S. J., & Pesut, B. (2022). Virtual care prior to and during COVID-19: Cross-sectional survey of rural and urban adults. *JMIR Formative Research*, 6(8), e37059.

Saad, A., Hossain, M. S., Muhammad, G., Alhamid, H., Guizani, M., & Al-Dhelaan, A. (2020). The Road to 6G: A Comprehensive Survey [Fourthquarter.]. *IEEE Communications Surveys and Tutorials*, 22(4), 2192–2235.

Saad, W., Bennis, M., & Chen, M. (2019). A Vision of 6G Wireless Systems: Applications, Trends, Technologies, and Open Research Problems. *IEEE Network*, 34(3), 134–142. DOI: 10.1109/MNET.001.1900287

Saarnisaari, H., ... (2020). "A 6G White Paper on Connectivity for Remote Areas." arXiv preprint arXiv:2004.14699.

Sahay, B. S., & Ranjan, J. (2008). Real time business intelligence in supply chain analytics. *Information Management & Computer Security*, 16(1), 28–48. DOI: 10.1108/09685220810862733

Sah, H. N. (2017). A Brief History of Mobile Generations and Satellite Wireless Communication System. [JETIR]. *Journal of Emerging Technologies and Innovative Research*, 4, 1211–1216.

Salahdine, F., Zhang, N., & Han, T. (2023). 5G, 6G, and Beyond: Recent advances and future challenges. *Annales des Télécommunications*, 78(9–10), 525–549. DOI: 10.1007/s12243-022-00938-3

Salameh, A. I., & El Tarhuni, M. (2022). From 5G to 6G—Challenges, Technologies, and Applications. *Future Internet*, 14(4), 117. DOI: 10.3390/fi14040117

Sánchez, J. A. H., Casilimas, K., & Rendón, Ó. M. C. (2022). Deep reinforcement learning for resource management on network slicing: A survey. *Sensors (Basel)*, 22(8), 3031. DOI: 10.3390/s22083031 PMID: 35459015

Saranovac, L., & Glisic, S. (2020). 5G impact on e-commerce: Opportunities and challenges. *Journal of Systems and Software*, 163, 110555.

Sarstedt, M., Hair, J. F.Jr, Cheah, J. H., Becker, J. M., & Ringle, C. M. (2019). How to Specify, Estimate, and Validate Higher-Order Constructs in PLS-SEM. *Australasian Marketing Journal*, 27(3), 197–211. DOI: 10.1016/j.ausmj.2019.05.003

Sarwar, B., Karypis, G., Konstan, J., & Riedl, J. (2002), Recommendation systems for large e-commerce:Scalable neighborhood formation using clustering. In *Proceedings of the fifth international conference on computer and information technology*

Sato, N., Hong, T., Hirose, K., & Yoshida, S. (2019). Terahertz wavefront sensing and real-time adaptation using a binary-phase spatial light modulator. *Journal of Infrared, Millimeter and Terahertz Waves*, 40, 823–834.

Scully, M. O., Chen, Y., & Yen, T.-H. (2016). Quantum optical tests of complementarity. *Proceedings of the National Academy of Sciences of the United States of America*, 113, 532–538.

Setthasuravich, P., & Kato, H. (2020). The Mediating Role of the Digital Divide in Outcomes of Short-Term Transportation Policy in Thailand. *Transport Policy*, 97(March), 161–171. DOI: 10.1016/j.tranpol.2020.07.008

Shafie, A. E. I., Khafagy, M. G., & Sultan, A. (2014). Optimization of a relay-assisted link with buffer state information at the source. *IEEE Communications Letters*, 18(12), 2149–2152. DOI: 10.1109/LCOMM.2014.2363115

Shafi, M., Molisch, A. F., Smith, P. J., Haustein, T., Zhu, P., Silva, P. D., & Wunder, G. (2017). 5G: A tutorial overview of standards, trials, challenges, deployment, and practice. *IEEE Journal on Selected Areas in Communications*, 35(6), 1201–1221. DOI: 10.1109/JSAC.2017.2692307

Shah, S. K., & Pate, V. A. (2016). Tea production in India: challenges and opportunities. *Journal of Tea Science Research, 6*.

Shahraki, A., Abbasi, M., Piran, M.J., Chen, M., & Cui, S. (2021). A Comprehensive Survey on 6G Networks: Applications, Core Services, Enabling Technologies, and Future Challenges. *ArXiv*, abs/2101.12475.

Sharma, C. K., & Barua, P. (2017). Small tea plantation and its impact on the rural landscape of contemporary Assam. *International Journal of Rural Management*, 13(2), 140–161. DOI: 10.1177/0973005217725454

Sharma, P. (2013). Evolution of mobile wireless communication networks-1G to 5G as well as future prospective of next generation communication network. *International Journal of Computer Science and Mobile Computing*, 2(8), 47–53.

Sharma, S. K., & Sharma, N. (2018). An Overview of 5G Technology. *International Journal of Engineering Research & Technology (Ahmedabad)*, 6(17), 1–3. https://www.ijert.org/research/an-overview-of-5g-technology-IJERTCONV6IS17001.pdf

Sharma, S., & Shrivastava, S. (2021). A review on artificial intelligence based edge computing in 6G wireless networks. *Wireless Personal Communications*. Advance online publication.

Shekhar, R., & Ashreetha, B. (2023). Performance Analysis of Energy Efficiency and Security Solutions of Internet of Things Protocols. *IJEER*, 11(2), 442–450. DOI: 10.37391/ijeer.110226

Shelton, T., Zook, M., & Wiig, A. (2015). The 'actually existing smart city'. *Cambridge Journal of Regions, Economy and Society*, 8(1), 13–25. DOI: 10.1093/cjres/rsu026

Sheng, Z., Cui, L., Nasir, A. A., Wang, R., & Fang, Y. (2023). URLLC-oriented secure communication for UAV relay-assisted network. *Physical Communication*, 59, 102063. DOI: 10.1016/j.phycom.2023.102063

Shen, X., Gao, J., Li, M., Zhou, C., Hu, S., He, M., & Zhuang, W. (2023). Toward immersive communications in 6G. *Frontiers of Computer Science*, 4, 1068478. Advance online publication. DOI: 10.3389/fcomp.2022.1068478

Shen, X., Lee, J.-S., Kim, K.-J., & Kwon, T. (2020). Artificial intelligence in edge computing: A review and research directions. *IEEE Access : Practical Innovations, Open Solutions*, 8, 47538–47556.

Shin, T., & Baek, S. J. (2019). Cooperative buffer-aided relaying using full-duplex relays with flow control. *IEEE Transactions on Vehicular Technology*, 68(2), 1824–1838. DOI: 10.1109/TVT.2018.2886753

Shirvanimoghaddam, M., Nguyen, T. M., Chatzinotas, S., Ottersten, B., & Juntti, M. (2018). Terahertz band communications: Applications, research challenges, and standardization activities. *IEEE Journal on Selected Areas in Communications*, 36, 2524–2538.

Shmueli, G., & Koppius, O. (2011). Predictive Analytics in Information Systems Research. *Management Information Systems Quarterly*, 35(3), 553–572. DOI: 10.2307/23042796

Shruthi, M., & Naveen, K. B. (2020). "Internet of Things: A Survey on Devices, Ecosystem, Components and Communication Protocols," *2020 4th International Conference on Electronics, Communication and Aerospace Technology (ICECA)*, pp. 611-616.

Shukla, S., Khare, V., Garg, S., & Sharma, P. (2013). Comparative Study of 1G, 2G, 3G and 4G. *Journal of Engineering, Computers & Applied Sciences (JEC&AS)*, *2*(4), 55–63.

Siddig, A. A. M., & Salleh, M. F. M. (2017). Buffer-aided relay selection for cooperative relay networks with certain information rates and delay bounds. *IEEE Transactions on Vehicular Technology*, 66(11), 10499–10514. DOI: 10.1109/TVT.2017.2738036

Silva, M. M., & Guerreiro, J. (2020). On the 5G and beyond. *Applied Sciences (Basel, Switzerland)*, 10(20), 7091. DOI: 10.3390/app10207091

Singh, A., Sharma, S., Deka, K., & Bhatia, V. (2023). DL-based OTFS signal detection in presence of hardware impairments. *IEEE Wireless Communications Letters*, 12(9), 1533–1537. DOI: 10.1109/LWC.2023.3281790

Singh, A., Sharma, S., Sharma, M., & Deka, K. (2023). Low complexity deep-decoder for OTSM with hardware impairments. *IEEE Communications Letters*, 27(12), 3240–3244. DOI: 10.1109/LCOMM.2023.3327249

Singh, P. (2024). Adoption of Blockchain Technology for Data Management of Human Resource Demands in Organizational Enterprises. *International Journal of Data Informatics and Intelligent Computing*, 3(2), 1–15.

Singh, P. R., Gupta, R., & Kumar, A. (2023). 6G networks for artificial intelligence-enabled smart cities applications: A scoping review. *Telematics and Informatics Reports*, 9, 100044. DOI: 10.1016/j.teler.2023.100044

Singh, S. K., Singh, R., & Kumbhani, B. (2020). The Evolution of Radio Access Network Towards Open-RAN: Challenges and Opportunities. *2020 IEEE Wireless Communications and Networking Conference Workshops (WCNCW)*, 1–6. DOI: 10.1109/WCNCW48565.2020.9124820

Sivanesan, R. (2013). Tea industry in India–Analysis of import and export of tea. *International Journal of Business and Management Invention*, 2(8), 9–15.

Smith, A., & Jones, M. (2022). Emerging Job Roles in 6G. *Future Jobs Journal*, 21(5), 175–190.

Smith, A., Jones, M., & Lee, K. (2022). Network Slicing and Edge Computing for 6G. *Communications Systems Review*, 52(4), 233–248.

Smith, J., & Jones, L. (2022). Interdisciplinary Collaboration in Telecommunications. *Collaboration and Technology Journal*, 15(2), 101–115.

Sohaili, N. L. M., Aminudin, N., & Amir, A. F. (2022). Bleisure Motivation Towards Leisure Attraction: A Systematic Literature Review. *Social Sciences*, 12(11), 910–922.

Soldani, D., & Manzalini, A. (2015). Horizon 2020 and Beyond: On the 5G Operating System for a True Digital Society. *IEEE Vehicular Technology Magazine*, 10(1), 32–42. DOI: 10.1109/MVT.2014.2380581

Solelhac, A. (2021). *Mountain resort marketing and management*. Routledge. DOI: 10.4324/9781003039396

Souza, J. H. I. D., Croisfelt, V., Kotaba, R., Abrão, T., & Popovski, P. (2024). Uplink multiplexing of eMBB/URLLC services assisted by reconfigurable intelligent surfaces. *IEEE Communications Letters*, 28(9), 2206–2210. DOI: 10.1109/LCOMM.2024.3426613

Speretta, M., & Gauch, S. (2005), Personalized search based on user search histories, in: The 2005 IEEE/WIC/ACM International Conference on Web Intelligence (WI'05), pp. 622–628. DOI: 10.1109/WI.2005.114

Stachniak, Z. (2015). old DOI doi - Clones: Soviet Computer Hobby Movement of the 1980s. *IEEE Annals of the History of Computing*, 1. https://doi.org/DOI: 10.1109/MAHC.2015.43 - good

Stojmenović, M., & Lin, X. (2014). Application of ambient backscatter for RF-powered cognitive radio networks. *IEEE Communications Magazine*, 52, 140–145.

Stone, P., Brooks, R., Brynjolfsson, E., Calo, R., Etzioni, O., Hager, G., & Teller, A. (2022). Artificial intelligence and life in 2030: the one hundred year study on artificial intelligence. *arXiv preprint arXiv:2211.06318.*

Subash, T. (2015). Tourism in India: Potentials, Challenges and Opportunities. *IJRAR-International Journal of Research and Analytical Reviews*, 2(4), 8–18.

Sujood, S., Bano, N., & Siddiqui, S. (2022). Consumers' intention towards the use of smart technologies in tourism and hospitality industry: A deeper insight into the integration of TAM, TPB and trust. *Journal of Hospitality and Tourism Insights*, 7(3), 1412–1434. DOI: 10.1108/JHTI-06-2022-0267

Sulaiman, R. B., & Kareem, A. (2020). A review on concepts and technologies of 6G cellular network and future scope. SSRN *Electronic Journal*. https://doi.org/ DOI: 10.2139/ssrn.3731869

Sundbo, J. (2008). *Creating Experiences In the Experience Economy*. Edward Elgar Publishing. DOI: 10.4337/9781848444003

Sun, S., Rappaport, T. S., Shafi, M., & MacCartney, G. R. (2016). MIMO techniques in millimeter wave wireless communications. *IEEE Transactions on Communications*, 64, 3201–3239.

Suram, S., & Miryala, R. K. (2023). Wearable technologies in human resource management: A comprehensive analysis of adoption and its impact. ANU Journal of Commerce and Management, 15(2), 89-96. https://doi.org/DOI: 10.6084/m9.figshare.25484752

Su, X., Liu, X., Motlagh, N. H., Cao, J., Su, P., Pellikka, P., Liu, Y., Petäjä, T., Kulmala, M., Hui, P., & Tarkoma, S. (2021). Intelligent and Scalable Air Quality Monitoring With 5G Edge. *IEEE Internet Computing*, 25(2), 35–44. DOI: 10.1109/MIC.2021.3059189

Suzuki, L. R. (2017). Smart Cities IoT: Enablers and Technology Road Map. In Rassia, S., & Pardalos, P. (Eds.), *Smart City Networks. Springer Optimization and Its Applications* (Vol. 125). Springer. DOI: 10.1007/978-3-319-61313-0_10

Taboada, I., & Shee, H. (2020). Understanding 5G technology for future supply chain management. *International Journal of Logistics*, 24(4), 392–406. DOI: 10.1080/13675567.2020.1762850

Tang, K., Chen, Y. L., & Hu, H. W. (2008). Context-based market basket analysis in a multiple-store environment. *Decision Support Systems*, 45(1), 150–163.

Tang, Y., Dananjayan, S., Hou, C., Guo, Q., Luo, S., & He, Y. (2021). A survey on the 5G network and its impact on agriculture: Challenges and opportunities. *Computers and Electronics in Agriculture*, 180, 105895. DOI: 10.1016/j.compag.2020.105895

Tashakkori, A., & Teddlie, C. (1998). *Mixed Methodology: Combining Qualitative and Quantitative Approaches*. SAGE.

Tashakkori, A., & Teddlie, C. (2003). *Handbook of mixed methods in the social and behavioural sciences*. Sage.

Tataria, H., Shafi, M., Molisch, A. F., Dohler, M., Sjöland, H., & Tufvesson, F. (2021). 6G Wireless Systems: Vision, Requirements, Challenges, Insights, and Opportunities. *Proceedings of the IEEE*, 109(7), 1166–1199. DOI: 10.1109/JPROC.2021.3061701

Taylor, R., & Green, S. (2023). Educating the Quantum Workforce. *Journal of Quantum Education*, 8(1), 45–58.

Tech Teamz. (2023). *Advances of 6G technology in electric cars - TechTeamZ*. Tech Teamz. https://techteamz.io/advances-of-6g-technology-in-electric-cars/

Tedeschi, P., Sciancalepore, S., & Pietro, R. D. (2020). Security in energy harvesting networks: A survey of current solutions and research challenges. *IEEE Communications Surveys and Tutorials*, 22(4), 2658–2693. DOI: 10.1109/COMST.2020.3017665

Teevan, J., Dumais, S. T., & Liebling, D. J. (2008), To personalize or not to personalize: Modeling queries with variation in user intent, in: Proc. Of the 31st Annual Int. ACM SIGIR Conf. on Research and Development in Information Retrieval, ACM, New York, NY, USA, pp. 163–170.

Teoh, Y. K., Gill, S. S., & Parlikad, A. K. (2023). IoT and Fog-Computing-Based Predictive Maintenance Model for Effective Asset Management in Industry 4.0 Using Machine Learning. *IEEE Internet of Things Journal*, 10(3), 2087–2094. DOI: 10.1109/JIOT.2021.3050441

Thantharate, A., Tondwalkar, A., Beard, C., & Kwasinski, A. (2022). Eco6G: Energy and cost analysis for network slicing deployment in beyond 5G networks. *Sensors (Basel)*, 22(22), 8614. DOI: 10.3390/s22228614 PMID: 36433208

Thüemmler, C., Paulin, A., Jell, T., & Lim, A. K. (2017). Information technology – next generation: the impact of 5G on the evolution of health and care services. *Advances in Intelligent Systems and Computing*, 811-817. https://doi.org/DOI: 10.1007/978-3-319-54978-1_100

Tian, M.-W., Wang, L., Yan, S., Tian, X., Liu, Z., & Rodrigues, J. J. P. C. (2019). Research on Financial Technology Innovation and Application Based on 5G Network. *IEEE Access : Practical Innovations, Open Solutions*, 7, 138614–138623. DOI: 10.1109/ACCESS.2019.2936860

Tian, Z., Gong, Y., Chen, G., & Chambers, J. A. (2017). Buffer-aided relay selection with reduced packet delay in cooperative networks. *IEEE Transactions on Vehicular Technology*, 66(3), 2567–2575. DOI: 10.1109/TVT.2016.2573378

Tong-On, P., Siripipatthanakul, S., & Phayaphrom, B. (2021). The implementation of business intelligence using data analytics and its effects towards on performance in the hotel industry in Thailand. *International Journal of Behavioral Analytics*, 1(2).

Trappe, W. (2015). The challenges facing physical layer security. *IEEE Communications Magazine.IEEE Communications Magazine*, 53(6), 16–20. DOI: 10.1109/MCOM.2015.7120011

Tsaih, R. H., & Hsu, C. C. (2018). Artificial intelligence in smart tourism: A conceptual framework.

Tsai, W., Bai, X., & Huang, Y. (2014). Software-as-a-service (SaaS): Perspectives and challenges. *Science China. Information Sciences*, 57(5), 1–15. DOI: 10.1007/s11432-013-5050-z

Uusitalo, M. A., Rugeland, P., Boldi, M., Strinati, E. C., Demestichas, P., Ericson, M., & Zou, Y. (2021). 6G vision, value, use cases and technologies from European 6G flagship project hexa-x. *IEEE Access : Practical Innovations, Open Solutions*, 9, 160004–160020. DOI: 10.1109/ACCESS.2021.3130030

Valcic, M., & Domsic, L. (2012). Information technology for management and promotion of sustainable cultural tourism. *Informatica (Slovenia)*, 36(2), 131–136. Retrieved from https://dblp.uni-trier.de/db/journals/informaticaSI/informaticaSI36.html#ValcicD12

Van Dijk, J. (2020). The Digital Divide. John Wiley & Sons. Docomo, N T T. 2020. "Inc. White Paper—5G Evolution and 6g; NTT Docomo." Inc.: Tokyo, Japan.

Vargo, S. L., & Lusch, R. F. (2004). Evolving to a new dominant logic for marketing. *Journal of Marketing*, 68(1), 1–17. DOI: 10.1509/jmkg.68.1.1.24036

Vargo, S. L., & Lusch, R. F. (2006). Service-dominant logic: What it is, what it is not, what it might be. In Lusch, R. F., & Vargo, S. L. (Eds.), *The service-dominant logic of marketing: Dialog, debate, and directions* (pp. 43–56). ME Sharpe.

Vega, M. T., Liaskos, C., Abadal, S., Papapetrou, E., Jain, A., Mouhouche, B., Kalem, G., Ergüt, S., Mach, M., Sabol, T., Cabellos-Aparicio, A., Grimm, C., De Turck, F., & Famaey, J. (2020). Immersive Interconnected Virtual and Augmented Reality: A 5G and IoT Perspective. *Journal of Network and Systems Management*, 28(4), 796–826. DOI: 10.1007/s10922-020-09545-w

Velásquez, K., Abreu, D. P., Curado, M., & Monteiro, E. (2022). Resource orchestration in 5G and beyond: Challenges and opportunities. *Computer Communications*, 192, 311–315. DOI: 10.1016/j.comcom.2022.06.019

Venkatesh, M., & Raj, D. J. (2016). Impact of tourism in India. *International Journal of Scientific Engineering and Applied Science*, 2(1), 167–184.

Vetrivel, S. C., Sowmiya, K. C., & Sabareeshwari, V. (2024). Digital Twins: Revolutionizing Business in the Age of AI. In Ponnusamy, S., Assaf, M., Antari, J., Singh, S., & Kalyanaraman, S. (Eds.), *Harnessing AI and Digital Twin Technologies in Businesses* (pp. 111–131). IGI Global., DOI: 10.4018/979-8-3693-3234-4.ch009

Vignaroli, L., Gramaglia, M., Fuentes, M., Casella, A., Odarchenko, R., Natale, L., & D'Andria, F. (2020, December). The touristic sector in the 5G technology era: The 5G-TOURS project approach. In *2020 IEEE Globecom Workshops (GC Wkshps* (pp. 1-6). IEEE.

Vignaroli, L., Gramaglia, M., Fuentes, M., Casella, A., Одарченко, P., Natale, L., & D'Andria, F. (2020). *The touristic sector in the 5G technology era: the 5G-tours project approach. 2020 IEEE Globecom Workshops*. GC WKSHPS., DOI: 10.1109/GCWkshps50303.2020.9367418

Viswanathan, H., & Mogensen, P. E. (2020). Communications in the 6G Era. *IEEE Access : Practical Innovations, Open Solutions*, 8, 57063–57074. DOI: 10.1109/ACCESS.2020.2981745

Vodafone. (2020). Vodafone and University Hospital Düsseldorf: First 5G remote surgery live. Retrieved from https://www.vodafone.com/news/press-release/vodafone-and-university-hospital-dusseldorf-first-5g-remote-surgery-live

Von Hippel, E. (1994). The Role of Technical Skills in Innovation: A Perspective from the Field of Engineering. *Research Policy*.

Vu, D., Dao, N., Won, D., & Cho, S. (2023). Potential enabling technologies for 6g mobile communication networks: a recent review. 2023 *Fourteenth International Conference on Ubiquitous and Future Networks (ICUFN)*.https://doi.org/DOI: 10.1109/ICUFN57995.2023.10199909

Walia, S., Kour, P., Choudhary, P., & Jasrotia, A. (2023). COVID-19 and the bleisure travellers: An investigation on the aftermaths and future implications. *Tourism Recreation Research*, 48(5), 657–667. DOI: 10.1080/02508281.2021.1946653

Walters, L. O., & Kritzinger, P. S. (2000). Cellular networks: past, present and future. *XRDS: Crossroads. The ACM Magazine for Students*, 7(2), 4–ff35.

Wang, C.-X., You, X., Gao, X., Zhu, X., Li, Z., Zhang, C., Wang, H., Huang, Y., Chen, Y., Haas, H., Thompson, J. S., Larsson, E. G., Di Renzo, M., Tong, W., Zhu, P., Shen, X., Poor, V. H., & Hanzo, L. (2023). On the Road to 6G: Visions, Requirements, Key Technologies, and Testbeds. *IEEE Communications Surveys and Tutorials*, 25(2), 905–974. DOI: 10.1109/COMST.2023.3249835

Wang, D., Bai, B., Zhao, W., & Han, Z. (2019). A survey of optimization approaches for wireless physical layer security. *IEEE Communications Surveys and Tutorials*, 21(2), 1878–1911. DOI: 10.1109/COMST.2018.2883144

Wang, D., Ren, P., & Cheng, J. (2018). Cooperative secure communication in two-hop buffer-aided networks. *IEEE Transactions on Communications*, 66(3), 972–985. DOI: 10.1109/TCOMM.2017.2776114

Wang, H., Al-Abdullah, M., Mao, B., & Giupponi, L. (2019). Physical layer security in millimeter wave networks: Challenges and opportunities. *IEEE Communications Magazine*, 57, 112–118.

Wang, H., Jiang, Z. H., Yu, C., Hou, D., Wang, H., Guo, C., & Zhou, J. (2021). The role of millimeter-wave technologies in 5G/6G wireless communications. *IEEE Journal of Microwaves*, 1(1), 101–122. DOI: 10.1109/JMW.2020.3035541

Wang, T., Wen, C.-K., Gao, F., Jiang, T., Jin, S., & Wang, H. (2017). Deep learning for wireless physical layer: Opportunities and challenges. *China Communications*, 14(11), 92–111. DOI: 10.1109/CC.2017.8233654

Wang, W., Kumar, N., Chen, J., Gong, Z., Kong, X., Wei, W., & Gao, H. (2020). Realising the potential of the Internet of Things for smart tourism with 5G and AI. *IEEE Network*, 34(6), 295–301. DOI: 10.1109/MNET.011.2000250

Wang, X., Guo, Y., & Gao, Y. (2024). Unmanned Autonomous Intelligent System in 6G Non-Terrestrial Network. *Information (Basel)*, 15(1), 38. DOI: 10.3390/info15010038

Wang, Y. (2022). Research on the Influence of Service Quality of Hotel Intelligent System on Customer Satisfaction Based on Artificial Intelligence Evaluation. *Mathematical Problems in Engineering*, 2022, 1–9. Advance online publication. DOI: 10.1155/2022/3832935

Wei, W. (2019). Research progress on virtual reality (VR) and augmented reality (AR) in tourism and hospitality. *Journal of Hospitality and Tourism Technology*, 10(4), 539–570. DOI: 10.1108/JHTT-04-2018-0030

White, S. K. (2023, August 31). *Boeing CIO Susan Doniz leads with curiosity and empathy*. CIO. https://www.cio.com/article/650681/boeing-cio-susan-doniz-leads-with-curiosity-and-empathy.html

White, G., Black, A., & Thompson, L. (2023). Updating Academic Curricula for 6G. *Educational Technology Review*, 19(2), 133–149.

Wicke, W., Zlatanov, N., Jamali, V., & Schober, R. (2017). Buffer-aided relaying with discrete transmission rates for the two-hop half-duplex relay network. *IEEE Transactions on Wireless Communications*, 16(2), 967–981. DOI: 10.1109/TWC.2016.2635142

Wilson, K., Miller, J., & Davis, T. (2023). Fostering Industry-Academia Collaboration for 6G. *Collaboration in Education Journal*, 14(3), 122–138.

Wu, J., Li, J., Guan, Z., Dong, M., & Ota, K. (2018). Big Data Analysis-Based Secure Cluster Management for Optimized Control Plane in Software-Defined Networks. *IEEE Transactions on Network and Service Management*, 15(1), 27–38. DOI: 10.1109/TNSM.2018.2799000

Wu, Q., Ding, Z., Lin, H., & Bhargava, V. K. (2016). Device-to-device communication in 5G cellular networks: Challenges, solutions, and future directions. *IEEE Access : Practical Innovations, Open Solutions*, 4, 6623–6639.

Wu, Q., & Zhang, R. (2020). Towards smart and reconfigurable environment: Intelligent reflecting surface aided wireless network. *IEEE Communications Magazine*, 58(1), 106–112. DOI: 10.1109/MCOM.001.1900107

Xia, B., Fan, Y., Thompson, J., & Poor, H. (2008). Buffering in a three-node relay network. *IEEE Transactions on Wireless Communications*, 7(11), 4492–4496. DOI: 10.1109/T-WC.2008.070867

Xiao, R. (2024). Human Resource Management Analysis in the AR/VR Industry: A Case Study on Meta Company. *Highlights in Business. Economics and Management*, 24, 462–469. DOI: 10.54097/90pavb05

Xie, D. (2020). Ping An Good Doctor launches first 5G telemedicine service in China. Retrieved from https://www.pingan.com/news/ping-an-good-doctor-launches-first-5g-telemedicine-service-in-china

Xu, H., Klaine, P. V., Onireti, O., Cao, B., Imran, M. A., & Zhang, L. (2020). Blockchain-enabled resource management and sharing for 6G communications. *Digital Communications and Networks*, 6(3), 261–269. DOI: 10.1016/j.dcan.2020.06.002

Xu, P., Chen, G., Yang, Z., & Lei, H. (2020). Buffer-state-based probabilistic relay selection for cooperative networks with delay constraints. *IEEE Wireless Communications Letters*, 9(11), 1855–1859. DOI: 10.1109/LWC.2020.3006079

Xu, T., Deeprasert, J., & Kanchanapoom, K. (2023). Employability Strategy by Focusing on Soft Skill Development of Animation Students in the Metaverse Job Market. *Educational Administration: Theory and Practice*, 30(2). Advance online publication. DOI: 10.52152/kuey.v30i2.1052

Xu, W., & Yang, S. (2023). Terahertz Waves for 6G Wireless Networks. *International Journal of Wireless Communications*, 27(3), 156–170.

Yadav, R., Dave, M., & Tiwari, P. (2019). Secure communication using 5G and IoT: A survey. *Wireless Personal Communications*, 109, 1109–1124.

Yadav, S., Kwon, J., & Jang, Y.-M. (2020). Machine learning in 6G networks: Current status and future directions. *IEEE Open Journal of the Communications Society*, 1, 145–152.

Yang, G., Yang, Y., & Li, K.. (2020). A Survey on Advanced Techniques for Spectrum Sharing in 5G and Beyond Mobile Networks. *IEEE Wireless Communications*, 27(1), 17–25.

Yang, H., Wu, K., Zhao, J., Niyato, D., Alphones, A., & Xiong, Z. (2019). Artificial-Intelligence-Enabled Intelligent 6G Networks. *IEEE Network*, 34(6), 272–280. DOI: 10.1109/MNET.011.2000195

Yan, J., Zheng, H., Zhang, Y., & Xing, C. (2019). THz-based ultra-high data rate wireless communication: Fundamental aspects, channel modeling, and applications. *IEEE Transactions on Terahertz Science and Technology*, 9, 542–563.

Yan, Z., Zhang, P., Cheng, Z., & Zhang, Y. (2017). Deep learning for image-based detection of unmanned aerial vehicles. *IEEE Transactions on Vehicular Technology*, 66, 8134–8145.

Yousefi, A., & Tang, J. (2012). *E-commerce: Consumer Online Shopping in Canada. Contemporary Research on E-business Technology and Strategy*. Springer Berlin Heidelberg.

You, X., Wang, C.-X., Huang, J., Gao, X., Zhang, Z., Wang, M., Huang, Y., Zhang, C., Jiang, Y., Wang, J., Zhu, M., Sheng, B., Wang, D., Pan, Z., Zhu, P., Yang, Y., Liu, Z., Zhang, P., Tao, X., & Liang, Y.-C. (2021). Towards 6G wireless communication networks: Vision, enabling technologies, and new paradigm shifts. *Science China. Information Sciences*, 64(1), 1–74. DOI: 10.1007/s11432-020-2955-6

Yuan, Y., Zou, J., Cui, Y., Li, X., Mu, J., & Han, K. (2023). Orthogonal time frequency space and predictive beamforming-enabled URLLC in vehicular networks. *IEEE Wireless Communications*, 30(2), 56–62. DOI: 10.1109/MWC.005.2200408

Yuen, S. C., Yaoyuneyong, G., & Johnson, E. (2011). Augmented reality: An overview and five directions for AR in education. *Journal of Educational Technology Development and Exchange*, 4(1). Advance online publication. DOI: 10.18785/jetde.0401.10

Yu, J., Mohan, S., Putthividhya, D. P., & Wong, W. K. (2014), Latent dirichlet allocation based diversified retrieval for e-commerce search, in: *Proceedings of the 7th ACM International Conference on Web Search and Data Mining*, ACM, New York, NY, USA, pp. 463–472. DOI: 10.1145/2556195.2556215

Yu, K., Yu, J., & Dong, A. (2022). Cooperative communication and mobility for securing URLLC of future wireless networks. *IEEE Transactions on Vehicular Technology*, 71(5), 5331–5342. DOI: 10.1109/TVT.2022.3151063

Zhang, B., Wang, K., Yang, K., & Zhang, G. (2022). IRS-assisted short packet wireless energy transfer and communications. *IEEE Wireless Communications Letters*, 11(2), 303–307. DOI: 10.1109/LWC.2021.3126596

Zhang, H., Xiao, Y., Wu, D., & Li, Y. (2020). Data-driven autonomous security resource management for 5G and beyond mobile networks. *IEEE Transactions on Mobile Computing*, 19, 145–160.

Zhang, L., Liang, Y.-C., & Niyato, D. (2019). 6G Visions: Mobile ultra-broadband, super internet-of-things, and artificial intelligence. *China Communications*, 16(8), 1–14. DOI: 10.23919/JCC.2019.08.001

Zhang, L., Li, Y., Qian, Y., & Shen, X. (2016). End-to-end security provisioning in mobile edge computing with NOMA. *IEEE Network*, 30, 46–51. DOI: 10.1109/MNET.2016.7474343

Zhang, R., Cui, Y., & Cui, Y.. (2021). AI-Empowered 6G: A Technology Review and Framework. *IEEE Communications Magazine*, 59(6), 26–32. DOI: 10.1109/MCOM.2012.6211482

Zhang, S., Xu, S., & Xiang, C. (2020). 6G: Connecting Everything by 1000 Times Price Reduction. *IEEE Open Journal of Vehicular Technology*, 1, 107–115. DOI: 10.1109/OJVT.2020.2980003

Zhang, S., & Zhang, R. (2021). Intelligent reflecting surface aided multi-user communication: Capacity region and deployment strategy. *IEEE Transactions on Communications*, 69(9), 5790–5806. DOI: 10.1109/TCOMM.2021.3079128

Zhang, S., Zhou, W., Cai, X., & Wang, Y. (2019). Green 5G enabling technologies: An overview. *IET Communications*, 13(2), 135–143. DOI: 10.1049/iet-com.2018.5448

Zhang, S., & Zhu, D. (2020). Towards artificial intelligence enabled 6G: State of the art, challenges, and opportunities. *Computer Networks*, 183, 107556. DOI: 10.1016/j.comnet.2020.107556

Zhang, X., Li, Y., & Wang, J. (2023). Artificial Intelligence and Machine Learning for 6G Networks. *Journal of Telecommunications Technology*, 45(3), 123–134.

Zhang, Y., Cheng, W., & Zhang, W. (2024). Multiple access integrated adaptive finite blocklength for ultra-low delay in 6G wireless networks. *IEEE Transactions on Wireless Communications*, 23(3), 1670–1683. DOI: 10.1109/TWC.2023.3290936

Zhang, Y., Yao, H., Shu, L., Liu, Z., & Li, D. (2018). Data-driven smart healthcare systems: A review. *IEEE Transactions on Industrial Informatics*, 14, 2230–2239.

Zhang, Z., Patra, B. G., Yaseen, A., Zhu, J., Sabharwal, R., Roberts, K., & Wu, H. (2023). Scholarly recommendation systems: A literature survey. *Knowledge and Information Systems*, 65(11), 4433–4478.

Zhang, Z., Xiao, Y., Ma, Z., Xiao, M., Ding, Z., Lei, X., Karagiannidis, G. K., & Fan, P. (2019). 6G Wireless Networks: Vision, Requirements, Architecture, and Key Technologies. *IEEE Vehicular Technology Magazine*, 14(3), 28–41. DOI: 10.1109/MVT.2019.2921208

Zhou, B., Cui, Y., & Tao, M. (2015). Stochastic throughput optimization for two-hop systems with finite relay buffers. *IEEE Transactions on Signal Processing*, 63(20), 5546–5560. DOI: 10.1109/TSP.2015.2452225

Zhou, Y., Yeoh, P., Chen, H., Li, Y., Schober, R., Zhuo, L., & Vucetic, B. (2018). Improving physical layer security via a UAV friendly jammer for unknown eavesdropper location. *IEEE Transactions on Vehicular Technology*, 67(11), 11280–11284. DOI: 10.1109/TVT.2018.2868944

Zlatanov, N., Schober, R., & Popovski, P. (2013). Buffer-aided relaying with adaptive link selection. *IEEE Journal on Selected Areas in Communications*, 31(8), 1530–1542. DOI: 10.1109/JSAC.2013.130816

Zsarnoczky, M. (2017). How does artificial intelligence affect the tourism industry? *Vadyba*, 31(2), 85–90.

About the Contributors

Ramesh Kumar Miryala holds Ph.D. in Business Management from Osmania University. Currently, he serves Mahatma Gandhi University as the Director of Internal Quality Assurance Cell, Director of Automation, and Chairperson of the Board of Studies in Business Management. Dr. Miryala has a distinguished academic career, having previously held positions such as the Controller of Examinations, Director of Admissions, Director of College Development Council, Principal of University College of Commerce & Business Management and Head, Department of Business Management at the same university. His areas of specialization include Organizational Behavior, Marketing, and Service Management. Dr. Miryala's contributions to academia have been significant, and he continues to be a driving force behind the academic excellence at Mahatma Gandhi University.

Young-Chan Lee is a Professor in the Department of Information Management at Dongguk University, Korea. He received a Ph.D. in Management Science in the Department of Business Administration from Sogang University, Korea. He is a scholar who has cited more than 2,500 research papers published in international journals in Google Scholar Search. His research appears in good-quality international journals such as Behavior and Information Technology, International Journal of Bank Marketing, Human Factors and Ergonomics in Manufacturing & Service Industries, Information Systems Management, Knowledge-Based Systems, Expert Systems with Applications, and so on. And he also has published tens of books, won many research funds, and participated in various governmental projects. He was awarded 'Most Cited Paper of Expert Systems with Applications (Elsevier) in 2006-2008' and 'Top Downloaded Paper of Human Factors and Ergonomics in Manufacturing & Service Industries (Wiley) in 2018-2019'. He was also awarded Best Paper of International Conference on Electronic Commerce (ICEC) in 2022, Korea Association of Information Systems in 2021, 2009, and 2010 successively, and Korea Academic Society of Business Administration in 2016. From October

2012 to February 2016, he served as Dean of the Office of International Affairs at Dongguk University. In 2016, he established the Institute for Global Studies and has served as the director, making great efforts in convergence research covering humanities, society, and science and technology. Specialties: His research interests include data mining and machine learning for business analytics, knowledge management, social media, systems thinking and system dynamics, multi-criteria decision making, innovation management, Fintech etc.

Marica Mazurek The author´s research interest is competitiveness of tourism destinations, place marketing, place branding. She participated in the academic exchanges in Canada at Brock University and the University of Waterloo. She is a member of the Central European Association of Canadian Studies and the author of several academic Scopus publications and book chapters in Emerald or Cambridge Scholar Publishing and the author of the book Models of Branding and their Application. She is a founding member of EATSA (European Asian Tourism Studies Association). The PhD. thesis defence was at the Zilina University, Slovakia and her contemporary affiliation is with the University of Zilina as a researcher and editorial board member of the Scopus Journal Communications (Scopus journal).

Madhavaiah Chendragiri is an Associate Professor of Marketing and the Head of the Department of Management, Pondicherry University, Karaikal Campus, Karaikal, India. He has a doctorate in management from Sri Venkateswara University, Tirupati, India. He has 22 years of teaching and 19 years of research experience. His areas of research are service quality measurement, technology adoption behaviour of customers, brand equity, and consumer behaviour in luxury fashion brands. He published a book on Measuring Service Quality in Banking Sector (2014) for Lambert Publishers, Germany. He has published more than 65 research articles in refereed national and international journals, including the Journal of Financial Economic Policy (Emerald Publications), Journal of Science and Technology Policy Management (Emerald Publications), Public Transport (Springer Nature), International Journal of Social Economics (Emerald Publications), Journal of Indian Business Research (Emerald Publications), Asia-Pacific Business Review (Sage Publications), and Vision – Journal of Business Perspective (Sage Publications). He is a winner of the Best Teacher Award at Pondicherry University. He organized national-level management meetings, national conferences, and research methodology workshops.

***

Swathi A is a fervent researcher pursuing a regular PhD at the Central University of Kerala. With an unwavering passion for exploration, she immerses herself in the realms of academia, driven by the desire to unravel new knowledge. Her journey began with completing an MBA at the same esteemed institution, where she honed her skills and cultivated a deep appreciation for the rigors of scholarly inquiry. Guided by curiosity and fueled by dedication, Swathi traverses the intricate pathways of research, eager to contribute meaningfully to her field. In each discovery lies the essence of her academic odyssey, a testament to her relentless pursuit of understanding.

Anurag A. S. obtained his BE degree in Electrical and Electronics from Anna University in 2013, followed by an MBA degree in Operations from Indira Gandhi National Open University in 2017. Accumulating nearly a decade of practical working experience in both the Engineering and Management fields, Anurag has garnered invaluable insights within various organizations across two culturally diverse countries. His extensive professional journey has deepened his understanding of the global significance of management functions. Currently pursuing a Ph.D. at the Central University of Kerala within the Management Studies Department, Anurag's research revolves around the intersection of Management and Artificial Intelligence. Fueled by a unique blend of engineering and business acumen, Anurag is wholeheartedly committed to unraveling the intricacies of how technology, specifically Artificial Intelligence, can enhance and streamline various facets of management in the contemporary business landscape.

Shambhavi Agnihotri is a research scholar in the Department of Tourism Management,Indira Gandhi National Tribal University,Amarkantak, Madhya Pradesh,India

Suryanarayana Alamuri is a former Dean, Faculty of Management, Osmania University, HYDERABAD-500007 (Telangana State) INDIA

Aluvala Ravi is a distinguished Professor of Business Management at Mahatma Gandhi University, Nalgonda, Telangana. He currently serves as Registrar (In-Charge). His notable previous positions include Officer on Special Duty (OSD) to the Vice-Chancellor, Chairperson of the Board of Studies, Head of the Department of Business Management, Principal of the University College of Commerce and Business Management, and Director of the Academic Audit Cell. With over 20 years of experience in teaching and research, and 5 years in industry, Dr. Ravi's expertise spans Business Management, Human Resource Management, Organizational Behavior, Performance Management, and more. He has published 60 research articles, guided 8 Ph.D. Scholars, and authored /edited 10 books. Dr. Ravi has organized 20 academic events, delivered 45 guest lectures, and given 5 keynote addresses. He received the State Level Best Teacher Award from the Government of Telangana in 2023. He has served on various state-level committees and professional bodies, including as a Life Member of the National Human Resource Development Network and a member of the Academic Senate and Board of Studies at Mahatma Gandhi University and Palamuru University.

Deepti Bhatt is a research scholar at Amity School of Hospitality, Amity University Noida. She is working in the capacity of Assistant Professor and Training and Placement Coordinator in Quantum University, Roorkee. She has an experience of 7 yrs. academic and Industry collectively. She has pursued her Masters from IHM Bangalore and Graduation from IHM Bhopal. Her Specialization is Front Office, Room Division, and Human resource Management. She is the reviewer of an International Journal published by Emerald. Her work published in Scopus Indexed Journal i.e. African Journal of Hospitality Tourism and leisure.

Jayeeta Bhattacharjee is a PhD Research Scholar in Assam University, Silchar. Completed MBA (Hospitality and Tourism Management) from Assam University, Silchar. Gold Medalist in MBA (Hospitality and Tourism Management). Has done 2 published chapters in books and participated in 1 national seminar.

D.Srinivas distinguished Faculty of Marketing and Communication at SR UNIVERSITY, Warangal,Telangana. He rececive MBA from Kakatiya University, Warangal and Ph.D degree from Kakatiya University,Warangal..He is author of many books. He also published many articles in National and International journals. His research covers Product strategy, Digital marketing, E-Business and IoT For Businees Operations. He is a active member of AIMA and HMA. He has delivered many guest lectures and organized FDP's across the country

Sunandita Debnath Presently, I serve as an Assistant Professor at IIIT Vadodara since 2021. Prior to this role, I held a faculty position at SENSE in VIT Chennai from June 2020 to February 2021. My academic contributions include seven publications in esteemed SCI journals alongside numerous conference presentations. My research concentrates on sensor networks and wireless communications, with a particular emphasis on devising sensor channel models applicable across varied propagation mediums. This aspect is pivotal for evaluating the service quality of deployed sensor networks. Additionally, my research extends to magnetic induction communication for underwater communication modalities.

Jayaprakashnarayana G is an assistant professor in the Department of Tourism Management Indira Gandhi National Tribal University Amarkantak, Madhya Pradesh.He published various research papers in national and international journals.

Dankan Gowda V is currently working as an Assistant Professor in the Department of Electronics and Communication Engineering at BMS Institute of Technology and Management in Bangalore. Previously, he worked as a Research Fellow at ADA DRDO and as a Software Engineer at Robert Bosch. With a total experience of 14 years, including teaching and industry, Dr. Gowda has made significant contributions to both academia and research. He has published over 120 research papers in renowned international journals and conferences. In recognition of his innovative work, Dr. Gowda has been granted six patents, including four from Indian authorities and two international patents. His research interests primarily lie in the fields of IoT and Signal Processing, where he has conducted workshops and handled industry projects. Dr. Gowda is passionate about teaching and strives to create an engaging and effective learning environment for his students. He consistently seeks opportunities to enhance his knowledge and stay updated with the latest advancements in his field.

Ashraf Hossain (Senior Member, IEEE) earned the Ph.D. degree in Electronics & Electrical Communication Engg., from Indian Institute of Technology (IIT), Kharagpur, India in 2011. He received the M.Tech. and B.Tech. degrees in Radio Physics & Electronics from the Institute of Radio Physics & Electronics, University of Calcutta, Kolkata, India in 2004 and 2002, respectively. He served Dept. of Electronics & Communication Engg., Aliah University, Kolkata and Haldia Institute of Technology as Asst. Professor. He is currently working as an Assistant Professor in the Dept. of Electronics & Communication Engg., National Institute of Technology (NIT), Silchar, Assam, India. His research interests include wireless sensor network, communication theory and systems. He is a Senior member of IEEE and IE (India). E-mail: ashrafiit@gmail.com

K.N. Lokesh Kumar is currently working as assistant professor in the Department of Hospitality and Tourism Management, Assam University (A Central University), Silchar, Assam. He has about 11 years of teaching experience. He completed his MBA from Prestigious Indian institute of tourism and travel management, Gwalior (An autonomous institute under Ministry of Tourism, Govt. of India). He is a UGC, NET& JRF holder in Tourism Management, Gold Medallist & S.K. Mishra merit awardee in MBA (Tourism & Travel). His areas of interest include tour operations, Travel agency, Airline operations and Management, Destination Marketing and Human resource in tourism. He has to his credit 05 publications and also 06 Published chapters in books and 12 participated national and international conferences. He developed E-content for Entrepreneurship in Tourism, Research Methods in Tourism to Centre for Distance Education, Acharya Nagarjuna University, developed the E-Content for the Diploma in Travel and Tourism Management (TTM) course which is being offered by the MOOCs Platform ANU and also edited the self-learning material for Board of Open Schooling & Skill Education (BOSSE), Sikkim. In addition, he also completed Basic Mountaineering course from Atal Bihari Vajpayee Institute of Mountaineering and Allied Sports (ABVIMAS), Manali, Himachal Pradesh. With regards to tourism activities and other related assignments he had travelled extensively in States of India and abroad to countries like Thailand and Taiwan.

Levent Karadağ He works as Assoc. Prof. at Mugla Sıtkı Kocman University-Datça Kazım Yılmaz Vocational School. He holds a bachelor's degree in tourism management. He did his master's degree in Tourism Management program on "Consumer decision-making process in tourism: A research on family groups who visit Turkey". He completed his PhD on "The importance of touristic product diversification in the development of sustainable tourism strategies: A sample of Muğla" in the Department of Business Administration Program.

Narendra Kumar is currently working as Assistant Professor at Amity Institute of Travel & Tourism, Amity University, Noida. With 23 years of experience in travel & tourism industry and academia, he has made significant contributions to the area of tourism education. Dr. Kumar has a strong interest in Business travel, tourist behaviour and visitor experiences has published extensively on related topics. In addition to his academic pursuits, Dr. Kumar is actively involved in helping young graduates of tourism and hospitality domain to help them in establishing their own start-ups. He has been invited to speak at numerous conferences and seminars on tourism and related fields and his current research focuses on student mentoring. Dr. Kumar's previous works include one edited book on "Tourist Behaviour" published by Apple Academic Press, USA; one book on "Digital Marketing" and one on "Ethical Aspects of Business in Tourism". Recent works from Dr. Kumar includes two books from IGI Global Publishing themed at Bleisure Tourism.

Rajeev Kumar received his Ph.D. and M. Tech. degrees in Dept. of Electronics & Communication Engineering (ECE) from National Institute of Technology (NIT), Silchar, Assam, India in 2019 and 2015, respectively. He obtained B.Tech. degree in Dept. of ECE from West Bengal University of Technology (WBUT), Kolkata, West Bengal, India in 2012. He held post as an Assistant Professor in the Dept. of ECE, Indian Institute of Information Technology (IIIT) Sri City, Chittoor, Andhra Pradesh,

India. His current research interests include Cooperative Communications, Wireless sensor Network, Green Technology, Advanced Protocol design, IoT applications for next-generation radio networks, and Optimization Theory. He is Reviewer of IEEE Transactions on Communications, IEEE Access, IET Communications, Wireless Personal Communications. Dr. Kumar is also, selected as an Exemplary Reviewer of National Conference on Communication (NCC) at IIT Hyderabad for 2017. He is also Reviewer of 26th (Virtual) Annual International Conference on ADCOM at NIT Silchar for 2020. E-mail ID: er.rajeevkr1989@gmail.com Alternate E-mail ID: rajeevece.nits@gmail.com

Johnpaul M is currently an Assistant Professor in the Department of Management Studies at the Central University of Kerala. Dr. John earned a Bachelor's Degree in Economics from the prominent Loyola College in Chennai, as well as a Master's Degree in Business Administration and MA Economics from Osmania University in Hyderabad. He earned his Ph.D. and cleared the AP SET and UGC NET in Management. From 2009 until June 2022, he was an Assistant Professor at Mahatma Gandhi University's Department of Management Studies in Nalgonda. He additionally served as a Hostel Warden, Coordinator for the Academic Audit Cell, NSS Programme Officer, and in a variety of other academic and administrative posts at MGU. To his credit, he has over 13 years of teaching experience. He has published 15 research papers in national and international journals. He has delivered 17 papers at national and international conferences. He has participated in various Faculty Development Programs. He has written one edited book, "Responsible Marketing for Sustainable Business". His teaching topics include marketing, marketing research, strategic management, international business strategy, and managerial economics.

P. VidhyaPriya is a dedicated academician with a strong foundation in both computer science and business management. She holds an MBA from Avinashilingam Deemed University, Coimbatore, and earned her Ph.D. from Bharathiar University, Coimbatore, in 2010. Before transitioning to academia, she gained valuable industry experience working with a textile firm for over three years. In 2000, she joined Kongu Engineering College, where she has since built an illustrious career in research and teaching. With over 24 years of academic experience, Dr. VidhyaPriya's research interests lie in corporate finance, corporate governance, and behavioral finance. She has published 57 research papers in reputed national and international journals, including Scopus-indexed publications. Throughout her career, she has actively engaged with the academic community by attending numerous Faculty Development Programs (FDPs) at leading B-Schools, including IIM Bangalore, IIT Madras, Pondicherry University, and Mangalore University. Dr.VidhyaPriya has also presented papers at 30 national and international conferences and has organized multiple conferences, seminars, FDPs, workshops, and training programs. She has successfully completed two ICSSR-sponsored projects focused on start-up entrepreneurship and the Stand-Up India Scheme for entrepreneurs.

KDV Prasad is working as Faculty (Research) in the Symbiosis Institute of Business Management. Hyderabad. He hold masters in computers and masters in software engineering from BITS Pilani. He possess PhD in Business Management and PhD in Business Administration. Dr Prasad publushed over 200 journal papers in national and international journals. Husband Research interests are general Management, behavioral finance. Human resources and marketing

S. Rajadurai is a Senior Lecturer in the Department of English Language Teaching,University of Colombo,Sri Lanka.

Rohith Ramesh earned his B.Tech degree in Computer Science and Engineering from LBS College of Engineering, Kasaragod, in 2018, and subsequently completed an MBA degree at the National Institute of Technology, Karnataka, in 2022. With substantial practical experience in marketing and teaching roles, Rohith has gained invaluable insights across various organizations. His extensive professional journey has deepened his understanding of the global significance of management functions. Currently pursuing a Ph.D. in Management Studies at the Central University of Kerala, Rohith focuses his research on marketing and branding. Combining engineering expertise with business acumen, Rohith is deeply dedicated to advancing the field of management, particularly in the realm of branding.

Manoj Kumar S B I am working as an associate professor in BGS Institute of Technology and having 13.5 Years of experience. I have Published 23 research papers and 2 Book chapters. I have a lifetime member for 4 professional bodies. Under my supervision KSCST 43rd and 44th series SPP projects were selected and received 22 thousand rupees. I was filed and published 4 National Patents, Currently I am working as PG Coordinator from Jan 2017 to till date, Project Coordinator, Internship Coordinator, Industrial visit coordinator, Innovation and science club coordinator.

SC Vetrivel is a faculty member in the Department of Management Studies, Kongu Engineering College (Autonomous), Perundurai, Erode Dt. Having experience in Industry 20 years and Teaching 19 years. Awarded with Doctoral Degree in Management Sciences in Anna University, Chennai. He has organized various workshops and Faculty Development Programmes. He is actively involved in research and consultancy works. He acted as a resource person to FDPs & MDPs to various industries like, SPB ltd, Tamilnadu Police, DIET, Rotary school and many. His areas of interest include Entrepreneurship, Business Law, Marketing and Case writing. Articles published more than 100 International and National Journals. Presented papers in more than 30 National and International conferences including IIM Bangalore, IIM Kozhikode, IIM Kashipur and IIM Indore. He was a Chief Co-ordinator of Entrepreneurship and Management Development Centre (EMDC) of Kongu Engineering College, he was instrumental in organizing various Awareness Camps, FDP, and TEDPs to aspiring entrepreneurs which was funded by NSTEDB – DST/GoI

Swati Sharma is the Associate Professor at the Amity Institute of Travel and Tourism at Amity University, Noida, and Uttar Pradesh. She has completed her Ph.D. in Tourism from Kurukshetra University in Tourism. She is UGC-NET Qualified and MBA in Hospitality Management. She has a vast teaching experience of around 17 years. Her Research Interest area are branding and Destination Marketing. Apart from this she has Published 15 Research papers in Scopus Indexed journal and International Journals.

Qazi Saima Syed is a research scholar at the Department of Management Studies, Islamic University of Science and Technology. My research interests span across omnichannel consumption behaviour, retailing, and technology acceptance behaviour.

V. Sathiyamoorthi is currently working as Lecturer in Computer Engineering Department at Government polytechnic college, Dharmapuri,Tamil Nadu, India. He was born on June 21, 1983, at Omalur in Salem District, Tamil Nadu, India. He received his Bachelor of Engineering degree in Information Technology from Periyar University, Salem with First Class. He obtained his Master of Engineering degree in Computer Science and Engineering from Anna University, Chennai with Distinction and secured 30th University Rank.He received his Ph.D degree from Anna University, Chennai in Web Mining. His areas of specialization include Web Usage Mining, Data Structures, Design and Analysis of Algorithm and Operating System. He has published many papers in International Journals and conferences. He has published many books and book chapters in various renowned international publishers. He has also participated in various National level Workshops and Seminars conducted by various reputed institutions.

V.P. Arun is a driven and accomplished professional with a diverse educational background and extensive hands-on experience across various industries. Graduating with honors, Arun earned his Master of Business Administration (M.B.A) with a specialization in Human Resources and Marketing from the renowned Sona School of Management in Salem in 2018, where he excelled academically with an impressive 8.3 Cumulative Grade Point Average (CGPA).Throughout his academic journey, Arun displayed an unwavering commitment to learning and personal growth, actively seeking opportunities to expand his knowledge and skills beyond the confines of traditional education. He sought practical experiences to complement his theoretical understanding, such as a 45-day summer internship focused on conducting a feasibility study for R-Doc Sustainability in the market. Additionally, Arun broadened his horizons through a 7-day industrial visit to Malaysia and Singapore, immersing himself in diverse cultural and professional environments. Arun's academic pursuits were further enriched by his involvement in hands-on projects, including a comprehensive study on Employee Job Satisfaction at Roots Cast Private Limited. His professional trajectory includes serving as a Growth Officer at Parle Agro Private Limited, where he played a crucial role in driving sales growth and market expansion. Arun is a dedicated professional, currently employed at JKKN Engineering College in the Management Studies Department. He holds expertise in his field and has contributed significantly to academia through the publication of two journal articles. Arun's work focuses on advancing knowledge and understanding in management studies, making valuable contributions to the academic community.

Prashant Kumar Yadav is currently pursuing Ph.D. in Tourism Management from the Department of Tourism Management, Indira Gandhi National Tribal University, Amarkantak, Madhya Pradesh. He has done Integrated B.Com.-LLB (Hons.), MBA in Tourism & Travel Management and qualified UGC-NET with JRF in 2022. He has presented six research papers in National and International Seminars/Conferences as well as published 3 research papers in UGC-CARE Listed and Scopus Index Journals. He has written 2 chapters & published in National and International Edited Books. He has participated in Various CBP, FDP and Research Methodology Workshops.

Index

Symbols

A

B

C

D

E

F

G

H

I

L

M

www.ingramcontent.com/pod-product-compliance
Lightning Source LLC
LaVergne TN
LVHW080343011026
849161LV00009B/2175